Fundamentals of Engineering Elasticity

Fundamentals of Engineering Elasticity

Second Edition

S. F. BORG
Professor Emeritus of Civil Engineering
Stevens Institute of Technology
Hoboken, NJ 07030, USA

World Scientific
Singapore • New Jersey • London • Hong Kong

Published by

World Scientific Publishing Co. Pte. Ltd.
5 Toh Tuck Link, Singapore 596224
USA office: 27 Warren Street, Suite 401-402, Hackensack, NJ 07601
UK office: 57 Shelton Street, Covent Garden, London WC2H 9HE

British Library Cataloguing-in-Publication Data
A catalogue record for this book is available from the British Library.

Printed in 1962 by Litton Educational Pub., Inc.

FUNDAMENTALS OF ENGINEERING ELASTICITY
(Revised 2nd Printing)

ISBN-13 978-981-02-0164-7
ISBN-10 981-02-0164-8
ISBN-13 978-981-02-0165-4 (pbk)
ISBN-10 981-02-0165-6 (pbk)

IN MEMORY OF AUDREY

PREFACE TO THE FIRST EDITION

This book is intended for sophomore and junior students in engineering curricula.

It may be trite to repeat that we are living in an era of "exploding technology," but for engineers this statement does perhaps bear repetition. In any event, it is this fact which has led the author to write this textbook. It is his conviction that the properly educated engineer and engineering scientist of the present and future must have a grasp of the fundamentals of engineering.

If the modern engineering curriculum is to include all of the groundwork knowledge in the many different fields that the well-prepared engineering student must know, then it will be impossible to cover, in the four-year engineering curriculum, the subject material which in the past was included in structural engineering. This broad field included strength of materials (or mechanics of materials), elementary structural analysis, and elementary structural design.

How then are we to prepare the student for truly professional engineering practice in the present and future?

It would seem that there is only one method that will fit within the framework of a four-year undergraduate curriculum, and this is to present all subjects from their truly fundamental points of view. We must go back to the origins of all fields, re-exploring the assumptions, hypotheses and approximations that ultimately led to the development of the engineering forms of the various subjects.

In the subject now called "strength of materials," going back to the origins means taking as our starting point the subject matter of the mathematical theory of elasticity. Otherwise stated, the presentation of material in this text is based on the assumption that the fountainhead, or essential source, of all knowledge of structural theory and practice is the mathematical theory of elasticity.

The mathematical theory of elasticity may be called the parent, and the engineering elasticity (or strength of materials) the offspring, even though the latter developed at an earlier time than the former, because the mathematical and scientific justification of the theories developed in the strength of materials is only to be found in the body and extensions of the mathematical theory of elasticity. The engineering elasticity is essentially an approximation or simplification of the more exact theory, so that the practicing engineer could utilize the predictions and results of mathematical elasticity. Textbooks on strength of materials have presented the engineering form of the mathematical theory of elasticity, but rarely, if ever, has the relationship between parent and offspring been made clear. Understanding the relationship would seem to be essential if the student is to have a good understanding of the coverage presented. To present the subject in such a manner and with such goals is what the present textbook attempts to do for the student

Because this book is written for the beginning student of structural engineering, we start with an explanatory chapter in which the basic philosophy behind the writing of the book is outlined. The purpose of the book is stated clearly and, as additional background, a brief historical summary of the field is presented. This will further help to explain to the student the purpose and intent of the text.

Tensor, the family name of the quantities of mathematical physics and, hence, of engineering, is a concept all modern engineers should understand and be familiar with. The tensor is introduced most naturally in this book because straightforward matrix-tensor statements are especially adaptable to discussions of engineering elasticity. For this reason, matrix-tensor arguments are utilized wherever possible. A basic, elementary treatment of matrix-tensor theory, suitable for our purposes, is given in Chapter 2. In this chapter also the elements of finite difference calculus are dealt with, because this also is a valuable tool that engineers should be familiar with and because elementary finite difference methods fit naturally into the framework of the coverage of the book.

The equations of the linearized theory of elasticity are presented next, and then it is shown how the ordinary, simplified structures of everyday engineering usage are analyzed using outgrowths and simplified forms of the basic, more exact theories. It is clearly shown just what approximations and assumptions are introduced in the basic theory of the engineering analysis — and it is shown why they are introduced.

In the remaining chapters the analysis of the key structural units (the bar, beam, etc.) is presented, first in the more exact linearized theory of elasticity solutions, and then in the approximate engineering solutions.

Extensions of the engineering analysis (the shear-moment curve construction, conjugate beam method, etc.) are introduced, where applicable, to indicate the directions in which this field has advanced.

Many problems dealing with the text material have been included, because an undergraduate engineering or science student can truly master his subject only when he can solve quantitative problems in connection with it.

In summary, then, the book has been written because the author believes the student in engineering must have a training and background in the field covered by the text. It is also the author's sincere hope that practicing engineers in the fields of applied mechanics and structural engineering will find the book worth while.

S. F. Borg
May 1961

PREFACE TO THE SECOND EDITION

The second edition follows the format of the first one. Several typographical errors have been corrected and several new topics or extensions of the original material have been included in an Appendix following Chapter 13 at the end of the book.

Overall however, the two fundamental premises have been adhered to, namely:

1. In order to attain a real understanding of the subject which we call "strength of materials" or "mechanics of materials" one must go back to the beginnings of these fields — the linearized mathematical theory of elasticity. Hence the title of the book stressing the words *engineering elasticity.*
2. The field of engineering elasticity is a good one to utilize in introducing to the undergraduate engineering student the important and useful topic of tensors. And for the engineer the matrix representation of the tensor is the easiest to visualize and to understand. Hence the use of the matrix-tensor notation.

The author wishes to express sincere thanks to his publishers, World Scientific Publishing Co., for reprinting the text and for their continual help and encouragement in seeing this task through to completion.

S. F. Borg
March 3, 1990

CONTENTS

Chapter 1

INTRODUCTION AND HISTORICAL BACKGROUND

1-1 Introduction

In this book we shall be concerned with the connections between the mathematical theory of elasticity and the engineering elasticity. In the normal engineering educational curriculum these two subjects are usually taught in two different courses. The first, the mathematical theory of elasticity, is generally a graduate course, taught at the highest mathematical level. The second is generally taught to sophomores as a course in "strength of materials" or "mechanics of materials."

Both subjects are primarily concerned with the stresses and deformations of solids, and therefore, are very intimately connected. The parent subject, if we may use this term, is the mathematical theory of elasticity. This is the "exact" formulation of the behavior of materials — subject to certain fundamental hypotheses — when the materials are subjected to forces or other disturbing effects. The subject starts with the exact laws of physics or of nature. Hence, solutions which are obtained in this field are (subject to the hypotheses and assumptions) capable of predicting exactly how the particular material, or part, will react to the imposed conditions. This is the ultimate aim of the engineering scientific method — to be able to predict, quantitatively, how an object or materials will react when subjected to a given set of conditions.

Unfortunately, the number of "exact" solutions obtainable, even for the mathematical theory developed under rather severely restricting assumptions of linearity, homogeneity and isotropy, is not large. Some fundamental solutions have been obtained, and they are of considerable historical and practical interest. But many more practical problems and structures exist than there are exact solutions available. It is unfortunate, too, that these structures which can not be solved "exactly" are part of our present civilization; also, more and more of them are being invented and constructed every day. The engineer who is responsible for the analysis and design of these structures must have some theory — albeit an approximate one — upon which his analysis is based.

This approximate theory — which is called "strength of materials," or "mechanics of materials" — is the second part of our study, or the second subject considered in this text. We shall call it "engineering elasticity" because it is, in the last analysis, the form which the more exact mathematical theory of elasticity has taken in order that its results may be utilized by engineers. And this is true even though, historically, the engineering elasticity developed before the mathematical theory of elasticity.

We are not implying that the subject of engineering elasticity is at a "lower level" than the subject of mathematical elasticity. Indeed, some of the developments in the field of engineering elasticity may be as complicated (mathematically) as any topics which appear in the mathematical elasticity field. We are saying, however, that in general, the jumping off place, the starting point, for many developments in engineering elasticity has been the more exact mathematical theory. The more exact theory, very often simplified and approximated, has been found capable of explaining the behavior of many of the more complicated structures.

It must also be emphasized at this point that the true test of any structural theory, exact or approximate, is a comparison of its predictions with the actual behavior of the structures. Hence, although most exact theories can be simplified by virtue of approximating assumptions, the results of these simplified theories are worthless unless they can predict — with accuracy sufficient for engineering purposes — the behavior of the structure.

Our purpose then is to present first, in sufficient detail, the fundamental, exact theory, that is, the basic equations and relations of the mathematical theory of elasticity. We shall derive these equations in the simplest manner (after pointing out that more rigorous and more sophisticated derivations are described in the literature). Wherever possible, a physical interpretation of a result will be given. This is especially necessary in a book of this kind, which is being written for use by engineering undergraduates.

We shall present this basic theory in terms of a notation which will probably be new to the student: the matrix-tensor notation. This is done not to confuse the student, but rather as a means of introducing him to this extremely important discipline in modern applied mathematics and modern engineering. It represents a new notation — a new language, one might say — and as such it is not easy to overcome at the beginning. It requires some application, some memorizing by the student. But it is not really difficult, and when the student has become familiar with it he will find that he actually enjoys using the notation. In addition, he will find that it offers economies in expression; equations can be written in fewer symbols using the tensor notation.

In Chapter 2 this subject of matrix-tensor analysis is presented in sufficient detail to give the student all the background necessary for this text and for many additional purposes as well. We shall also discuss in the next chapter the introductory material to finite difference calculus. Here again we have a discipline or tool in applied mathematics which is extremely useful and entirely understandable to any student who has taken the first course in calculus. We shall use this technique for solving differential equations in later sections of the book.

The fundamental or exact linearized mathematical theory of elasticity which we shall present will also be given in the language of the modern scientist, namely, in the form of partial differential equations. Here also it must be assumed that, whereas the student may not have had a thorough course in partial differential equations, he is, however, familiar with the notation and also the meaning of partial differential quantities. Very little more than this is expected of the student. Those solutions of partial differential equations which are worked out are of a most elementary type — generally in the form of ordinary differential equations. However, it is deemed essential that the student be introduced to the basic equations in their proper form; this means in their form as partial differential equations.

So much for the fundamental relations of the exact linearized theory of elasticity. Then, having obtained the relations, we shall show in later chapters how various simple solutions to these relations may be obtained. The solutions are exact, subject to the assumptions and hypotheses used in deriving the original fundamental relations. As such they are extremely important and extremely useful. They represent positions from which we may move in various directions, making approximations, simplifying assumptions, or otherwise altering the results. But, as noted before, *in every case an approximation or a simplification is worthless unless it predicts actual behavior — at least with sufficient accuracy for use in engineering analysis.*

This is the essence of the subject of engineering elasticity — the second field covered in this text. That is, we are very interested in the alterations made to the exact solutions in order that they may be used to design more complicated structures than the original theory could be applied to. And as pointed out before, these altered solutions are the material covered in the so-called courses of "strength of materials" or "mechanics of materials." This is the subject which in this text we call "engineering elasticity."

By using the term engineering elasticity we also wish to emphasize that the results used in this field follow directly from the more exact results obtained in the linearized theory of elasticity. We emphasize this point repeatedly in the text, and we show in the text how we go from the exact to the engineering solution. We indicate just what assumptions are made, what simplifications are introduced, and what degree of approximation is involved. We consider, in this manner, the tension-compression structure, the beam, the torsion structure, the column, and briefly, some special structures. *In each case we start with the more exact and proceed to the more approximate.*

It is felt that the student studying this book may be interested in a brief historical survey covering the combination of the two fields. This is therefore presented in the next section.

1-2 Brief Historical Survey

It is generally agreed that Galileo, in 1636, reported the first quantitative investigation in the field of structural engineering. Before his time the strengths and deformations of structures were determined primarily by trial and error. A structure was built. If it stood up, well and good. If not, then the next structure was made stronger where the first one failed, and so on.

Galileo studied the behavior of beams. He tested beams under various conditions and derived an expression for the distribution of stresses across a cross section. Although this relation was not quite correct, he did arrive at some general properties concerning the strengths of geometrically similar beams. Furthermore, his work ultimately led to further investigations in the field of structures.

The next great advance was made by Robert Hooke, who in 1678 announced his general law of the proportionality between stress and strain.

The derivation of the engineering form of the beam relations as we know them today is due to Edme Mariotte, Jacques Bernoulli and Leonhard Euler. Mariotte in 1680 experimentally determined that the neutral axis, or axis of zero strain, is at the centroid of the beam. Bernoulli introduced the assumption that a plane section before bending remains a plane section after bending, and Euler studied the relations governing the elastic curve or deformed position of the beam. The formulation of the problem of the beam as given by these three men is called the Bernoulli-Euler theory of beams.

Charles Augustin Coulomb, in a paper published in 1776, presented a complete engineering solution for the beam problem essentially as we solve it today. He has been called the father of mechanics of materials.

The first textbook on strength of materials as we know it today was published by Navier in 1826. It included many of the results of Coulomb, Bernoulli and Euler on beam analysis, as well as extensions of their work by Navier.

The foundations of the mathematical theory of elasticity were developed later than those in the strength of materials. The originators of this theory may be taken as Cauchy, Poisson and Navier. Navier presented his formulation of the theory in 1821, Cauchy presented his in 1822, and Poisson his in 1828. Great names associated with the early development of this subject are George Green, Lagrange, Lamé, Saint-Venant, and

Boussinesq, and many other great mathematicians to this day have been associated with this field.

For our purposes, the important fact which emerges from the above is that from its beginnings the subject of structural analysis seemed to separate naturally into two fields — the theory of elasticity and the strength of materials. The first was generally considered the province of the mathematician and was admittedly more theoretical and mathematical than the second. It became an important, active field in applied mathematics, generally ignored by the engineer. Strength of materials, on the other hand, tended more toward the practical. If the mathematical theory of elasticity could not give the answer, then an approximation to it was looked for. If necessary, experiments or tests were performed, and these were frequently looked upon as sources for developing further engineering theories.

Through the years, the two fields tended to drift apart, occasionally approaching each other and even overlapping, but more often ignoring each other. In recent years, however, the engineer more and more is finding that his problems require solutions only attainable by the theory of elasticity. His new structures, new techniques of manufacture, new uses to which structures are being put — all, very frequently, require a solution that is given only by the most exact theory available, and this means the mathematical theory of elasticity.

So it is that the engineer of today, and certainly the engineer of the future, will of necessity require some knowledge of the powers and limits of the mathematical theory of elasticity, and an essential element of this is a knowledge and awareness of the place of strength of materials with relation to the more exact theory. It is an introduction to this knowledge that this text hopes to give the student of engineering.

Chapter 2

MATHEMATICAL PRELIMINARIES: THE ELEMENTS OF MATRIX-TENSOR THEORY AND OF THE FINITE DIFFERENCE METHOD

2-1 Introduction

In the presentation to be used in later portions of this text, recourse will be made to matrix-tensor methods and also to finite difference methods. In this chapter, introductory treatments of both topics will be given. We shall begin with a discussion of matrix algebra which will include coverage of the basic matrix operations. Following this, a brief treatment will be given of tensor analysis — from the matrix point of view. In particular, in this chapter we shall describe the tensor of zero, first and second order, since these are the quantities which will be used in later chapters.

Finally, we shall describe the elements of finite difference methods, with special emphasis placed on the subject from the point of view of the beam — since this is the main object of study in this text.

The finite difference method is discussed in this book for two reasons.

1. The finite difference technique is one which has applications in all of mathematical physics, and hence, in engineering. Our particular applications to beams (as shown in later chapters) are relatively simple ones and — although typical of finite difference methods generally — because they are applied to beams, it is a discussion that readers of this book should find particularly easy to follow.

2. The modern electronic computer permits us to solve problems involving many simultaneous equations. Before the advent of the electronic computer, the very weaknesses of existing methods of solutions limited the setup of many problems to a relatively small number of simultaneous equations. The computer fits in with the finite difference technique, since this method essentially approximates differential equations by linear algebraic simultaneous equations — and in many problems the number of these simultaneous equations is such that only electronic computers can give solutions to the problem. Thus, it is deemed desirable to introduce the modern engineer and engineering student to one of the important steps in the formulation of many current problems for electronic computer programing.

2-2 Matrix Algebra

A rectangular array of m rows and n columns of numbers or other quantities is called a *matrix*. We designate this matrix with a capital letter, as A, and show it in its expanded form as

$$A = \begin{pmatrix} a_{11} & a_{12} & a_{13} \\ a_{21} & a_{22} & a_{23} \end{pmatrix} \tag{2.1}$$

In the above expression, a_{ij} represents an *element* of the matrix. Note particularly that the subscripts of the elements carry a position significance. That is, the first subscript represents the row position of the element and the second subscript represents the column position.

A matrix is not a determinant. As a reminder of this, the enclosing bars are shown curved, as against the ordinary usage of straight bars for the determinant.

The number of rows in a matrix need not necessarily be the same as the number of columns. If the number of rows *does* equal the number of columns, then the matrix is a *square* matrix.

The elements of a matrix may or may not have any physical significance. For example, the elements may be pure numbers, as

$$\begin{pmatrix} 6 & 15 & 1028 \end{pmatrix} \tag{2.2}$$

or the elements may be components of a velocity vector, as

$$\begin{pmatrix} u \\ v \\ w \end{pmatrix} \tag{2.3}$$

Indeed, they could even be colors, as

$$\begin{pmatrix} \text{red} & \text{blue} & \text{green} \end{pmatrix} \tag{2.4}$$

or animals as

$$\begin{pmatrix} \text{cat} \\ \text{dog} \\ \text{hare} \end{pmatrix} \tag{2.5}$$

No significance must be attached to the use of a row form for the numbers and colors and a column form for the velocity and animals in the above matrices.

The elements of a matrix may also be complex quantities, or equations, or, in fact, any quantity whatever.†

†Some recent developments in chemistry and chemical engineering utilize matrices in the formulation and solution of chemical equations in the usual chemical symbolic notation. See "Matrix Algebra For Calculating Multicomponent Mixtures," by Fred Ordway, Paper No. 74, March 1960, Portland Cement Association and National Bureau of Standards.

Two matrices A and B are equal only if each has the same number of rows and the same number of columns, and if corresponding elements are equal. Thus, given

$$A = \begin{pmatrix} a_{11} & a_{12} & a_{13} & a_{14} \\ a_{21} & a_{22} & a_{23} & a_{24} \end{pmatrix} \tag{2.6}$$

$$B = \begin{pmatrix} b_{11} & b_{12} & b_{13} & b_{14} \\ b_{21} & b_{22} & b_{23} & b_{24} \end{pmatrix} \tag{2.7}$$

Then, if

$$a_{ij} = b_{ij} \tag{2.8}$$

it follows that

$$A = B \tag{2.9}$$

Thus, the simple algebraic equations

$$\left.\begin{aligned} a &= p + 2 \\ b &= q + 7 \\ c &= r - 2 \\ d &= s + 7 \end{aligned}\right\} \tag{2.10}$$

may be given in matrix form as

$$\begin{pmatrix} a & b \\ c & d \end{pmatrix} = \begin{pmatrix} p + 2 & q + 7 \\ r - 2 & s + 7 \end{pmatrix} \tag{2.11}$$

The *zero* matrix, 0, has all elements equal to zero.

The *unit* matrix E_n is an n by n square matrix whose left to right diagonal elements equal unity and whose off-diagonal elements are equal to zero. That is, in

$$E_n, \begin{cases} a_{ij} = 1 & \text{if} \quad i = j \\ a_{ij} = 0 & \text{if} \quad i \neq j \end{cases} \tag{2.12}$$

and, as an example,†

$$E_3 = \begin{pmatrix} 1 & 0 & 0 \\ 0 & 1 & 0 \\ 0 & 0 & 1 \end{pmatrix} \tag{2.13}$$

A square matrix is *symmetrical* if

$$a_{ij} = a_{ji} \tag{2.14}$$

An example of a symmetrical matrix is the following:

$$\begin{pmatrix} 1 & 2x^2 & 3 - y \\ 2x^2 & e^s & 0 \\ 3 - y & 0 & q^n \end{pmatrix} \tag{2.15}$$

†It may be shown (see Eq. 2.32) that the product of any matrix and E_n is just equal to the original matrix. In analogy to the algebraic product of any quantity and unity, we may therefore call the matrix E_n the "unit matrix."

A square matrix is *anti-symmetrical* or *skew-symmetric* if

$$a_{ij} = -a_{ji} \tag{2.16}$$

An example of a skew-symmetric matrix is

$$\begin{pmatrix} 0 & e^q & 9-x \\ -e^q & 0 & 1 \\ -(9-x) & -1 & 0 \end{pmatrix} \tag{2.17}$$

Note that in a skew-symmetric matrix the main diagonal (left to right) elements must be zero, for only then will $a_{ij} = -a_{ji}$ be true for these elements.

The *transpose* of a matrix A is shown as A^* and is obtained by interchanging the rows and columns of A. Thus, if

$$A = \begin{pmatrix} 3z & y^2 & t^n \\ 2-x & 3^n & e^s \end{pmatrix} \tag{2.18}$$

then

$$A^* = \begin{pmatrix} 3z & 2-x \\ y^2 & 3^n \\ t^n & e^s \end{pmatrix} \tag{2.19}$$

We may obtain the sum of two matrices A and B only if A and B have the same number of rows and the same number of columns. The sum $A + B$ is then a matrix C of the same number of rows as A (and B) and the same number of columns as A (and B) and with

$$c_{ij} = a_{ij} + b_{ij} \tag{2.20}$$

For example, the algebraic equations

$$\left.\begin{aligned} c_{11} &= a_{11} + b_{11} \\ c_{12} &= a_{12} + b_{12} \\ c_{21} &= a_{21} + b_{21} \\ c_{22} &= a_{22} + b_{22} \end{aligned}\right\} \tag{2.21}$$

are equivalent to

$$\begin{pmatrix} c_{11} & c_{12} \\ c_{21} & c_{22} \end{pmatrix} = \begin{pmatrix} a_{11} & a_{12} \\ a_{21} & a_{22} \end{pmatrix} + \begin{pmatrix} b_{11} & b_{12} \\ b_{21} & b_{22} \end{pmatrix} \tag{2.22}$$

The student should note that, consistent with this definition of addition, it follows that KA, or any constant times a matrix, is a new matrix, each of whose elements is K times the elements of matrix A.

It may be shown (the student should verify this and the following statement for simple matrices) that the sum of two matrices is commutative, that is,

$$A + B = B + A \tag{2.23}$$

Also it may be shown that the addition of matrices is associative, that is,

$$(A + B) + C = A + (B + C) \tag{2.24}$$

The difference of two matrices A and B is defined similarly. Thus,

$$C = A - B \tag{2.25}$$

with

$$c_{ij} = a_{ij} - b_{ij} \tag{2.26}$$

Any square matrix may be given as the sum of a symmetrical and anti-symmetrical matrix, for, if A is a square matrix, then obviously

$$A = \frac{A + A^*}{2} + \frac{A - A^*}{2} \tag{2.27}$$

The first term on the right is a symmetrical matrix and the second term is an anti-symmetrical matrix. This may be verified for a 2×2 matrix as follows: if

$$A = \begin{pmatrix} a_{11} & a_{12} \\ a_{21} & a_{22} \end{pmatrix} \tag{2.28}$$

then

$$A^* = \begin{pmatrix} a_{11} & a_{21} \\ a_{12} & a_{22} \end{pmatrix} \tag{2.29}$$

so that

$$\frac{A + A^*}{2} = \begin{pmatrix} a_{11} & \dfrac{a_{12} + a_{21}}{2} \\ \dfrac{a_{21} + a_{12}}{2} & a_{22} \end{pmatrix} \tag{2.30}$$

and

$$\frac{A - A^*}{2} = \begin{pmatrix} 0 & \dfrac{a_{12} - a_{21}}{2} \\ \dfrac{a_{21} - a_{12}}{2} & 0 \end{pmatrix} \tag{2.31}$$

The *product* of two matrices is obtained as follows: given two matrices A and B such that the number of rows in B equals the number of columns in A, then the product AB is given by C, in which the element C_{ij} is obtained by multiplying each element of the i-th row of A by the corresponding element of the j-th column of B and adding. For example,

$$C = A \quad B \tag{2.32}$$

$$= \begin{pmatrix} a_{11} & a_{12} & a_{13} \\ a_{21} & a_{22} & a_{23} \end{pmatrix} \begin{pmatrix} b_{11} \\ b_{21} \\ b_{31} \end{pmatrix} \tag{2.33}$$

$$= \begin{pmatrix} a_{11}b_{11} + a_{12}b_{21} + a_{13}b_{31} \\ a_{21}b_{11} + a_{22}b_{21} + a_{23}b_{31} \end{pmatrix} \tag{2.34}$$

Note in the above product that C is a 2 by 1 matrix. In general, if

$$C = AB \tag{2.35}$$

and if $A = A_{ij}$ and $B = B_{jk}$ then $C = C_{ik}$ which is obtained by cancelling the common subscript. It will be obvious from the above example, that in general

$$AB \neq BA \tag{2.36}$$

that is, the position of a matrix in a matrix multiplication is not immaterial.

As another example of matrix multiplication, the student should satisfy himself that the algebraic equations

$$\left.\begin{aligned} a &= 2e + 3g \\ b &= 2f + 3h \\ c &= -e + 7g \\ d &= -f + 7h \end{aligned}\right\} \tag{2.37}$$

are equivalent to the matrix equation

$$\begin{pmatrix} a & b \\ c & d \end{pmatrix} = \begin{pmatrix} 2 & 3 \\ -1 & 7 \end{pmatrix} \begin{pmatrix} e & f \\ g & h \end{pmatrix} \tag{2.38}$$

It may be shown (the student should verify this and the following statement using simple matrices) that the product of matrices is associative, that is,

$$(AB)C = A(BC) \tag{2.39}$$

Also, the product of matrices is distributive, that is,

$$A(B + C) = AB + AC \tag{2.40}$$

The following expression is the statement of the very important *transpose product rule*,

$$(AC)^* = C^*A^* \tag{2.41}$$

The student should verify this for a simple case.

The *cofactor* matrix of any square matrix A, with n rows and columns (denoted by co A) is the matrix obtained by replacing each element of A by its cofactor, the cofactor of a_{ij} being the product of the determinant of the matrix with $n - 1$ rows and columns obtained by erasing the i-th row and j-th column of A, by $(-1)^{(i+j)}$. Thus, if A is given by

$$A = \begin{pmatrix} a_{11} & a_{12} \\ a_{21} & a_{22} \end{pmatrix} \tag{2.42}$$

then

$$\text{co } A = \begin{pmatrix} a_{22} & -a_{21} \\ -a_{12} & a_{11} \end{pmatrix} \tag{2.43}$$

If A is any square matrix of n columns and rows, then the *inverse* matrix, denoted by A^{-1}, is the matrix such that

$$AA^{-1} = E_n \tag{2.44}$$

in which E_n is the matrix equivalent of the *arithmetic unity*, and A^{-1} is given by

$$A^{-1} = \frac{(\text{co } A)^*}{\text{determinant of } A} \tag{2.45}$$

The *derivative* of a matrix with respect to a variable, say y, is obtained by differentiating each element separately. Thus

$$\frac{d}{dy}\begin{pmatrix} y^3 & e^y & y^{-1} \\ 3 & 2y & y^2 \end{pmatrix} = \begin{pmatrix} 3y^2 & e^y & -y^{-2} \\ 0 & 2 & 2y \end{pmatrix} \tag{2.46}$$

The integral of a matrix is defined in a similar manner.

The above represents the elements of matrix algebra sufficient for our purposes. It may also be noted that the matrix operations described herein are of interest, not only in connection with tensor analysis, but also because they occur in some electronic computer operations. Examples of this will be given in a later chapter.

2-3 Scalar, Vector and Tensor Analysis

The scalar, vector and tensor are quantities of fundamental importance in engineering and physics. We shall define these quantities with reference to their behaviour under a transformation of axes.† This precise definition, we shall see, has an intimate connection with the matrix analysis discussed in the preceding section.

2-4 The Scalar

A scalar quantity is one whose value does not change when we rotate the coordinate axes of the system in question about its origin.

For example, in a given room, at a particular point (say at a distance x, y, z from some origin and orthogonal set of axes), the temperature may be 70°F. If now we rotate the axes about the origin, so that the point is at x', y', z' from the origin, the temperature is still 70°F; that is, it is independent of the orientation of the axes, and this is the definition of a

†There are in general two types of axial transformations — translational (in which the origin moves but the axes remain parallel to their original directions) and rotational. The translational transformation is of no interest in our present study; hence, the behaviour which we shall consider will be due only to a rotational transformation of axes about a stationary origin.

scalar quantity. Thus, pure members are scalars, pressure is a scalar, work is a scalar, and so on for other physical quantities. For reasons which will be clear shortly, we also may call a scalar a *tensor of zero order*.

2-5 The Vector

The transformation law satisfied by the vector quantity may be obtained as follows (we do this first for the two dimensional case; the three dimensional vector will then be obtained as a generalization).

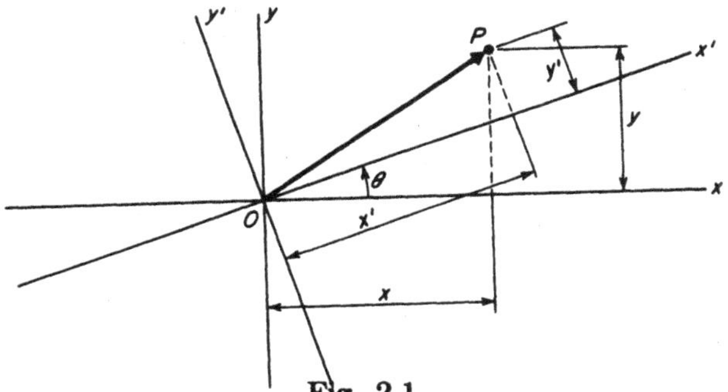

Fig. 2.1

In Fig. 2.1 let oP be a vector, whose coordinates in the oxy system are (x,y). Now assume the axes are rotated through an angle θ, about o, to position $ox'y'$. The coordinates of the vector in the $ox'y'$ system are (x',y') and, furthermore, as may be verified,

$$x' = x \cos\theta + y \sin\theta \tag{2.47}$$

$$y' = -x \sin\theta + y \cos\theta \tag{2.48}$$

or, in our matrix notation,

$$\begin{pmatrix} x' \\ y' \end{pmatrix} = \begin{pmatrix} \cos\theta & \sin\theta \\ -\sin\theta & \cos\theta \end{pmatrix} \begin{pmatrix} x \\ y \end{pmatrix} \tag{2.49}$$

For example, consider a force vector $\bar{F}$, of value 100 lb, directed as shown in Fig. 2.2.

It is obvious that for the oxy system

$$\bar{F} = \begin{pmatrix} F_x \\ F_y \end{pmatrix} = \begin{pmatrix} 80 \\ 60 \end{pmatrix} \tag{2.50}$$

Now assume the axis rotated to $ox'y'$ as shown in Fig. 2.3.

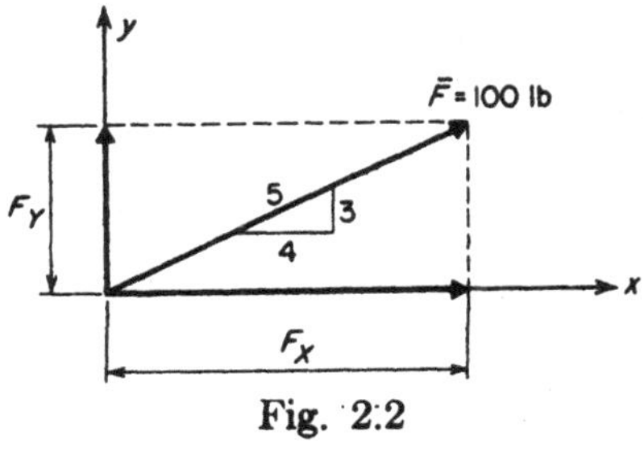

Fig. 2.2

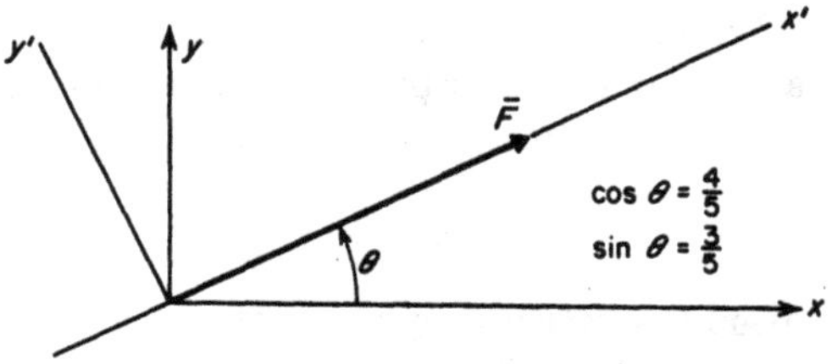

Fig. 2.3

It is equally obvious that

$$\bar{F} = \begin{pmatrix} F_{x'} \\ F_{y'} \end{pmatrix} = \begin{pmatrix} 100 \\ 0 \end{pmatrix} \tag{2.51}$$

We verify that this is as it should be by checking Eqs. 2.47 and 2.48;

$$F_{x'} = F_x \cos\theta + F_y \sin\theta \tag{2.52}$$

$$F_{y'} = -F_x \sin\theta + F_y \cos\theta \tag{2.53}$$

or,

$$F_{x'} = \tfrac{4}{5}(80) + \tfrac{3}{5}(60) = 64 + 36 = 100 \tag{2.54}$$

$$F_{y'} = -\tfrac{3}{5}(80) + \tfrac{4}{5}(60) = -48 + 48 = 0 \tag{2.55}$$

which is as required.

The above relation, Eq. 2.49, is the general transformation law for vectors in two dimensions, that is,

$$V' = RV \tag{2.56}$$

in which V' is the vector referred to the $ox'y'$ system of coordinates, V is the vector referred to the oxy system of coordinates, and R is the so-called *rotation matrix*,† which we may show as

$$R = \begin{pmatrix} l_{11} & l_{12} \\ l_{21} & l_{22} \end{pmatrix} \tag{2.57}$$

In this representation of R, each element represents the cosine of the angle between the axes corresponding to the subscripts — in which the first subscript is the primed axis and the second is the unprimed — so that, for example,

$$l_{21} = \text{cosine of the angle between the } y' \text{ and the } x \text{ axes} \tag{2.58}$$

Note that neither x, y nor x', y' in Eqs. 2.47 and 2.48 are scalars, that is, the value of each component depends upon the orientation of the axes.

†For obvious reasons, R is also called the *direction cosine matrix*.

But there is a scalar connected with the vector, this being the squared length, that is,

$$x^2 + y^2 = x'^2 + y'^2 = \text{independent of axial orientation.} \tag{2.59}$$

For example, for the force vector $\overline{F}$ considered above in Eq. 2.50,

$$F_x^2 + F_y^2 = 80^2 + 60^2 = 10{,}000 \tag{2.60}$$

and (see Eq. 2.51)

$$F_{x'}^2 + F_{y'}^2 = 100^2 + 0^2 = 10{,}000 \tag{2.61}$$

One may show quite easily that the equation

$$V' = RV \tag{2.62}$$

is a general one which also holds true in a space of three dimensions. In this,

$$l_{ij} = \text{cosine of angle between the } i' \text{ and } j \text{ axes.} \tag{2.63}$$

For example, consider the force vector in space, shown in Fig. 2.4.

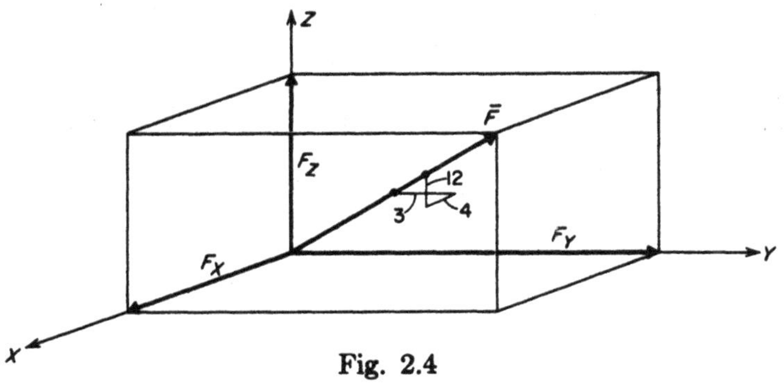

Fig. 2.4

If $\overline{F} = 260$ lb, then

$$\overline{F} = \begin{pmatrix} F_x \\ F_y \\ F_z \end{pmatrix} = \begin{pmatrix} 80 \\ 60 \\ 240 \end{pmatrix} \tag{2.64}$$

Now suppose that the axes are rotated so that x' coincides with $\overline{F}$, and

y' and z' are perpendicular to it and to each other. Then, in the direction cosine table, the elements l_{11}, l_{12} and l_{13}, given in the first row are

$$R = \begin{pmatrix} l_{11} & l_{12} & l_{13} \\ l_{21} & l_{22} & l_{23} \\ l_{31} & l_{32} & l_{33} \end{pmatrix} = \begin{pmatrix} \frac{4}{13} & \frac{3}{13} & \frac{12}{13} \\ - & - & - \\ - & - & - \end{pmatrix} \tag{2.65}$$

It follows that $F_{x'}$ is given by

$$F_{x'} = l_{11}F_x + l_{12}F_y + l_{13}F_z \tag{2.66}$$

$$= \tfrac{4}{13}(80) + \tfrac{3}{13}(60) + \tfrac{12}{13}(240) \tag{2.67}$$

$$= 260 \text{ lb} \tag{2.68}$$

as required.

The vector is also called a *tensor of the first order.*

2-6 The Tensor

The quantity which we shall call a *tensor* is, in reality, a *tensor of the second order*. As pointed out previously, scalars and vectors are also part of the tensor family — membership in the family being restricted to physical quantities which satisfy certain transformation laws. Practically all of the quantities of engineering and physical interest are members of this family. From this point on, we shall reserve the name "tensor," or "matrix," for the "tensor of second order" only, and we shall call the first order tensor, a vector — and the zero order tensor, a scalar. But the student should understand and appreciate the very intimate connection between all of these fundamental physical quantities.†

The transformation law for the tensor is obtained in the following manner. Consider two vectors, U and V, which are represented by

$$U = \begin{pmatrix} u_x \\ u_y \\ u_z \end{pmatrix} \tag{2.69}$$

$$V = \begin{pmatrix} v_x \\ v_y \\ v_z \end{pmatrix} \tag{2.70}$$

Then, we define the tensor of the second order as the quantity

$$T = UV^* \tag{2.71}$$

†In this text we shall consider only such quantities as exist in a three-dimensional physical space. However, the general tensor concept and the associated transformation laws are perfectly general and permit logical, straightforward extensions to spaces of N dimensions. We can, for example, define an N-dimensional vector by means of Eq. 2.62, with V and V' containing N components, and R being an N by N "direction cosine matrix." This extension applies as well to the scalar and tensor.

or, by simply performing the indicated operation,

$$T = \begin{pmatrix} u_x v_x & u_x v_y & u_x v_z \\ u_y v_x & u_y v_y & u_y v_z \\ u_z v_x & u_z v_y & u_z v_z \end{pmatrix} \tag{2.72}$$

This being so (see Eq. 2.56), it follows that

$$U' = RU \tag{2.73}$$

and (see Eq. 2.56)

$$V' = RV \tag{2.74}$$

so that (see Eq. 2.41), by taking the transpose of both sides of Eq. 2.74, we obtain

$$V'^* = V^* R^* \tag{2.75}$$

and therefore

$$U'V'^* = RUV^*R^* \tag{2.76}$$

or (see Eqs. 2.39 and 2.71)

$$T' = RTR^* \tag{2.77}$$

and this is the transformation law that is satisfied by the tensor of the second order.†

It will be instructive to obtain the expanded form of T' in terms of the elements of T for both the three-dimensional and the two-dimensional form of T. Thus, if

$$T = \begin{pmatrix} T_{11} & T_{12} & T_{13} \\ T_{21} & T_{22} & T_{23} \\ T_{31} & T_{32} & T_{33} \end{pmatrix} \tag{2.78}$$

then

$$T' = \begin{pmatrix} l_{11} & l_{12} & l_{13} \\ l_{21} & l_{22} & l_{23} \\ l_{31} & l_{32} & l_{33} \end{pmatrix} \begin{pmatrix} T_{11} & T_{12} & T_{13} \\ T_{21} & T_{22} & T_{23} \\ T_{31} & T_{32} & T_{33} \end{pmatrix} \begin{pmatrix} l_{11} & l_{21} & l_{31} \\ l_{12} & l_{22} & l_{32} \\ l_{13} & l_{23} & l_{33} \end{pmatrix} \tag{2.79}$$

where

$$T' = \begin{pmatrix} T'_{11} & T'_{12} & T'_{13} \\ T'_{21} & T'_{22} & T'_{23} \\ T'_{31} & T'_{32} & T'_{33} \end{pmatrix} \tag{2.80}$$

†The student should verify at this point that the unit matrix E_n, of Eqs. 2.12 and 2.13, is a tensor of the second order which does not change under the above transformation, which defines the second order tensor. To make this verification, use the results of Prob. 7 at the end of this chapter. Hence, this unit matrix may be called the "scalar tensor." See Ref. 4 and Prob. 15 at the end of this chapter for additional identity and similar relations with reference to tensors.

Performing the multiplications indicated, we obtain

$$\left.\begin{aligned}
T'_{11} &= T_{11}l_{11}^2 + T_{12}l_{12}l_{11} + T_{13}l_{13}l_{11} + T_{21}l_{11}l_{12} + T_{22}l_{12}^2 \\
&\quad + T_{23}l_{13}l_{12} + T_{31}l_{11}l_{13} + T_{32}l_{12}l_{13} + T_{33}l_{13}^2 \\
T'_{12} &= T_{11}l_{21}l_{11} + T_{12}l_{22}l_{11} + T_{13}l_{23}l_{11} + T_{21}l_{21}l_{12} + T_{22}l_{22}l_{12} \\
&\quad + T_{23}l_{23}l_{12} + T_{31}l_{21}l_{13} + T_{32}l_{22}l_{13} + T_{33}l_{23}l_{13} \\
T'_{13} &= T_{11}l_{31}l_{11} + T_{12}l_{32}l_{11} + T_{13}l_{33}l_{11} + T_{21}l_{31}l_{12} + T_{22}l_{32}l_{12} \\
&\quad + T_{23}l_{33}l_{12} + T_{31}l_{31}l_{13} + T_{32}l_{32}l_{13} + T_{33}l_{33}l_{13} \\
T'_{21} &= T_{11}l_{11}l_{21} + T_{12}l_{12}l_{21} + T_{13}l_{13}l_{21} + T_{21}l_{11}l_{22} + T_{22}l_{12}l_{22} \\
&\quad + T_{23}l_{13}l_{22} + T_{31}l_{11}l_{23} + T_{32}l_{12}l_{23} + T_{33}l_{13}l_{23} \\
T'_{22} &= T_{11}l_{21}^2 + T_{12}l_{22}l_{21} + T_{13}l_{23}l_{21} + T_{21}l_{21}l_{22} + T_{22}l_{22}^2 \\
&\quad + T_{23}l_{23}l_{22} + T_{31}l_{21}l_{23} + T_{32}l_{22}l_{23} + T_{33}l_{23}^2 \\
T'_{23} &= T_{11}l_{31}l_{21} + T_{12}l_{32}l_{21} + T_{13}l_{33}l_{21} + T_{21}l_{31}l_{22} + T_{22}l_{32}l_{22} \\
&\quad + T_{23}l_{33}l_{22} + T_{31}l_{31}l_{23} + T_{32}l_{32}l_{23} + T_{33}l_{33}l_{23} \\
T'_{31} &= T_{11}l_{11}l_{31} + T_{12}l_{12}l_{31} + T_{13}l_{13}l_{31} + T_{21}l_{11}l_{32} + T_{22}l_{12}l_{32} \\
&\quad + T_{23}l_{13}l_{32} + T_{31}l_{11}l_{33} + T_{32}l_{12}l_{33} + T_{33}l_{13}l_{33} \\
T'_{32} &= T_{11}l_{21}l_{31} + T_{12}l_{22}l_{31} + T_{13}l_{23}l_{31} + T_{21}l_{21}l_{32} + T_{22}l_{22}l_{32} \\
&\quad + T_{23}l_{23}l_{32} + T_{31}l_{21}l_{33} + T_{32}l_{22}l_{33} + T_{33}l_{23}l_{33} \\
T'_{33} &= T_{11}l_{31}^2 + T_{12}l_{32}l_{31} + T_{13}l_{33}l_{31} + T_{21}l_{31}l_{32} + T_{22}l_{32}^2 \\
&\quad + T_{23}l_{33}l_{32} + T_{31}l_{31}l_{33} + T_{32}l_{32}l_{33} + T_{33}l_{33}^2
\end{aligned}\right\} \quad (2.81)$$

For the two-dimensional case, this becomes

$$\begin{pmatrix} T'_{11} & T'_{12} \\ T'_{21} & T'_{22} \end{pmatrix} = \begin{pmatrix} \cos\theta & \sin\theta \\ -\sin\theta & \cos\theta \end{pmatrix} \begin{pmatrix} T_{11} & T_{12} \\ T_{21} & T_{22} \end{pmatrix} \begin{pmatrix} \cos\theta & -\sin\theta \\ \sin\theta & \cos\theta \end{pmatrix} \quad (2.82)$$

which, upon expansion, becomes

$$T' = \begin{pmatrix} \begin{aligned} &T_{11}\cos^2\theta + T_{12}\sin\theta\cos\theta \\ &+ T_{21}\sin\theta\cos\theta + T_{22}\sin^2\theta \end{aligned} & \begin{aligned} &-T_{11}\sin\theta\cos\theta + T_{12}\cos^2\theta \\ &- T_{21}\sin^2\theta + T_{22}\sin\theta\cos\theta \end{aligned} \\ \\ \begin{aligned} &-T_{11}\sin\theta\cos\theta - T_{12}\sin^2\theta \\ &+ T_{21}\cos^2\theta + T_{22}\sin\theta\cos\theta \end{aligned} & \begin{aligned} &T_{11}\sin^2\theta - T_{12}\sin\theta\cos\theta \\ &- T_{21}\sin\theta\cos\theta + T_{22}\cos^2\theta \end{aligned} \end{pmatrix} \quad (2.83)$$

We emphasize that these are perfectly general relations which hold for all tensors of the second order. We shall see later how Eq. 2.83 is the basis of the Mohr circle construction. But before we do this we shall discuss two very important properties of tensors: diagonalization and invariance. These general properties of tensors will be given without proof. The proofs may be found in Refs. 1 and 2.

PROPERTY 1 — DIAGONALIZATION OF SYMMETRICAL TENSORS

Given any symmetrical tensor, A, whose elements in an $oxyz$ system of coordinates are given by

$$A = \begin{pmatrix} a_{11} & a_{12} & a_{13} \\ a_{21} & a_{22} & a_{23} \\ a_{31} & a_{32} & a_{33} \end{pmatrix} \tag{2.84}$$

then there is an orthogonal set of axes $ox'y'z'$ (called the *principal axes*) with respect to which the tensor may be given in diagonal form, as

$$A' = \begin{pmatrix} a'_{11} & 0 & 0 \\ 0 & a'_{22} & 0 \\ 0 & 0 & a'_{33} \end{pmatrix} \tag{2.85}$$

This property holds also for the 2×2 (that is, the two-dimensional) tensor. An example of diagonalization will be presented in the next chapter in connection with the inertia tensor.

PROPERTY 2 — INVARIANTS OF THE TENSOR

For any 3×3 tensor A, whose elements in the $oxyz$ system are given by

$$A = \begin{pmatrix} a_{11} & a_{12} & a_{13} \\ a_{21} & a_{22} & a_{23} \\ a_{31} & a_{32} & a_{33} \end{pmatrix} \tag{2.86}$$

there are three *invariants* (or scalars), which are quantities whose values do not change when the coordinate axes are changed to $ox'y'z'$ (a rotation of axes), for which the tensor becomes

$$A' = \begin{pmatrix} a'_{11} & a'_{12} & a'_{13} \\ a'_{21} & a'_{22} & a'_{23} \\ a'_{31} & a'_{32} & a'_{33} \end{pmatrix} \tag{2.87}$$

These invariants are I_1, the trace:

$$\left.\begin{aligned} I_1 &= a_{11} + a_{22} + a_{33} \\ &= a'_{11} + a'_{22} + a'_{33} \end{aligned}\right\} \tag{2.88}$$

and I_2, the sum of the principal two-rowed minors:

$$\left.\begin{aligned} I_2 &= \begin{vmatrix} a_{22} & a_{23} \\ a_{32} & a_{33} \end{vmatrix} + \begin{vmatrix} a_{11} & a_{13} \\ a_{31} & a_{33} \end{vmatrix} + \begin{vmatrix} a_{11} & a_{12} \\ a_{21} & a_{22} \end{vmatrix} \\ &= \begin{vmatrix} a'_{22} & a'_{23} \\ a'_{32} & a'_{33} \end{vmatrix} + \begin{vmatrix} a'_{11} & a'_{13} \\ a'_{31} & a'_{33} \end{vmatrix} + \begin{vmatrix} a'_{11} & a'_{12} \\ a'_{21} & a'_{22} \end{vmatrix} \end{aligned}\right\} \tag{2.89}$$

I_3, the determinant of the matrix:

$$\left.\begin{aligned} I_3 &= \begin{vmatrix} a_{11} & a_{12} & a_{13} \\ a_{21} & a_{22} & a_{23} \\ a_{31} & a_{32} & a_{33} \end{vmatrix} \\ &= \begin{vmatrix} a'_{11} & a'_{12} & a'_{13} \\ a'_{21} & a'_{22} & a'_{23} \\ a'_{31} & a'_{32} & a'_{33} \end{vmatrix} \end{aligned}\right\} \tag{2.90}$$

For the 2 × 2 tensor, there are two invariants:

$$I_1 = a_{11} + a_{22} = a'_{11} + a'_{22} \tag{2.91}$$

$$I_2 = \begin{vmatrix} a_{11} & a_{12} \\ a_{21} & a_{22} \end{vmatrix} = \begin{vmatrix} a'_{11} & a'_{12} \\ a'_{21} & a'_{22} \end{vmatrix} \tag{2.92}$$

Examples of invariants will be presented in the next chapter in connection with the discussion of the inertia tensor.

The complete discussion given above represents sufficient background in the field of matrix-tensor analysis for our present purposes and is, indeed, a sound foundation for additional advanced work in these areas.

In the next chapter we shall present illustrations of all of the above, by referring to tensors which occur especially in the field of elasticity and structures — the inertia tensor, the stress tensor and the strain tensor. Further discussions of these topics will be found in Refs. 1, 2, 3 and 4.

2-7 Brief Remarks on the Tensor Notations

In connection with the discussion of tensor analysis as presented herein, the following point should be noted. There are, in general usage, two methods for representing tensors,

1. The matrix notation;
2. The indicial-summation convention notation.

In this textbook we shall use method (1) exclusively, since it is the author's experience that the beginning student can visualize and grasp the meaning of tensors more easily when they are presented in this fashion. Both notations are in use; perhaps (1) is used more frequently by engineers and other applied scientists, whereas (2) is used most frequently by mathematicians and physicists.

The student who is interested in a more detailed discussion of the indicial-summation tensor notation will find it in References (10) and (15).

2-8 Finite Difference Method — Introduction

The finite difference method is a very useful technique of applied mathematics which enables one to solve, in an approximate, numerical manner, many differential equations not otherwise solvable. Generally, as will be seen, the finite difference method substitutes for a given differential equation and boundary conditions, a set of simultaneous equations. With the advent of the electronic computer and the consequent ability to solve very easily and quickly simultaneous equations with as many as 100 unknowns, the finite difference method has acquired additional advantages and power. It is now a recognized tool in all fields of engineering science and is especially useful in the field of engineering elasticity.

In this chapter we shall develop the general technique and then apply this to a simplified example. In later sections of the book the method will be used to solve specific problems pertaining to the topics under discussion.

2-9 Outline of the Finite Difference Method

In the finite difference method, we substitute "difference" (i.e., finite)

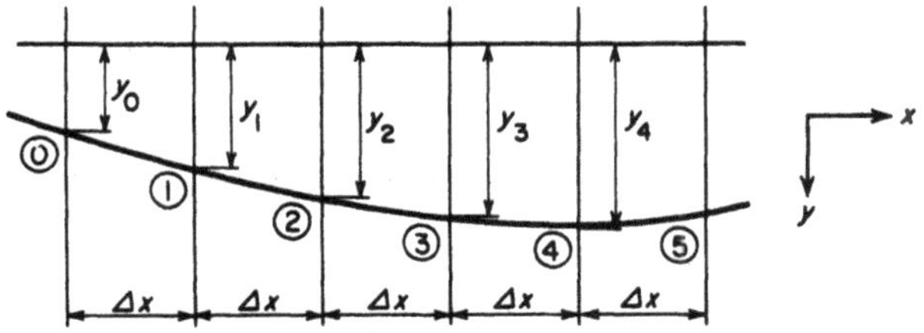

Fig. 2.5

quantities for differential quantities. Thus, in Fig. 2.5 the slope at point 1 , which is given exactly by

$$\text{slope} = \left(\frac{dy}{dx}\right)_1 \tag{2.93}$$

may be approximated by†

$$\text{slope} = \left(\frac{\Delta y}{\Delta x}\right)_1 = \frac{y_2 - y_1}{\Delta x} \tag{2.94}$$

†In this problem we have arbitrarily designated the grid spacing as Δx. In the general finite difference formulation of problems which may contain time as well as space variables, the grid spacing, or interval, in any or all of the coordinates is an extremely important quantity in the following respects:

1. The size of the grid spacing determines the number of equations or relations which must be solved, and hence is a direct measure of the complexity of the finite difference formulation.

2. The size of the grid spacing determines the "stability" of the solution, that is, determines whether the approximate solution (for the particular grid chosen) obtained by substituting *difference* for *differential* quantities is one which will approach the exact solution if the grid spacing is decreased. It is possible, in finite difference (and hence electronic computer) formulation, to so choose the grid intervals that the solution obtained is not a stable one in this sense, and hence is not a solution to the problem being considered. It is generally possible to obtain stable solutions if the intervals are made small enough — but this tends to add to the number of equations and hence to the complexity of the formulation.

The approximation for the second differential quantity, d^2y/dx^2, is obtained in the same way, namely,

$$\left(\frac{d^2y}{dx^2}\right)_1 = \left[\frac{d\left(\frac{dy}{dx}\right)}{dx}\right]_1 = \left[\frac{\Delta\left(\frac{\Delta y}{\Delta x}\right)}{\Delta x}\right]_1 \tag{2.95}$$

$$= \frac{\left(\frac{y_2 - y_1}{\Delta x}\right) - \left(\frac{y_1 - y_0}{\Delta x}\right)}{\Delta x} \tag{2.96}$$

$$= \frac{y_2 - 2y_1 + y_0}{(\Delta x)^2} \tag{2.97}$$

and in general, at points $n - 1$, n and $n + 1$, this gives

$$\left(\frac{d^2y}{dx^2}\right)_n = \frac{y_{n+1} - 2y_n + y_{n-1}}{(\Delta x)^2} \tag{2.98}$$

In exactly the same way, we may obtain the finite difference form for the operator, d^4y/dx^4. In this case we have, at point 2 , for example,

$$\left(\frac{d^4y}{dx^4}\right)_2 = \frac{d^2\left(\frac{d^2y}{dx^2}\right)}{dx^2} = \frac{\Delta\left[\frac{\Delta\left(\frac{\Delta^2y}{\Delta x^2}\right)}{\Delta x}\right]}{\Delta x} \tag{2.99}$$

$$= \frac{\left[\frac{\left(\frac{\Delta^2y}{\Delta x^2}\right)_3 - \left(\frac{\Delta^2y}{\Delta x^2}\right)_2}{\Delta x}\right] - \left[\frac{\left(\frac{\Delta^2y}{\Delta x^2}\right)_2 - \left(\frac{\Delta^2y}{\Delta x^2}\right)_1}{\Delta x}\right]}{\Delta x} \tag{2.100}$$

$$= \frac{(y_4 - 2y_3 + y_2) - (y_3 - 2y_2 + y_1) - (y_3 - 2y_2 + y_1) + (y_2 - 2y_1 + y_0)}{(\Delta x)^4} \tag{2.101}$$

or

$$\left(\frac{d^4y}{dx^4}\right)_2 = \frac{y_4 - 4y_3 + 6y_2 - 4y_1 + y_0}{(\Delta x)^4} \tag{2.102}$$

so that, in general, at any point, n,

$$\left(\frac{d^4y}{dx^4}\right)_n = \frac{y_{n+2} - 4y_{n+1} + 6y_n - 4y_{n-1} + y_{n-2}}{(\Delta x)^4} \tag{2.103}$$

The above relations are the only ones needed for our work. However, the manner in which they were obtained is typical and perfectly general for all differential equations, partial as well as ordinary.

In order to illustrate the application of the method to a simple problem, let us consider the equation

$$\frac{d^2y}{dx^2} = f(x) \quad \text{Valid between } 0 < x < 12, \text{ with boundary conditions}$$

$$y = 0 \begin{cases} \text{at } x = 0 \\ \text{at } x = 12 \end{cases} \tag{2.104}$$

This (as we shall see later) is actually the differential equation for a deflected beam, but for the present it need be considered as simply an ordinary differential equation. Assume further that $f(x)$ is known at all points x. Then we shall solve this differential equation for y at the interior points $x = 3$, $x = 6$ and $x = 9$. To illustrate graphically the procedure which is being used, we refer to Fig. 2.6, on which the values of $f(x)$ are shown at the points indicated.

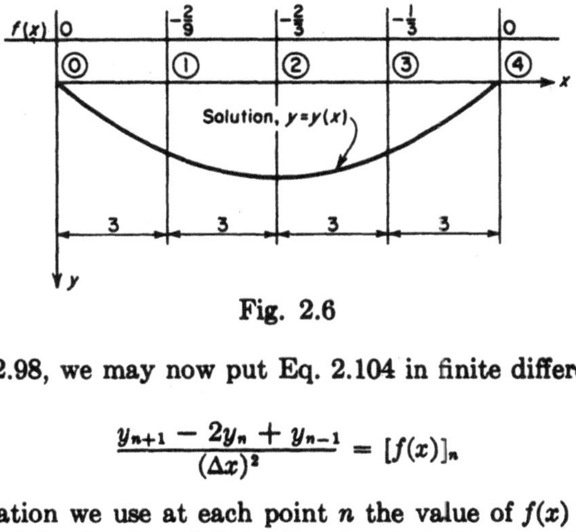

Fig. 2.6

Using Eq. 2.98, we may now put Eq. 2.104 in finite difference form, as follows:

$$\frac{y_{n+1} - 2y_n + y_{n-1}}{(\Delta x)^2} = [f(x)]_n \tag{2.105}$$

In this equation we use at each point n the value of $f(x)$ at this point instead of a weighted value of $f(x)$. See Art. 9.6 for a further discussion of this point.

Then, applying the relation of Eq. 2.105 to the points 1, 2 and 3 in turn, noting that $\Delta x = 3$, we obtain

$$\left.\begin{aligned} y_2 - 2y_1 + y_0 &= -2 \\ y_3 - 2y_2 + y_1 &= -6 \\ y_4 - 2y_3 + y_2 &= -3 \end{aligned}\right\} \tag{2.106}$$

These may be solved for y_1, y_2, and y_3 (noting that $y_0 = y_4 = 0$) without difficulty, giving

$$\left.\begin{aligned} y_1 &= 5\tfrac{1}{4} \\ y_2 &= 8\tfrac{1}{2} \\ y_3 &= 5\tfrac{3}{4} \end{aligned}\right\} \tag{2.107}$$

See Prob. 14 at the end of this chapter for a check on the solution given above.

The above discussion of finite difference methods represents sufficient background for our present purposes. Further discussion of this topic will be found in Refs. 5, 6 and 7.

2-10 Summary

In this chapter we presented introductory treatments of the elements of matrix-tensor algebra and of the finite difference method. In addition, we discussed the three tensors of engineering — the scalar (tensor of zero order), the vector (tensor of first order) and the tensor (tensor of second order) pointing out especially that all three are members of the same family in that all three are defined by reference to their behavior under a rotation of axes.

In the next chapter we shall give physical interpretations of tensors and their behavior by referring especially to the tensors of the mathematical theory of elasticity.

Problems

1. Given

$$A = \begin{pmatrix} 3 & z^2 & y \\ e^x & x & 3y \\ -y & -16 & 0 \end{pmatrix} \qquad B = \begin{pmatrix} 1 & 0 & 0 \\ 1 & 1 & 0 \\ 0 & 1 & 1 \end{pmatrix}$$

determine

(a) $A + B$
(b) $A - B$
(c) $\dfrac{\partial A}{\partial z}$
(d) AB
(e) BA. Is this equal to AB?

2. Show how the three equations,

$$\begin{aligned} x &= 3a + 2b - 4c \\ y &= 6a + 3c \\ z &= \sqrt{2}\, a + 8b - c \end{aligned}$$

can be represented by a matrix equation of the form

$$A = BC$$

3. Obtain the symmetrical and anti-symmetrical parts of

$$\begin{pmatrix} z^2 & e^y & t^x \\ 1 - 3y & x^x & 3 \\ 0 & 0 & 1 \end{pmatrix}$$

4. Given

$$A = \begin{pmatrix} 3 & 0 & 16 \\ 12 & 4 & 2 \\ 15 & 1 & 0 \end{pmatrix} \qquad C = \begin{pmatrix} x \\ y \\ z \end{pmatrix}$$

(a) determine

$$(AC)^*$$

(b) Determine, for the above,

$$C^*A^*$$

Compare the two results (see Eq. 2.41).

5. Given

$$A = \begin{pmatrix} 3 & 2 & 1 \\ 0 & -4 & 3 \\ 8 & 12 & -5 \end{pmatrix}$$

determine A^{-1}.

6. Using the transformation relations for the first order tensor, obtain the components of $\bar{V}$ if the axes are rotated to new positions x'-y', making an angle of 45° with x-y.

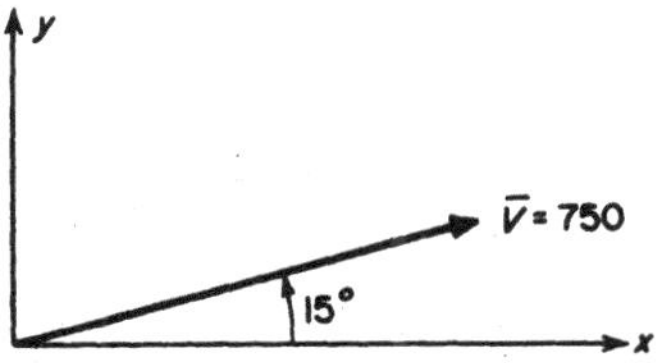

7. For the direction cosine matrix

$$R = \begin{pmatrix} l_{11} & l_{12} & l_{13} \\ l_{21} & l_{22} & l_{23} \\ l_{31} & l_{32} & l_{33} \end{pmatrix}$$

show that

(a) $l_{11}^2 + l_{12}^2 + l_{13}^2 = 1$ and similarly for the other two rows.
(b) $l_{11}^2 + l_{21}^2 + l_{31}^2 = 1$ and similarly for the other two columns.
(c) $l_{11}l_{21} + l_{12}l_{22} + l_{13}l_{23} = 0$ and similarly for the other paired rows.
(d) $l_{11}l_{12} + l_{21}l_{22} + l_{31}l_{32} = 0$ and similarly for the other paired columns.

Thus, $RR^* = E$ and $R^*R = E$.

8. Determine the two invariants for the stress tensor

$$T = \begin{pmatrix} \sigma_x & \tau_{xy} \\ \tau_{yx} & \sigma_y \end{pmatrix}$$

9. The two-dimensional strain tensor is

$$\eta = \begin{pmatrix} \dfrac{\partial u}{\partial x} & \dfrac{1}{2}\left(\dfrac{\partial u}{\partial y} + \dfrac{\partial v}{\partial x}\right) \\ \dfrac{1}{2}\left(\dfrac{\partial v}{\partial x} + \dfrac{\partial u}{\partial y}\right) & \dfrac{\partial v}{\partial y} \end{pmatrix}$$

Determine the two invariants for this tensor.

10. Given a grid as shown in the figure, show that, in finite difference notation, the Laplace equation

$$\nabla^2 w = \frac{\partial^2 w}{\partial x^2} + \frac{\partial^2 w}{\partial y^2}$$

becomes

$$\nabla^2 w_i = \frac{1}{h^2}(w_a + w_b + w_r + w_l - 4w_i)$$

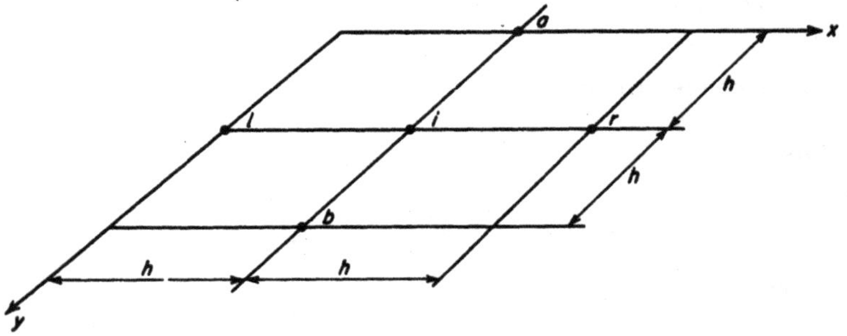

11. Given the grid shown, with equal increments in x and y (both equal to h), show that the *biharmonic* equation

$$\nabla^2\nabla^2 w = \nabla^4 w = \frac{\partial^4 w}{\partial x^4} + \frac{2\partial^4 w}{\partial x^2 \partial y^2} + \frac{\partial^4 w}{\partial y^4}$$

becomes, in finite difference notation,

$$\nabla^4 w_i = \frac{1}{h^4}[(w_{aa} + w_{bb} + w_{ll} + w_{rr}) + 2(w_{ar} + w_{al} + w_{br} + w_{bl}) - 8(w_a + w_b + w_l + w_r) + 20w_i]$$

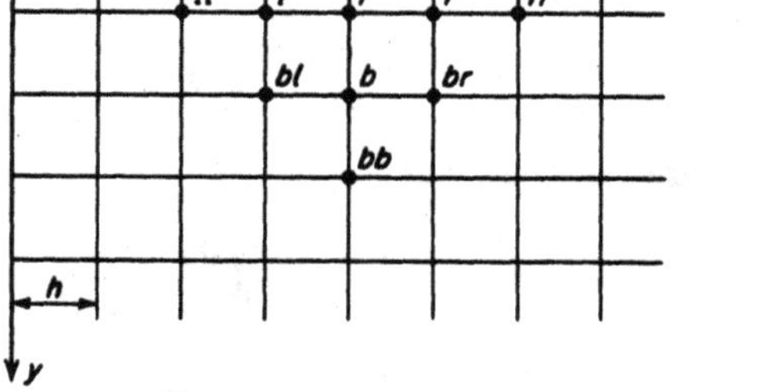

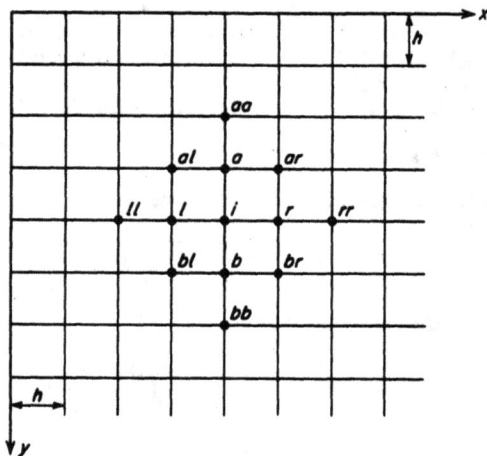

12. Derive the expression for $\frac{d^6 y}{dx^6}$ in finite difference notation. Hint: $\frac{d^6 y}{dx^6} = \frac{d^2}{dx^2}\left(\frac{d^4 y}{dx^4}\right)$ and use the results given in the text.

13. Derive the expression for $\frac{d^8y}{dx^8}$ in finite difference notation. Hint: $\frac{d^8y}{dx^8} = \frac{d^4}{dx^4}\left(\frac{d^4y}{dx^4}\right)$ and use the results given in the text.

14. Refer to the illustrative problem on the finite difference method, Eq. 2.104. To check the solution obtained in Eq. 2.107, assume

$$f(x) = c_1x + c_2x^2 + c_3x^3$$

and, using the values of $f(x)$ at $x = 3$, $x = 6$, $x = 9$, determine c_1, c_2 and c_3 in the above equation. Using this value of $f(x)$ in Eq. 2.104, obtain the corresponding solution $y(x)$ of the differential equation. Having this $y = y(x)$, check the solution obtained by the finite difference method as given in Eq. 2.107.

15. (a) If T_1 is a tensor (i.e., $T_1 = RT_1R^*$ and T_2 is a tensor), prove that $T_1 + T_2$ is a tensor. Hint: show $T_1 + T_2 = R(T_1 + T_2)R^*$.
(b) If A is a tensor and c is a constant, show that cA is a tensor.
(c) If B is a tensor, show that B^* is a tensor. Hint: use the results of Prob. 7.
(d) If C is a tensor and D is a tensor, show that CD is a tensor. Hint: use the results of Prob. 7.
(e) If G is a tensor, show that the symmetrical and the antisymmetrical parts of G are each a tensor.

16. For the square region shown in Prob. 10, with the origin at the center, obtain the finite difference solution to the Poisson equation $\nabla^2 w = 2$, subject to the boundary condition $w = 0$ everywhere on the boundary.

17. For the square region shown in Prob. 10, with origin as shown, obtain the finite difference solution to the Laplace equation $\nabla^2 w = 0$, subject to the boundary condition $w = xy + 3$ everywhere on the boundary.

18. Obtain the finite difference form for the Laplace equation $\nabla^2 w = 0$ for the rectangular net shown.

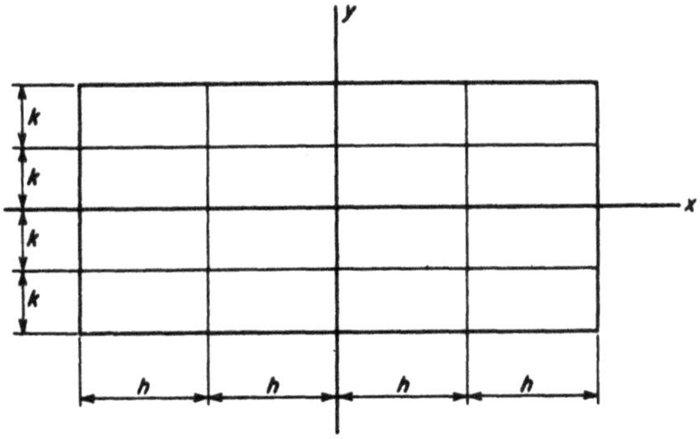

Chapter 3

THE TENSORS OF ELASTICITY — THE INERTIA, STRAIN AND STRESS TENSORS

3-1 Introduction

In this chapter we shall discuss in some detail the tensors which appear in the mathematical theory of elasticity — the inertia tensor, the strain tensor and the stress tensor. We shall emphasize the transformation, diagonalization and invariance properties of these tensors and shall discuss the graphical solution for the transformation relations for the two-dimensional tensor — the Mohr's circle. It will be emphasized that, although our illustration of this graphical solution will be given in connection with the inertia tensor, it holds for *all* tensors of the second order because it represents a solution of a general property of these tensors.

3-2 The Inertia Tensor

Let us consider the following expression, which occurs repeatedly in many problems in dynamics, and (as we shall see) also occurs in the present field of engineering elasticity:

$$I = r^2E - r^*r \tag{3.1}$$

in which

$$r = (x \quad y), \text{ the distance vector} \tag{3.2}$$

$$r^2 = x^2 + y^2, \text{ a scalar} \tag{3.3}$$

$$E = \text{unit matrix} \tag{3.4}$$

$$= \begin{pmatrix} 1 & 0 \\ 0 & 1 \end{pmatrix} \tag{3.5}$$

If we expand the above in the usual way, we find that

$$I = \begin{pmatrix} yy & -xy \\ -yx & xx \end{pmatrix} \tag{3.6}$$

See Fig. 3.1.

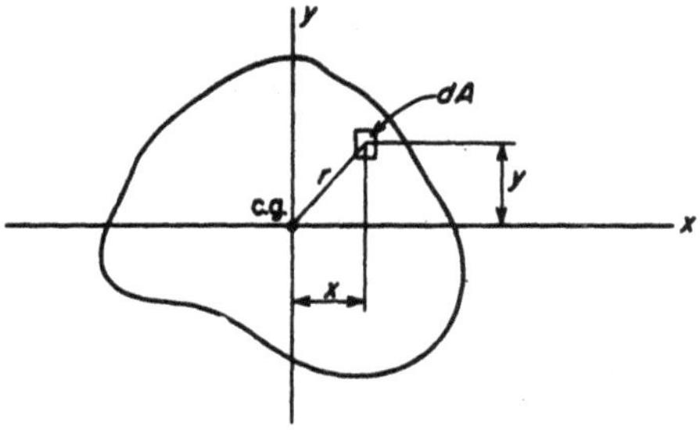

Fig. 3.1

If we multiply the right-hand side of Eq. 3.6 by dA and integrate over the entire area, then each element represents a moment of inertia (the off-diagonal elements being the "products of inertia") and we have the very important second order tensor, the *inertia tensor*, given by

$$I = \begin{pmatrix} I_{xx} & -I_{xy} \\ -I_{yx} & I_{yy} \end{pmatrix} \tag{3.7}$$

In this expression,

I_{xx} = area moment of inertia with respect to x-x axes.

I_{yy} = area moment of inertia with respect to y-y axes.

$I_{xy} = I_{yx}$ = area product of inertia with respect to x-y axes.

Note that I_{xx} and I_{yy} are *always* positive. $I_{xy} = I_{yx}$ may be positive or negative.

To illustrate a simple moment of inertia calculation, let us determine I_{xx} and I_{yy} for the rectangle about its centroidal axis; see Fig. 3.2.

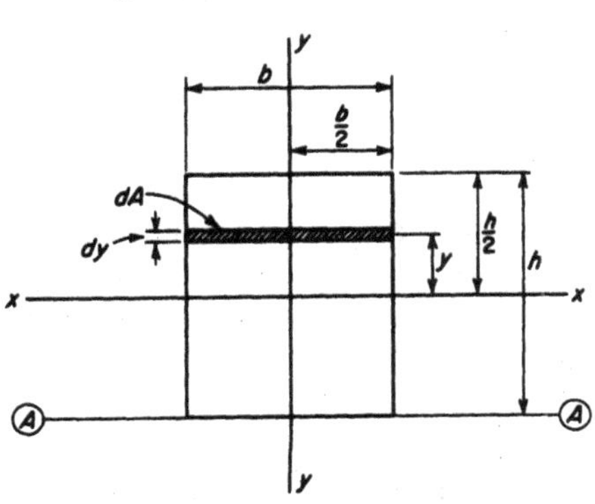

Fig. 3.2

Consider I_{xx} first. We have

$$I_{xx} = \int_{-h/2}^{+h/2} y^2 dA \tag{3.8}$$

$$= \int_{-h/2}^{+h/2} y^2 b dy \tag{3.9}$$

$$I_{xx} = \frac{bh^3}{12} \tag{3.10}$$

In the same way, we would find

$$I_{yy} = \int_{-b/2}^{+b/2} x^2 dA \tag{3.11}$$

$$= \int_{-b/2}^{+b/2} x^2 h dx \tag{3.12}$$

$$I_{yy} = \frac{hb^3}{12} \tag{3.13}$$

Also, it is obvious that $I_{xy} = 0$, since for every positive contribution to this term, there is an equal negative one.

If we wish we can determine I_{AA} where A-A is an axis through the base. In this case we have

$$I_{AA} = \int_0^h y^2 dA \tag{3.14}$$

$$= \int_0^h y^2 b dy \tag{3.15}$$

$$I_{AA} = \frac{bh^3}{3} \tag{3.16}$$

There are other moments of inertia besides *area moments of inertia*. For example, in certain fields the concept of *line moment of inertia* is important. This is essentially a one-dimensional counterpart of the area moment of inertia. Also we may speak of *mass moments of inertia*, which is a three-dimensional form of the inertia tensor.

In the example cited above we have I_{xx} and I_{yy}, the moments of inertia with respect to the x- and y-axes respectively. It is also possible to define a moment of inertia, I_r, the so-called *polar moment* of inertia, which is the moment of inertia with respect to an axis through the origin and normal to the area, that is,

$$I_r = \int_A r^2 dA \tag{3.17}$$

$$= \int_A (x^2 + y^2) dA \tag{3.18}$$

$$= \int_A x^2 dA + \int_A y^2 dA \tag{3.19}$$

or, in general,

$$I_r = I_{yy} + I_{xx} \tag{3.20}$$

Also, we may show that if x and y are the centroidal axes, then, given an axes x_1 parallel to x and at, say, a distance d from it (see Fig. 3.3), then

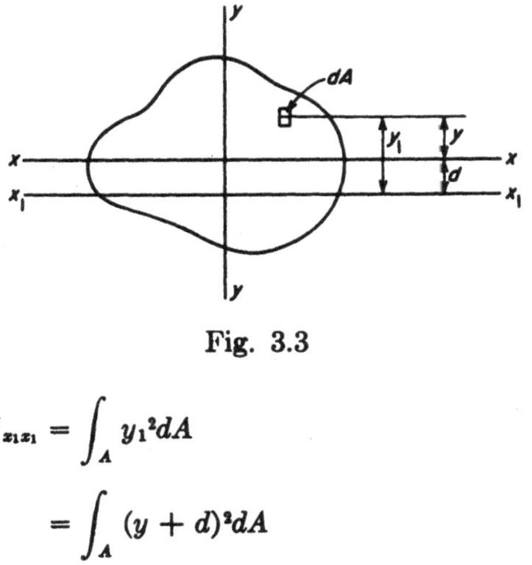

Fig. 3.3

$$I_{x_1x_1} = \int_A y_1^2 dA \tag{3.21}$$

$$= \int_A (y + d)^2 dA \tag{3.22}$$

$$= \int_A y^2 dA + 2d \int_A y dA + d^2 \int_A dA \tag{3.23}$$

and since the x-axis is a centroidal axis, the middle integral equals zero, or,

$$I_{x_1x_1} = I_{xx} + Ad^2 \tag{3.24}$$

which is the very important *parallel axis theorem*, stated in words, as follows:

"The moment of inertia of an area with respect to any axis exceeds its moment of inertia with respect to a parallel axis drawn through its centroid by the product of the area by the square of the distance between the parallel axes."

The student should verify the theorem for the rectangle example given earlier in the chapter.

A similar parallel axis theorem may be given for lines and for volumes, and also for products of inertia.

Going back to the tensor representation of the inertia, it will be instructive to determine I', the above tensor in an $ox'y'$ system which makes an angle θ with the oxy system. This is given at once (see Eq. 2.83) by

$$I' = \begin{pmatrix} I_{x'x'} & -I_{x'y'} \\ -I_{y'x'} & I_{y'y'} \end{pmatrix} \tag{3.25}$$

$$I' = \begin{pmatrix} I_{xx}\cos^2\theta + I_{yy}\sin^2\theta - 2I_{xy}\sin\theta\cos\theta & (I_{yy} - I_{xx})\sin\theta\cos\theta - (\cos^2\theta - \sin^2\theta)I_{xy} \\ (I_{yy} - I_{xx})\sin\theta\cos\theta - (\cos^2\theta - \sin^2\theta)I_{xy} & I_{xx}\sin^2\theta + I_{yy}\cos^2\theta + 2I_{xy}\sin\theta\cos\theta \end{pmatrix} \tag{3.26}$$

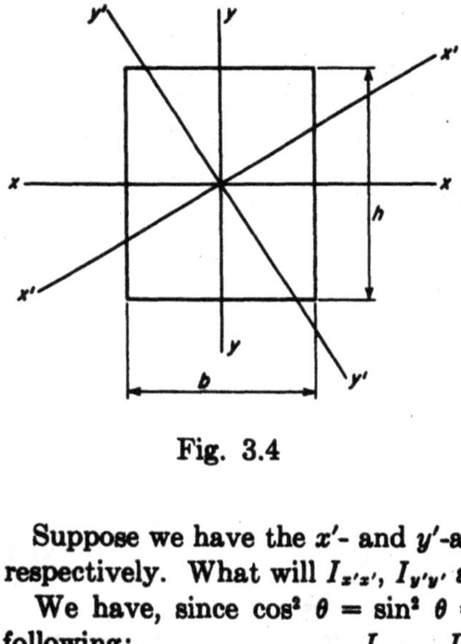

Fig. 3.4

Equation 3.26 represents the rotation of axes expressions for the area moment of inertia.

For example, for the rectangle of Fig. 3.2, repeated in Fig. 3.4, we had about the centroidal x- and y-axes,

$$I_{xx} = \frac{bh^3}{12} \tag{3.27}$$

$$I_{yy} = \frac{hb^3}{12} \tag{3.28}$$

$$I_{xy} = 0 \tag{3.29}$$

Suppose we have the x'- and y'-axes making 45° with the x- and y-axes, respectively. What will $I_{x'x'}$, $I_{y'y'}$ and $I_{x'y'}$ be given by?

We have, since $\cos^2\theta = \sin^2\theta = \sin\theta\cos\theta = \frac{1}{2}$, from Eq. 3.26, the following:

$$I_{x'x'} = \frac{I_{xx}}{2} + \frac{I_{yy}}{2} - I_{xy} = \frac{bh^3}{24} + \frac{hb^3}{24} \tag{3.30}$$

$$I_{y'y'} = \frac{I_{xx}}{2} + \frac{I_{yy}}{2} + I_{xy} = \frac{bh^3}{24} + \frac{hb^3}{24} \tag{3.31}$$

$$-I_{x'y'} = \frac{I_{yy} - I_{xx}}{2} = \frac{hb^3}{24} - \frac{bh^3}{24} \tag{3.32}$$

Let us now determine the invariants of the area moment of inertia tensor.

These are given at once (see Eqs. 2.91 and 2.92) by

$$\mathcal{I}_1 = I_{xx} + I_{yy} = I_{x'x'} + I_{y'y'} \tag{3.33}$$

and

$$\mathcal{I}_2 = I_{xx}I_{yy} - I^2_{xy} = I_{x'x'}I_{y'y'} - I^2_{x'y'} \tag{3.34}$$

The invariant $\mathcal{I}_1$ is usually discussed in elementary courses of statics. The invariant $\mathcal{I}_2$ is rarely mentioned; however, it is an invariant just as $\mathcal{I}_1$ is, and in some ways it is of more interest, particularly since it is an invariant which contains all elements of the inertia tensor.

We may verify the invariance relations for the rectangle considered above as follows:

$$I_{xx} + I_{yy} = \frac{bh^3}{12} + \frac{hb^3}{12} \tag{3.35}$$

and (from Eqs. 3.30 and 3.31)

$$I_{x'x'} + I_{y'y'} = \frac{bh^3}{12} + \frac{hb^3}{12} \tag{3.36}$$

which verifies the relation for $\mathcal{I}_1$

For $\mathcal{I}_2$ we have

$$I_{xx}I_{yy} - I^2_{xy} = \frac{b^4h^4}{144} \tag{3.37}$$

and

$$I_{x'x'}I_{y'y'} - I^2_{x'y'} = \left(\frac{bh^3}{24} + \frac{hb^3}{24}\right)^2 - \left(\frac{bh^3}{24} - \frac{hb^3}{24}\right)^2 \tag{3.38}$$

$$= \frac{b^4h^4}{144} \tag{3.39}$$

which verifies the relation for $\mathcal{I}_2$.

We may note also that since the tensor of Eq. 3.7 is a symmetric one, then by virtue of the theorem stated in Art. 2-6, it follows that this tensor may be put in diagonal form, that is, there is a set of axes, ox_py_p, the *principal axes*, such that the tensor becomes

$$I_p = \begin{pmatrix} I_{x_px_p} & 0 \\ 0 & I_{y_py_p} \end{pmatrix} \tag{3.40}$$

Thus, the products of inertia are zero about the principal axes.

It is clear, therefore, that for the rectangle of Fig. 3.2, the x- and y-axes are the principal axes. In fact, it must follow that axes of symmetry are *always* principal axes, since for an axis of symmetry I_{xy} will equal zero.

Also, we may show that $I_{x_px_p}$ and $I_{y_py_p}$ are either the maximum or minimum values of the moment of inertia about any set of orthogonal axes through 0. This is done as follows: assume $I_{xx} > I_{yy}$. This is no real restriction, since we can always label the axes so that this is so.† We have, by virtue of the invariance relations,

$$I_{x_px_p} + I_{y_py_p} = I_{xx} + I_{yy} \tag{3.41}$$

$$I_{x_px_p}I_{y_py_p} = I_{xx}I_{yy} - I^2_{xy} \tag{3.42}$$

where the left-hand sides are the principal axes values and the right-hand sides are the values for any other set of orthogonal axes through the origin. Then,

$$I_{x_px_p} = I_{xx} + I_{yy} - I_{y_py_p} \tag{3.43}$$

and, substituting this in the second of the above, we have

$$I^2_{y_py_p} - I_{y_py_p}(I_{xx} + I_{yy}) + (I_{xx}I_{yy} - I^2_{xy}) = 0 \tag{3.44}$$

Solving for $I_{y_py_p}$, we have

$$I_{y_py_p} = \frac{+(I_{xx} + I_{yy}) \pm \sqrt{(I_{xx} + I_{yy})^2 - 4(I_{xx}I_{yy} - I^2_{xy})}}{2} \tag{3.45}$$

$$= \frac{+(I_{xx} + I_{yy}) \pm [(I_{xx} - I_{yy}) + \epsilon]}{2} \tag{3.46}$$

in which ϵ is some positive quantity.
Hence

$$I_{y_py_p} = \begin{cases} I_{xx} + \dfrac{\epsilon}{2} \\ \text{or} \\ I_{yy} - \dfrac{\epsilon}{2} \end{cases} \tag{3.47}$$

In other words, $I_{y_py_p}$ is either greater or less than the moment of inertia about any other set of axes. This may also be demonstrated for $I_{x_px_p}$ in the same way. Hence, in order for the equality of the first invariance to be valid, it follows that $I_{x_px_p}$ and $I_{y_py_p}$ are the $\left\{\begin{matrix}\text{maximum}\\ \text{minimum}\end{matrix}\right\}$ or $\left\{\begin{matrix}\text{minimum}\\ \text{maximum}\end{matrix}\right\}$ values of the moment of inertia about any set of orthogonal axes through the origin.

Equation 3.26 has a most suggestive form. Whenever equations are given in terms of the circular functions (sine, cosine, etc.) one is led to consider the possibility of constructing a graphical solution of the equations using circular arcs. Such a construction does in fact exist for these equations and, therefore, exists also for all other tensors of the second order, in two dimensions. This construction is the Mohr circle, which we describe next.

We consider Eq. 3.26 and assume that for a given set of rectangular axes the moments of inertia I_{xx}, I_{yy} and I_{xy} are known. Then for any other rectangular set of axes making an angle θ with the initial set we have the

†If $I_{xx} = I_{yy}$ for a particular set of axes, the development which follows still holds. If $I_{xx} = I_{yy}$ for *all* axes, then obviously $I_{xx} = I_{yy} = I_{x_px_p} = I_{y_py_p}$, and this also follows from the analysis given above.

primed quantities as given by Eqs. 3.25 and 3.26. From these equations we obtain

$$I_{x'x'} = \frac{I_{xx} + I_{yy}}{2} + \frac{I_{xx} - I_{yy}}{2}\cos 2\theta - I_{xy}\sin 2\theta \tag{3.48}$$

$$I_{x'y'} = \frac{I_{xx} - I_{yy}}{2}\sin 2\theta + I_{xy}\cos 2\theta \tag{3.49}$$

or

$$\left(I_{x'x'} - \frac{I_{xx} + I_{yy}}{2}\right)^2 + I^2_{x'y'} = \left(\frac{I_{xx} - I_{yy}}{2}\right)^2 + I^2_{xy} \tag{3.50}$$

$$= \text{constant for any } \theta\dagger \tag{3.51}$$

But, Eq. 3.50 is the equation of a circle in an I plane, its center at $\left(\frac{I_{xx} + I_{yy}}{2}, 0\right)$, of radius equal to $\sqrt{\left(\frac{I_{xx} - I_{yy}}{2}\right)^2 + I^2_{xy}}$. This circle may most easily be constructed as shown in Fig. 3.5.

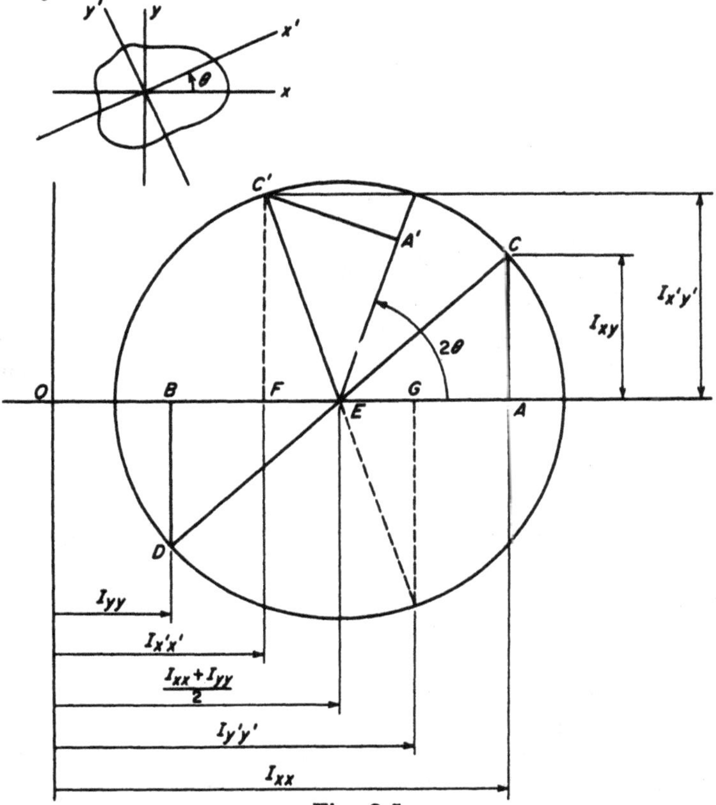

Fig. 3.5

†The student may verify this constancy with θ by considering the invariants of the tensor. See also Prob. 9 at the end of this chapter.

We choose the x direction such that $I_{xx} > I_{yy}$. Lay off abscissae equal to $I_{xx}(= 0A)$ and $I_{yy}(= 0B)$. Lay off ordinates (assumed positive) $I_{xy}(= AC, BD)$ as shown. From center E draw a circle through C and D. This is the Mohr circle.

Now consider ΔECA as a movable indicator and to find $I_{x'x'}$, $I_{y'y'}$, $I_{x'y'}$ at the angle θ, rotate ECA through 2θ in the same direction as θ to position $EA'C'$. Then $0F = I_{x'x'}$, $0G = I_{y'y'}$ and $C'F = I_{x'y'}$. In this way, a complete graphical solution of the inertia (and/other) tensors may be obtained. For example, the student should verify the following general relations, by referring to the Mohr circle:

1. Principal moments of inertia always occur on planes perpendicular to each other

2. $$(I_p)_{max} = \text{the maximum principal } I,$$

$$= \frac{I_{xx} + I_{yy}}{2} + (I_{xy})_{max} \tag{3.52}$$

and $$(I_p)_{min} = \frac{I_{xx} + I_{yy}}{2} - (I_{xy})_{max} \tag{3.53}$$

3. The plane of $(I_{xy})_{max}$ is at 45° to the principal planes.

In a similar way, other general relations concerning the elements of the second order tensor can be obtained from the Mohr circle.

We shall now discuss two other tensors of the second order which are of fundamental importance in the theory of elasticity and hence also in the engineering applications of elasticity. These are the strain tensor and the stress tensor.

3-3 The Strain Tensor

In this section we shall derive the expressions for the elements of the strain tensor. In our derivation, a basic postulate will be the assumption that the elongations or shortenings or other deformations are *small*. This assumption (and the companion assumption that gradiants or rates of change of deformations are also small) is basic to the so-called *small deformation theory of elasticity* with which we are concerned in this text.

The assumption that the deformations (or gradiants of deformations) are small means, mathematically, that the square of either of these quantities is negligibly small when compared to the first power of the quantities. Another way of stating this is that we are interested only in the *linear* strain terms, and for this reason, the theory of elasticity which is based upon this assumption is frequently called the *linearized theory of elasticity*.

Although in this text we shall be concerned only with the linearized terms of the strain tensor, for the interested reader we shall, in the next section, show the complete non-linearized form of the tensor. The derivation will not be given here. It can be found in Refs. 5, 10 and 15.

We are concerned in elasticity with deformable bodies, i.e., bodies which elongate or shorten or otherwise deform under the application of forces, loads or other effects. If we consider the elemental cube shown in Fig. 3.6, we see that there are basically two types of deformation to which this body can be subjected in that any deformation, no matter how complicated, can be given as a combination of these. They are

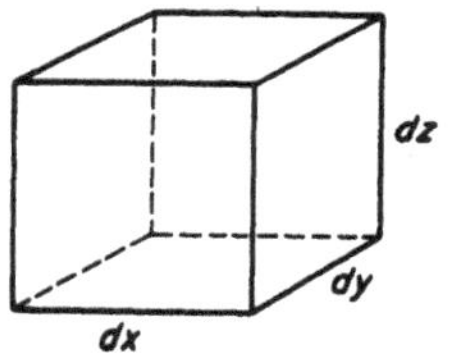

Fig. 3.6

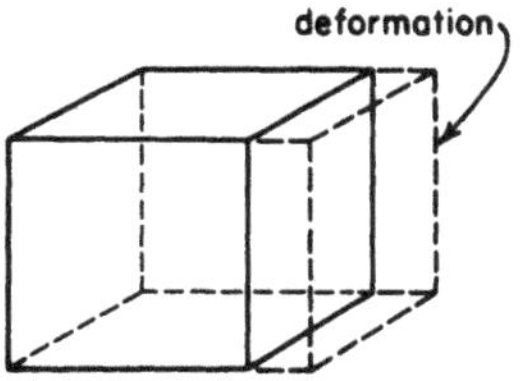

Fig. 3.7

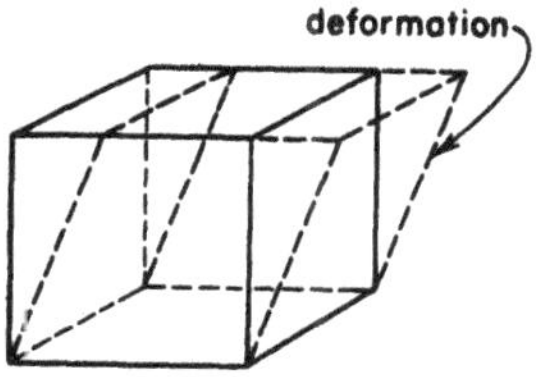

Fig. 3.8

1. A pure elongation or contraction; see Fig. 3.7.
2. A pure sliding or shearing action; see Fig. 3.8.

As a physical example of the first type of deformation, consider a bar subjected to a simple tension load; see Fig. 3.9.

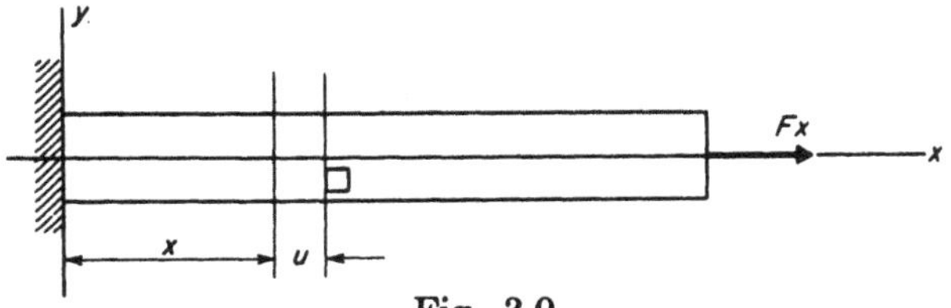

Fig. 3.9

Consider conditions at a plane originally at a distance x from the left end, which is fixed in a wall. At this point the deformation is u. Now let us consider a differential element of the body at this point originally at x.

We obtain a measure of this first type of deformation as follows: we consider the projection of the element on an x-y plane and consider first only deformations in the x direction. The deformations in the y and z directions then follow directly from this; u is the deformation in the x direction of the face closest to the origin. Then, since the deformation changes over the length, dx, of the element, we have for the far face, a movement given by $u + (\partial u/\partial x)\, dx$.

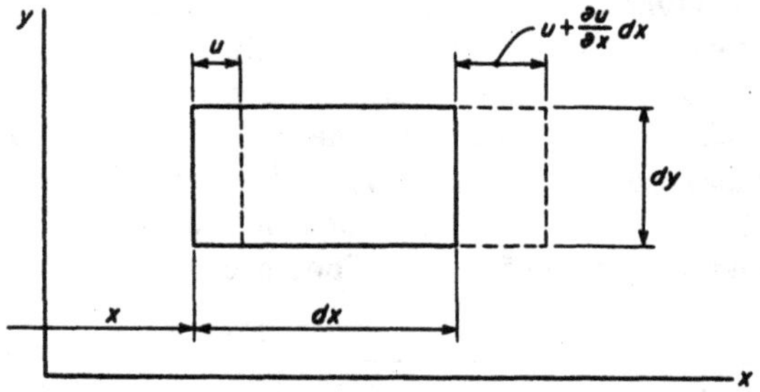

Fig. 3.10

This means that the

$$\text{net total elongation of element} = \left(u + \frac{\partial u}{\partial x}\, dx\right) - u \tag{3.54}$$

$$= \frac{\partial u}{\partial x}\, dx \tag{3.55}$$

and also, that

$$\text{unit elongation} = \frac{\text{total elongation}}{\text{original length}} \tag{3.56}$$

$$= \frac{\frac{\partial u}{\partial x}\, dx}{dx} \tag{3.57}$$

$$= \frac{\partial u}{\partial x} \tag{3.58}$$

In exactly the same way, we can find the unit elongations (or contractions) in the y and z directions. Thus, if v is the movement in the y direction, and w is the movement in the z direction, then

$$\text{unit elongation in the } y \text{ direction} = \frac{\partial v}{\partial y} \tag{3.59}$$

$$\text{unit elongation in the } z \text{ direction} = \frac{\partial w}{\partial z} \tag{3.60}$$

Note: although the word "elongation" is used in the above expressions, the statements also hold for "contractions", in which case the right-hand sides of Eqs. 3.59 and 3.60 are simply negative quantities.

Now let us consider the sliding or shearing deformations, and let us obtain a measure of this quantity. Once again we consider first a projection on the x-y plane of the $dx\,dy\,dz$ body, undergoing a sliding or shearing deformation; see Fig. 3.11.

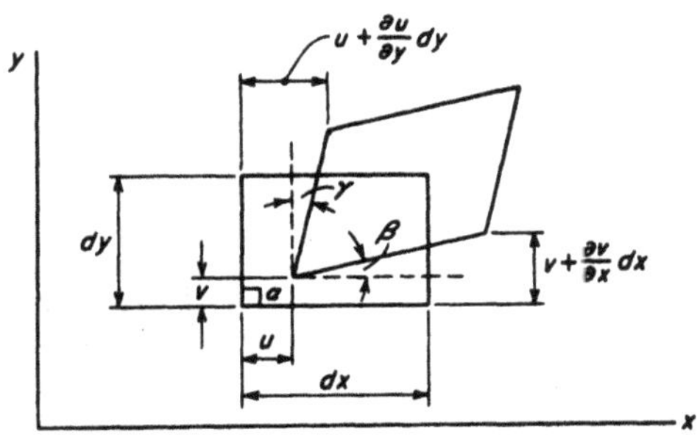

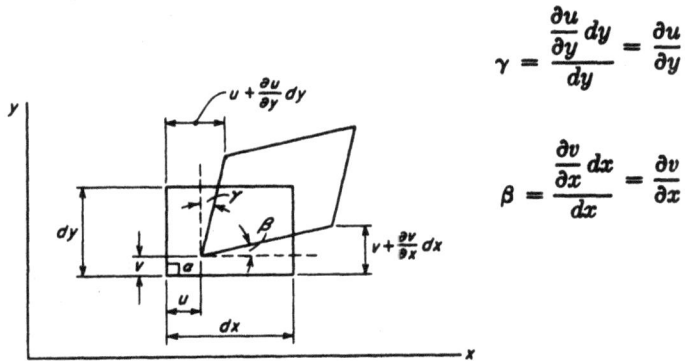

$$\gamma = \frac{\frac{\partial u}{\partial y}\,dy}{dy} = \frac{\partial u}{\partial y}$$

$$\beta = \frac{\frac{\partial v}{\partial x}\,dx}{dx} = \frac{\partial v}{\partial x}$$

Fig. 3.11

We take as a measure of the shearing or sliding deformation the change in the original right angle, α, of the element:

$$\text{shearing unit deformation in the } x\text{-}y \text{ plane} = \beta + \gamma \qquad (3.61)$$

$$= \frac{\partial v}{\partial x} + \frac{\partial u}{\partial y} \qquad (3.62)$$

In the same way, we have for the y-z and z-x planes,

$$\text{shearing unit deformation in the } y\text{-}z \text{ plane} = \frac{\partial w}{\partial y} + \frac{\partial v}{\partial z} \qquad (3.63)$$

$$\text{and shearing unit deformation in the } z\text{-}x \text{ plane} = \frac{\partial u}{\partial z} + \frac{\partial w}{\partial x} \qquad (3.64)$$

We repeat, for convenience, the six fundamental unit deformations of the theory of elasticity; we add, furthermore, some of the common European and American notations for these:

u = deformation in x direction
v = deformation in y direction
w = deformation in z direction

Unit elongation or contraction in x direction	$\frac{\partial u}{\partial x}$	e_x	e_{11}	e_{xx}
Unit elongation or contraction in y direction	$\frac{\partial v}{\partial y}$	e_y	e_{22}	e_{yy}
Unit elongation or contraction in z direction	$\frac{\partial w}{\partial z}$	e_z	e_{33}	e_{zz}
Unit shearing strain in x-y plane	$\frac{\partial u}{\partial y} + \frac{\partial v}{\partial x}$	γ_{xy}	e_{12}	e_{xy}
Unit shearing strain in y-z plane	$\frac{\partial v}{\partial z} + \frac{\partial w}{\partial y}$	γ_{yz}	e_{23}	e_{yz}
Unit shearing strain in z-x plane	$\frac{\partial w}{\partial x} + \frac{\partial u}{\partial z}$	γ_{zx}	e_{31}	e_{zx}
		American practice: see Ref. 9	Russian practice: see V.V. Novozhilov, *The Theory of Thin Shells*, P. Noordhoff, Groningen, The Netherlands, 1959	British practice: see Ref. 8

Now, it may be shown (see Refs. 4 and 10) that these unit strains are connected with the elements of the "strain tensor," which we shall label η, as follows:

$$\eta = \begin{pmatrix} \eta_{xx} & \eta_{xy} & \eta_{xz} \\ \eta_{yx} & \eta_{yy} & \eta_{yz} \\ \eta_{zx} & \eta_{zy} & \eta_{zz} \end{pmatrix} = \begin{pmatrix} \frac{\partial u}{\partial x} & \frac{1}{2}\left(\frac{\partial u}{\partial y} + \frac{\partial v}{\partial x}\right) & \frac{1}{2}\left(\frac{\partial u}{\partial z} + \frac{\partial w}{\partial x}\right) \\ \frac{1}{2}\left(\frac{\partial v}{\partial x} + \frac{\partial u}{\partial y}\right) & \frac{\partial v}{\partial y} & \frac{1}{2}\left(\frac{\partial v}{\partial z} + \frac{\partial w}{\partial y}\right) \\ \frac{1}{2}\left(\frac{\partial w}{\partial x} + \frac{\partial u}{\partial z}\right) & \frac{1}{2}\left(\frac{\partial w}{\partial y} + \frac{\partial v}{\partial z}\right) & \frac{\partial w}{\partial z} \end{pmatrix} \tag{3.65}$$

Note especially, that the off-diagonal elements of the strain tensor are *not* the shearing strains as defined above, but rather are one-half these strains. The $\frac{1}{2}$ values are essential — without these, the matrix is not a tensor, i.e., does not follow the required transformation law.

Note also that this tensor is a symmetrical one. Hence it may be put in diagonal (i.e., principal axis) form.

The two-dimensional form of the strain tensor, for the x-y system becomes (also shown is the American notation, which we will use in this text)

$$\begin{pmatrix} \eta_{xx} & \eta_{xy} \\ \eta_{yx} & \eta_{yy} \end{pmatrix} = \begin{pmatrix} e_x & \frac{1}{2}\gamma_{xy} \\ \frac{1}{2}\gamma_{yx} & e_y \end{pmatrix} = \begin{pmatrix} \frac{\partial u}{\partial x} & \frac{1}{2}\left(\frac{\partial u}{\partial y} + \frac{\partial v}{\partial x}\right) \\ \frac{1}{2}\left(\frac{\partial v}{\partial x} + \frac{\partial u}{\partial y}\right) & \frac{\partial v}{\partial y} \end{pmatrix} \tag{3.66}$$

Finally, as noted previously, because this is a tensor of the second order, it satisfies the transformation laws for this tensor (Eq. 2.83); furthermore, the Mohr circle construction of Fig. 3.5 applies to the elements of this tensor.

3-4 The Complete Non-Linear Strain Tensor

The non-linearized form of the strain tensor (i.e., the tensor containing quadratic terms as well as linear terms) is given by

$$\eta = \begin{pmatrix} \frac{\partial u}{\partial x} & \frac{1}{2}\left(\frac{\partial v}{\partial x} + \frac{\partial u}{\partial y}\right) & \frac{1}{2}\left(\frac{\partial w}{\partial x} + \frac{\partial u}{\partial z}\right) \\ \frac{1}{2}\left(\frac{\partial u}{\partial y} + \frac{\partial v}{\partial x}\right) & \frac{\partial v}{\partial y} & \frac{1}{2}\left(\frac{\partial w}{\partial y} + \frac{\partial v}{\partial z}\right) \\ \frac{1}{2}\left(\frac{\partial u}{\partial z} + \frac{\partial w}{\partial x}\right) & \frac{1}{2}\left(\frac{\partial v}{\partial z} + \frac{\partial w}{\partial y}\right) & \frac{\partial w}{\partial z} \end{pmatrix} +$$

$$\frac{1}{2}\begin{pmatrix} \left(\frac{\partial u}{\partial x}\right)^2 + \left(\frac{\partial v}{\partial x}\right)^2 + \left(\frac{\partial w}{\partial x}\right)^2 & \frac{\partial u}{\partial x}\frac{\partial u}{\partial y} + \frac{\partial v}{\partial x}\frac{\partial v}{\partial y} + \frac{\partial w}{\partial x}\frac{\partial w}{\partial y} & \frac{\partial u}{\partial x}\frac{\partial u}{\partial z} + \frac{\partial v}{\partial x}\frac{\partial v}{\partial z} + \frac{\partial w}{\partial x}\frac{\partial w}{\partial z} \\ \frac{\partial u}{\partial x}\frac{\partial u}{\partial y} + \frac{\partial v}{\partial x}\frac{\partial v}{\partial y} + \frac{\partial w}{\partial x}\frac{\partial w}{\partial y} & \left(\frac{\partial u}{\partial y}\right)^2 + \left(\frac{\partial v}{\partial y}\right)^2 + \left(\frac{\partial w}{\partial y}\right)^2 & \frac{\partial u}{\partial y}\frac{\partial u}{\partial z} + \frac{\partial v}{\partial y}\frac{\partial v}{\partial z} + \frac{\partial w}{\partial y}\frac{\partial w}{\partial z} \\ \frac{\partial u}{\partial x}\frac{\partial u}{\partial z} + \frac{\partial v}{\partial x}\frac{\partial v}{\partial z} + \frac{\partial w}{\partial x}\frac{\partial w}{\partial z} & \frac{\partial u}{\partial y}\frac{\partial u}{\partial z} + \frac{\partial v}{\partial y}\frac{\partial v}{\partial z} + \frac{\partial w}{\partial y}\frac{\partial w}{\partial z} & \left(\frac{\partial u}{\partial z}\right)^2 + \left(\frac{\partial v}{\partial z}\right)^2 + \left(\frac{\partial w}{\partial z}\right)^2 \end{pmatrix} \tag{3.67}$$

As pointed out in Art. 3-3, we shall neglect the quadratic (or higher order terms) since, for small deformations, these are negligible in comparison to the linear or first order terms.

3-5 The Stress Tensor

The last of the fundamental tensors of elasticity which we shall consider is the stress tensor. We define a stress as the force per unit area, and on the basis of the pure elongation (or shortening) strains and sliding (shearing) strains, we look for related forces and hence stresses. Thus, a type of force which would cause a pure elongation is the one shown in Fig. 3.9, a small portion of which is shown below:†

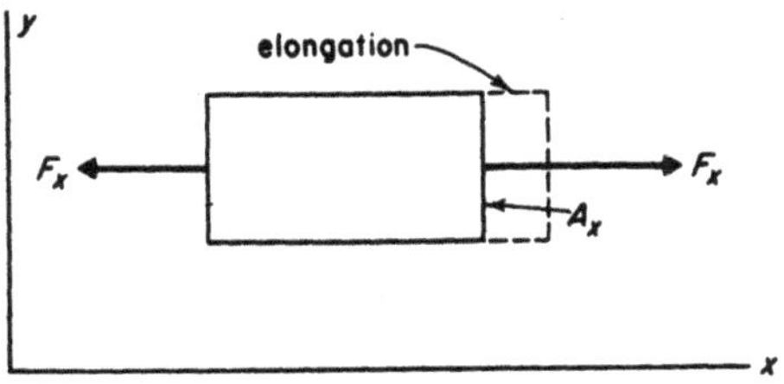

Fig. 3.12

†In this figure, it is assumed that the force F_x is applied at the centroid of the cross section, so that the stress σ_x is a uniform stress. Also, in the figure we do not show the lateral deformation which occurs due to the Poisson ratio effect. See Art. 4-5.

This is a force normal to the area, A_x, hence it is a normal force, and this leads to a normal stress, σ_x, defined by

$$\sigma_x = \frac{F_x}{A_x} \tag{3.68}$$

in which A_x is the area (normal to x-axis) on which the force F_x acts. In a similar manner we have

$$\sigma_y = \frac{F_y}{A_y} \tag{3.69}$$

and

$$\sigma_z = \frac{F_z}{A_z} \tag{3.70}$$

The stresses σ_x, σ_y and σ_z given above are the only normal stresses which can act on the faces of an elemental volume $dx\ dy\ dz$ having faces normal to the orthogonal axes x, y, z.

The shearing or sliding effect is obviously caused by a force acting as shown in Fig. 3.13. Defining the shearing stress τ_{yx} (note the subscripts — a stress in the x direction acting on a plane perpendicular to the y-axis) as a force divided by an area, we have

$$\tau_{yx} = \frac{F_x}{A_y} \tag{3.71}$$

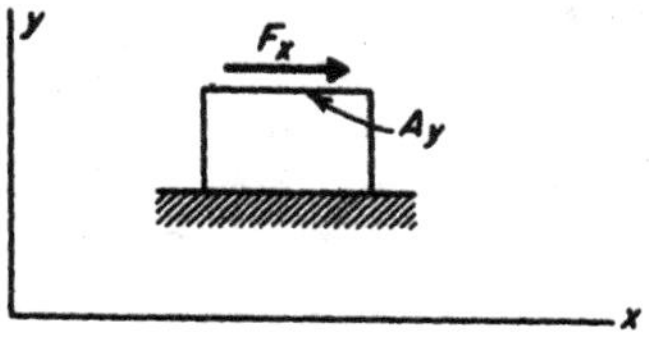

Fig. 3.13

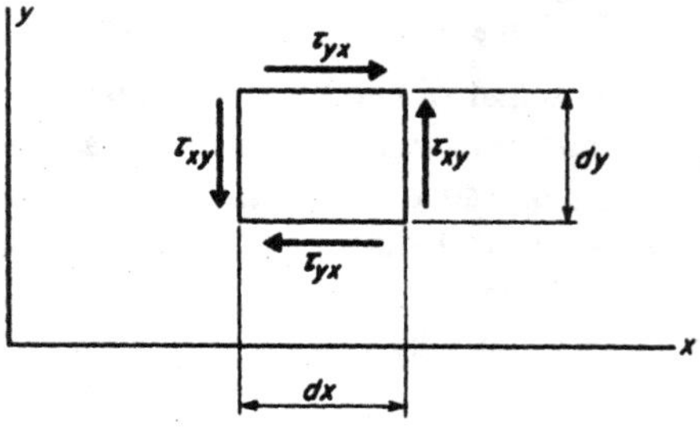

Fig. 3.14

and similarly

$$\tau_{zy} = \frac{F_y}{A_z} \tag{3.72}$$

$$\tau_{xz} = \frac{F_z}{A_x} \tag{3.73}$$

Corresponding to the stress τ_{yx}, we have a companion stress τ_{xy}; see Fig. 3.14,

It is clear that in order for the element shown to be in static equilibrium

$$\tau_{xy} = \tau_{yx} \tag{3.74}$$

since moment balance† requires that

$$(\tau_{xy})(dx)(dy) = (\tau_{yx})(dy)(dx) \tag{3.75}$$

In the same way it may be shown that there are stresses, τ_{yz} and τ_{zx}, which are given by the equalities

$$\tau_{zy} = \tau_{yz} \tag{3.76}$$

$$\tau_{xz} = \tau_{zx} \tag{3.77}$$

This equality of shear stresses on mutually perpendicular planes is a general property which is always true. We shall utilize this property of shear stresses in later portions of the book.

The nine stresses,‡

$$\left.\begin{array}{l} \sigma_x \\ \sigma_y \\ \sigma_z \\ \tau_{xy} = \tau_{yx} \\ \tau_{yz} = \tau_{zy} \\ \tau_{zx} = \tau_{xz} \end{array}\right\} \tag{3.78}$$

are the only possible independent stresses which can act on the faces of the body $dx\,dy\,dz$. This is so since a force in any direction on any of the faces can always be resolved into components corresponding to the normal and shear stresses. It may be shown that they are the elements of a tensor, T, the stress tensor (see Refs. 15 and 18 for the proof that this is a tensor). Note, this tensor is a *symmetrical* tensor. This is so because of the shear equalities shown in Eq. 3.78.

In the standard American notation this tensor is shown in matrix form as follows:

$$\begin{pmatrix} \sigma_x & \tau_{xy} & \tau_{xz} \\ \tau_{yx} & \sigma_y & \tau_{yz} \\ \tau_{zx} & \tau_{zy} & \sigma_z \end{pmatrix} \tag{3.79}$$

†We are perhaps anticipating ourselves somewhat at this point, since we have not as yet discussed "moment balance" (see Art. 4-2). However, it is advantageous for our purposes to prove the symmetry of the stress tensor at this point and so the equality of shears is proved here.

‡In this text we shall use the following sign convention for stresses: the normal stresses σ_x, σ_y, and σ_z are positive when they are tensile stresses. They are negative stresses when they are compressive stresses. A shear stress due to a force in the direction of positive axis, acting on an area whose outward normal is in the direction of positive axis, is a positive shear stress. Thus, all stresses in Fig. 3.14 are positive.

In typical Russian and British notations this tensor is shown as, respectively,

$$\begin{pmatrix} \sigma_{11} & \sigma_{12} & \sigma_{13} \\ \sigma_{21} & \sigma_{22} & \sigma_{23} \\ \sigma_{31} & \sigma_{32} & \sigma_{33} \end{pmatrix} \qquad \begin{pmatrix} x_x & x_y & x_z \\ y_x & y_y & y_z \\ z_x & z_y & z_z \end{pmatrix} \tag{3.80}$$

In Fig. 3.15 we show these stresses acting on a typical element. Note: the stresses shown are all positive stresses.

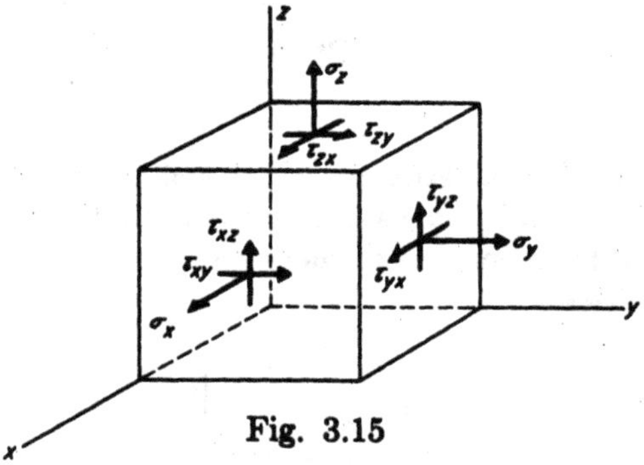

Fig. 3.15

In the two-dimensional x-y space, the stress tensor becomes

$$\begin{pmatrix} \sigma_x & \tau_{xy} \\ \tau_{yx} & \sigma_y \end{pmatrix} \tag{3.81}$$

and because this is a symmetrical tensor it may be put in diagonal form, i.e., it has principal stresses. Also, because it is a tensor of the second order, it may be shown graphically by means of the Mohr circle.

3-6 Summary

In this chapter we discussed the important tensors of elasticity — the inertia tensor, the strain tensor and the stress tensor. The physical significance of all elements of these tensors was described, the different invariance properties and transformation relations were discussed, and the graphical solution, the Mohr circle, of the transformation relations for the two-dimensional tensor was developed.

In the next chapter we discuss the connection between the mathematical theory of elasticity and the engineering form of elasticity, which we call "engineering elasticity."

Problems

1. Determine I_{xx}, I_{yy} and I_{xy} for the shapes shown; $h = 10$ in., $b = 8$ in., $t = 1$ in., $l = 5$ in.

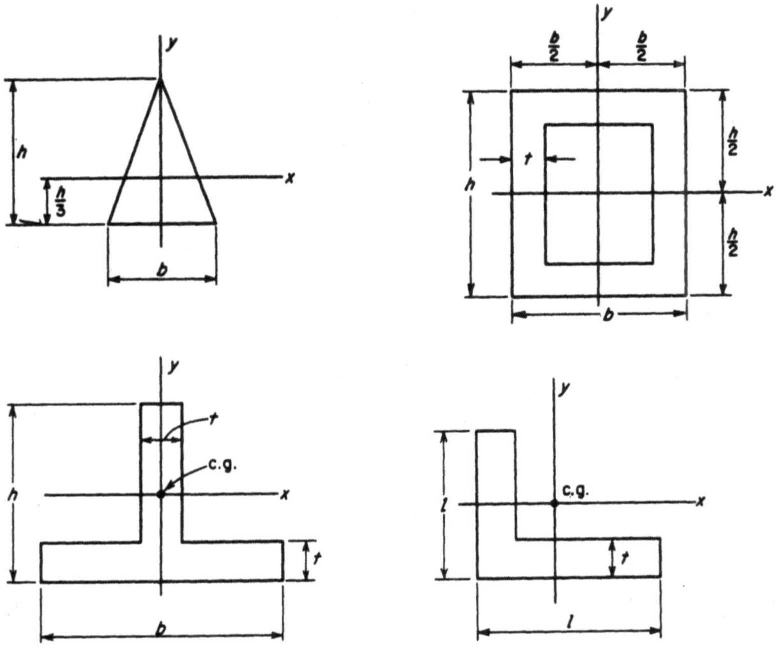

2. Determine the moment of inertia about the base for each figure in Prob. 1.

3. Determine $I_{x'x'}$, $I_{y'y'}$, $I_{x'y'}$ for each figure in Prob. 1, if x'-y' make 45° with x-y.

4. Given

$$\sigma_x = 5000 \text{ psi}$$
$$\sigma_y = 2000 \text{ psi}$$
$$\tau_{xy} = 1500 \text{ psi}$$

on a certain plane at a point in a solid. Draw Mohr's circle for the stress condition at this point. Determine

(a) $(\sigma_x)_{max}$

(b) $(\sigma_y)_{min}$

(c) $(\tau_{xy})_{max}$

(d) What angles do the principal planes make with the plane of given stresses?

5. Draw Mohr circles for each of the figures given in Problem 1. Determine $(I_{xy})_{max}$ from the construction. What are the values of $I_{x'x'}$ and $I_{y'y'}$ for the planes corresponding to $(I_{xy})_{max}$?

6. A fundamental tensor in small deformation plate theory is the curvature tensor,

$$\begin{pmatrix} \frac{\partial^2 w}{\partial x^2} & \frac{\partial^2 w}{\partial x \partial y} \\ \frac{\partial^2 w}{\partial y \partial x} & \frac{\partial^2 w}{\partial y^2} \end{pmatrix} = \begin{pmatrix} -\frac{1}{\rho_{xx}} & \frac{1}{\rho_{xy}} \\ \frac{1}{\rho_{yx}} & -\frac{1}{\rho_{yy}} \end{pmatrix}$$

(a) What are the invariants of this tensor?
(b) If, at a particular point,

$$\frac{1}{\rho_{xx}} = 0.01 \qquad \frac{1}{\rho_{yy}} = 0.02 \qquad \frac{1}{\rho_{xy}} = -0.01$$

draw the Mohr circle and determine the principal values of $1/\rho_{xx}$ and $1/\rho_{yy}$.
(c) What angle does the plane of principal curvature make with the plane of given curvatures in (b)?

7. We may define a *line* moment of inertia similarly to *area* moment of inertia by substituting length for area.
(a) Determine the moment of inertia of the line shown about the x-x axis.
(b) What is the moment of inertia of this line about its base?
(c) Assume axis x'-x' making 45° with the x-x axis, and axis y'-y' normal to this. Obtain $I_{x'-x'}$, $I_{y'-y'}$ and $I_{x'-y'}$. Do this by direct integration and also by using the transformation formulas.

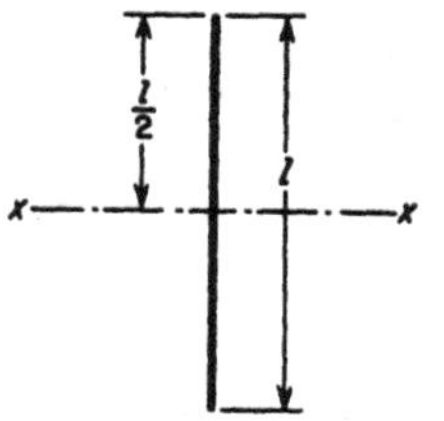

8. The three-dimensional (mass) moment of inertia tensor may be shown as (see Art. 3-2)

$$\begin{pmatrix} I_{xx} & -I_{xy} & -I_{xz} \\ -I_{yx} & I_{yy} & -I_{yz} \\ -I_{zx} & -I_{zy} & I_{zz} \end{pmatrix}$$

(a) Show typical terms, I_{xx}, I_{yz} of this tensor, in their integral forms. What are the invariants of this tensor?

9. Refer to Fig. 3.5 and
(a) verify Eq. 3.48 and 3.49;
(b) prove that the principal plane makes an angle θ with the planes of I_{xx}, I_{yy} and I_{xy} given by

$$\tan 2\theta = -\frac{I_{xy}}{\frac{1}{2}(I_{xx} - I_{yy})}$$

10. Prove that the linear form of the strain tensor is given by the symmetrical part of the tensor

$$\nabla W = \begin{pmatrix} \frac{\partial}{\partial x} \\ \frac{\partial}{\partial y} \\ \frac{\partial}{\partial z} \end{pmatrix} (uvw)$$

in which ∇ is the gradient vector and W is the deformation vector.

Chapter 4

THE CONNECTION BETWEEN THE LINEARIZED MATHEMATICAL THEORY OF ELASTICITY AND ENGINEERING ELASTICITY

4-1 Introduction

In this chapter we shall indicate in an abbreviated fashion the connection between the equations of the linearized mathematical theory of elasticity and the ordinary engineering applications of these equations.

To do this, we shall first discuss the basic equations and requirements of the linearized theory of elasticity as embodied in

1. the equilibrium equations;
2. the boundary conditions;
3. the compatibility equations;
4. the linearized relation between stress and strain, as given by Hooke's Law.

These shall be given in the three- and two-dimensional forms in both matrix-tensor notation and the usual expanded notation.

Following this, we shall discuss the manner in which these equations and requirements are simplified for use in engineering applications. Because the basic structure under consideration in this text is the beam, arguments, wherever possible, shall be illustrated by reference to the beam. It will be shown that all of the requirements of the more exact linearized theory of elasticity are approximated or otherwise simplified for use in the engineering applications.

Finally, a brief discussion of the St. Venant Principle is given.

4-2 The Equations of Equilibrium in the Linearized Theory of Elasticity

One major set of equations of the mathematical theory of elasticity is the stress equilibrium equations. If these equations are satisfied throughout the body, then we can be assured that the stresses are in conformance with the requirement that the body be in static equilibrium. Also, although in this text we are concerned primarily with *elastic* bodies, the equilibrium equations (and boundary conditions of Art. 4.3) hold not only for elastic bodies but for *all continuous bodies* of whatever material, be it elastic, viscous-fluid, or viscoelastic.

We are concerned in this text with the analysis of bodies in static equilibrium. This means simply that the bodies in question do not move under the applied forces or other disturbing agents.†

Now, any motion which a body may have can be analyzed in terms of the following two basic types of movement:

1. Rectilinear movement, or movement in a straight line.
2. Angular movement, or rotation about a line.

The most general motion for (1) will be movement in the three directions, x, y and z.

The most general motion for (2) will be rotation about the three mutually perpendicular axes, x, y and z.

Newton's Law relates the movements (1) and (2) to the forces and moments acting on the bodies and also to the resulting accelerations, linear and angular. The statement of his law for (1) is

$$F_x = ma_x \tag{4.1}$$

$$F_y = ma_y \tag{4.2}$$

$$F_z = ma_z \tag{4.3}$$

in which F_x, F_y, F_z are the forces in the x, y, and z directions, m is the mass, and a_x, a_y, and a_z are the corresponding accelerations. If the body is to be in static equilibrium (i.e., stationary) then it follows that the net forces F_x, F_y and F_z acting on the body must be zero. These are the relations which we shall use in determining the equilibrium equations of elasticity theory (see below).

Newton's Law for (2) is

$$M_x = I_x\alpha_x \tag{4.4}$$

$$M_y = I_y\alpha_y \tag{4.5}$$

$$M_z = I_z\alpha_z \tag{4.6}$$

in which M_x, M_y, M_z are the moments about the x-, y- and z-axes; I_x, I_y, I_z are the moments of inertia of the body about these axes and α_x, α_y, α_z are the corresponding angular accelerations. If the body is to be in static equilibrium (i.e., stationary) then it follows that the net moments M_x, M_y and M_z acting on the body must be zero. These are just the relations which were

†Actually our equations will also hold for bodies moving with *constant velocities*. However, we are not, in this text, concerned with questions of relativistic mechanics, and for simplicity the assumption of *no movement* will be adhered to. Note also, we distinguish between *motion* and *deformation*. The bodies we consider in this text *deform*, but they are not in *motion*.

used to prove the symmetry of the stress tensor in the last chapter (see Eq. 3.75).

Returning to the static equilibrium relations which are obtained by virtue of the absence of rectilinear motion, we proceed to obtain these as follows:

See Fig. 4.1,† which shows as elemental solid with corner closest to the origin at x, y and z and having faces of length dx, dy and dz. The stresses acting on this body are as shown. Note that on the faces closest to the origin the stresses of the tensor, T, act, where

$$T = \begin{pmatrix} \sigma_x & \tau_{xy} & \tau_{xz} \\ \tau_{yx} & \sigma_y & \tau_{yz} \\ \tau_{zx} & \tau_{zy} & \sigma_z \end{pmatrix} \tag{4.7}$$

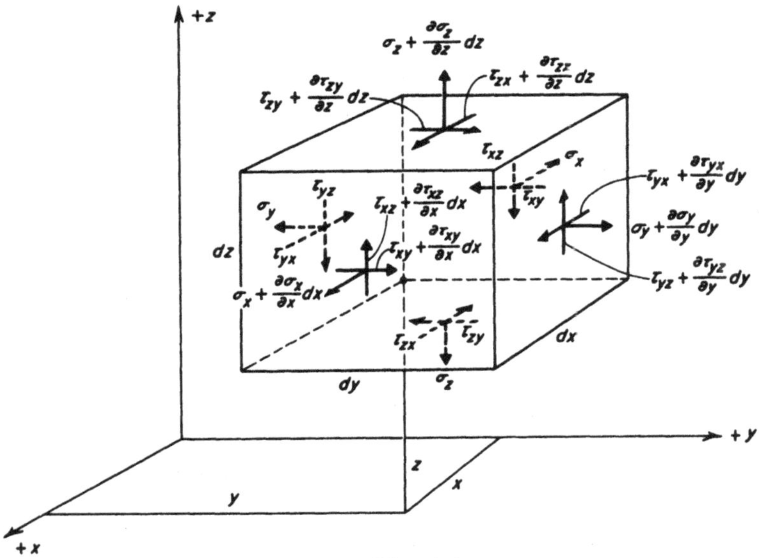

Fig. 4.1

whereas on the far faces we have stresses given by

$$T + dT = \begin{pmatrix} \sigma_x + \dfrac{\partial \sigma_x}{\partial x}\,dx & \tau_{xy} + \dfrac{\partial \tau_{xy}}{\partial x}\,dx & \tau_{xz} + \dfrac{\partial \tau_{xz}}{\partial x}\,dx \\ \tau_{yx} + \dfrac{\partial \tau_{yx}}{\partial y}\,dy & \sigma_y + \dfrac{\partial \sigma_y}{\partial y}\,dy & \tau_{yz} + \dfrac{\partial \tau_{yz}}{\partial y}\,dy \\ \tau_{zx} + \dfrac{\partial \tau_{zx}}{\partial z}\,dz & \tau_{zy} + \dfrac{\partial \tau_{zy}}{\partial z}\,dz & \sigma_z + \dfrac{\partial \sigma_z}{\partial z}\,dz \end{pmatrix} \tag{4.8}$$

†In this figure the stresses are shown increasing in the positive directions of the axes.

This is the standard method for indicating variations of quantities in partial differential notation. Thus, on face $x + dx$ we have acting the stress $\sigma_x + \frac{\partial \sigma_x}{\partial x} dx$, which may be described in words as "the stress σ_x plus the rate of change of stress σ_x in the x direction times the distance dx." The other faces and stresses are handled in a similar manner.

In addition to the stresses shown acting on the body, let us assume that there is acting a *body force* per unit mass. This is, for example, a force due to the weight of the body or a force due to magnetic effects or a similar force. This force per unit mass has components $\mathfrak{F}_x$, $\mathfrak{F}_y$, and $\mathfrak{F}_z$ in the x, y, and z directions respectively. Therefore the *total body force* on the body in, say, the x direction is given (ρ is the density) by

$$\rho \mathfrak{F}_x \, dx \, dy \, dz \tag{4.9}$$

Similar expressions apply in the y and z directions.

If now the sum of the forces (i.e., stresses times area and body force times mass) in the three directions x, y and z are determined for the body of Fig. 4.1, we must have†

$$\Sigma F_x = 0 \tag{4.10}$$

$$\Sigma F_y = 0 \tag{4.11}$$

$$\Sigma F_z = 0 \tag{4.12}$$

A typical statement of one of the above, say

$$\Sigma F_x = 0 \tag{4.13}$$

is given by the following (noting the positive and negative directions of forces as shown on the figure):

$$-\sigma_x(dy\, dz) + \left(\sigma_x + \frac{\partial \sigma_x}{\partial x} dx\right) dy\, dz - \tau_{yx}(dx\, dz) + \left(\tau_{yx} + \frac{\partial \tau_{yx}}{\partial y} dy\right) dx\, dz$$
$$- \tau_{zx}(dx\, dy) + \left(\tau_{zx} + \frac{\partial \tau_{zx}}{\partial z} dz\right) dx\, dy + \rho \mathfrak{F}_x \, dx\, dy\, dz = 0 \tag{4.14}$$

Now, collecting terms, cancelling, and dividing through by the common term $dx\, dy\, dz$, we obtain

$$\frac{\partial \sigma_x}{\partial x} + \frac{\partial \tau_{yx}}{\partial y} + \frac{\partial \tau_{zx}}{\partial z} + \rho \mathfrak{F}_x = 0 \tag{4.15}$$

†The other three equilibrium requirements (Eqs. 4.4, 4.5 and 4.6) were used to establish the symmetry of the stress tensor (see Art. 3-5).

The second and third of these equations are obtained in a similar manner, by summing forces in the y and z directions. These give

$$\frac{\partial \tau_{xy}}{\partial x} + \frac{\partial \sigma_y}{\partial y} + \frac{\partial \tau_{zy}}{\partial z} + \rho \mathfrak{F}_y = 0 \tag{4.16}$$

and

$$\frac{\partial \tau_{xz}}{\partial x} + \frac{\partial \tau_{yz}}{\partial y} + \frac{\partial \sigma_z}{\partial z} + \rho \mathfrak{F}_z = 0 \tag{4.17}$$

Thus, in the mathematical theory of elasticity, if a body is in static equilibrium, the stresses acting on the body must satisfy the following three equations:†

$$\frac{\partial \sigma_x}{\partial x} + \frac{\partial \tau_{yx}}{\partial y} + \frac{\partial \tau_{zx}}{\partial z} + \rho \mathfrak{F}_x = 0 \tag{4.18}$$

$$\frac{\partial \tau_{xy}}{\partial x} + \frac{\partial \sigma_y}{\partial y} + \frac{\partial \tau_{zy}}{\partial z} + \rho \mathfrak{F}_y = 0 \tag{4.19}$$

$$\frac{\partial \tau_{xz}}{\partial x} + \frac{\partial \tau_{yz}}{\partial y} + \frac{\partial \sigma_z}{\partial z} + \rho \mathfrak{F}_z = 0 \tag{4.20}$$

or, in matrix-tensor notation,

$$\begin{pmatrix} \frac{\partial}{\partial x} & \frac{\partial}{\partial y} & \frac{\partial}{\partial z} \end{pmatrix} \begin{pmatrix} \sigma_x & \tau_{xy} & \tau_{xz} \\ \tau_{yx} & \sigma_y & \tau_{yz} \\ \tau_{zx} & \tau_{zy} & \sigma_z \end{pmatrix} + \rho(\mathfrak{F}_x \mathfrak{F}_y \mathfrak{F}_z) = 0 \tag{4.21}$$

or,

$$\text{div } T + \rho \mathfrak{F} = 0 \tag{4.22}$$

in which T is the stress tensor, ρ is the material density, and $\mathfrak{F}$ is the body force per unit mass acting on the body. In the last equation, the operation "divergence of a tensor" is the operation defined in Eq. 4.21.

4-3 The Boundary Conditions in the Linearized Theory of Elasticity

A second major set of equations of the mathematical theory of elasticity are the boundary conditions. We shall give these in terms of stresses, that is, we shall derive the equations which must be satisfied by the stresses applied on the boundary or surface of the body.

†If the body is in non-uniform motion, we may still use these equations. In this case, however, the body force, $\mathfrak{F}$, will introduce the inertia effect or non-uniform motion effect in accordance with D'Alembert's Principle.

Let us consider a differential portion of the body adjacent to and containing the boundary surface (see Fig. 4.2).

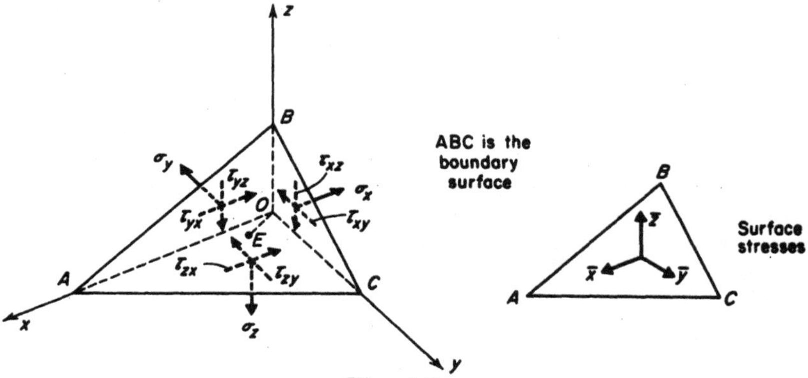

Fig. 4.2

Without limiting the generality, we may assume the normal distance from the origin O to the boundary surface, OE, is unity. On the boundary we have acting stresses, $\bar{x}$, $\bar{y}$ and $\bar{z}$ in the x, y, z directions. On the other faces of the elemental tetrahedron we have acting the stresses of the stress tensor, as shown.

Now if $\overline{N}$ is the unit normal vector (of length $OE = 1$) from the origin to the surface ABC, then $\overline{N}$ has components l, m, and n, which are the direction cosines of the normal, where

$$l = \cos\,(x,\overline{N}) \tag{4.23}$$

or, l is the cosine of the angle between the x-axis and the normal $\overline{N}$ — the other two cosines being defined similarly.

We can show that the area, A_x of face OBC normal to the x-axis is given by

$$A_x = lA \tag{4.24}$$

where A is the area of the boundary surface ABC; similarly,

$$A_y = mA \tag{4.25}$$

$$A_z = nA \tag{4.26}$$

in the following manner:

We give the expression for the volume of the tetrahedron considering each of the four faces, in turn, as the base, noting that

$$\text{volume} = \tfrac{1}{3}(\text{base})(\text{height}). \tag{4.27}$$

Thus,

$$\text{vol.} = \tfrac{1}{3}(A)\,1 \tag{4.28}$$

and

$$\text{vol.} = \tfrac{1}{3}(A_x)(OA) \tag{4.29}$$

and

$$\text{vol.} = \tfrac{1}{3}(A_y)(OC) \tag{4.30}$$

and

$$\text{vol.} = \tfrac{1}{3}(A_z)(OB) \tag{4.31}$$

so that

$$A_x = A\left(\frac{1}{OA}\right) = lA \tag{4.32}$$

Similarly,

$$A_y = mA \tag{4.33}$$

$$A_z = nA \tag{4.34}$$

The stresses acting on the faces A_x, A_y and A_z, when multiplied by the corresponding areas, are equivalent to forces. Noting once more the requirement from Newton's Law that if the body is to be in static equilibrium

$$\Sigma F_x = 0 \tag{4.35}$$

$$\Sigma F_y = 0 \tag{4.36}$$

$$\Sigma F_z = 0 \tag{4.37}$$

we obtain (in the x direction; the other two relations are obtained in a similar manner)

$$\bar{x}A = \sigma_x(lA) + \tau_{yx}(mA) + \tau_{zx}(nA) \tag{4.38}$$

from which, finally, we obtain the three equations which are the boundary conditions that the stresses must satisfy

$$\bar{x} = l\sigma_x + m\tau_{yx} + n\tau_{zx} \tag{4.39}$$

$$\bar{y} = l\tau_{xy} + m\sigma_y + n\tau_{zy} \tag{4.40}$$

$$\bar{z} = l\tau_{xz} + m\tau_{yz} + n\sigma_z \tag{4.41}$$

In matrix-tensor notation these three equations are given by†

$$(\bar{x}\;\bar{y}\;\bar{z}) = (l\;m\;n)\begin{pmatrix} \sigma_x & \tau_{xy} & \tau_{xz} \\ \tau_{yx} & \sigma_y & \tau_{yz} \\ \tau_{zx} & \tau_{zy} & \sigma_z \end{pmatrix} \tag{4.42}$$

†In Eqs. 4.21 and 4.42, we give the equilibrium equations and also the boundary conditions in terms of stresses. Since the stresses are related to the strains through Hooke's Law, it is obvious that these equations can also be given in terms of the strains. We shall not obtain this form of the equations in this text. The student who is interested in seeing these developments will find them in Ref. 9.

4-4 The Strain Compatibility Conditions in the Linearized Theory of Elasticity

A third major set of equations of the mathematical theory of elasticity are the "strain compatibility conditions." The strain compatibility conditions also can be obtained in several different ways. Two of these different ways are shown in Refs. 4 and 15. In this text we shall obtain them in a simpler manner. We shall differentiate the various elements of the strain tensor and shall form identities resulting from these differentiations. The resulting identities will be the strain compatibility conditions.

We have, for the strain tensor,

$$\eta = \begin{pmatrix} e_x & \frac{1}{2}\gamma_{xy} & \frac{1}{2}\gamma_{xz} \\ \frac{1}{2}\gamma_{yx} & e_y & \frac{1}{2}\gamma_{yz} \\ \frac{1}{2}\gamma_{zx} & \frac{1}{2}\gamma_{zy} & e_z \end{pmatrix} = \begin{pmatrix} \frac{\partial u}{\partial x} & \frac{1}{2}\left(\frac{\partial u}{\partial y} + \frac{\partial v}{\partial x}\right) & \frac{1}{2}\left(\frac{\partial u}{\partial z} + \frac{\partial w}{\partial x}\right) \\ \frac{1}{2}\left(\frac{\partial v}{\partial x} + \frac{\partial u}{\partial y}\right) & \frac{\partial v}{\partial y} & \frac{1}{2}\left(\frac{\partial v}{\partial z} + \frac{\partial w}{\partial y}\right) \\ \frac{1}{2}\left(\frac{\partial w}{\partial x} + \frac{\partial u}{\partial z}\right) & \frac{1}{2}\left(\frac{\partial w}{\partial y} + \frac{\partial v}{\partial z}\right) & \frac{\partial w}{\partial z} \end{pmatrix} \tag{4.43}$$

Now

$$\frac{\partial^2 e_x}{\partial y^2} = \frac{\partial^2}{\partial y^2}\left(\frac{\partial u}{\partial x}\right) = \frac{\partial^3 u}{\partial y^2\, \partial x} \tag{4.44}$$

and

$$\frac{\partial^2 e_y}{\partial x^2} = \frac{\partial^2}{\partial x^2}\left(\frac{\partial v}{\partial y}\right) = \frac{\partial^3 v}{\partial x^2\, \partial y} \tag{4.45}$$

Also,

$$\frac{\partial^2 \gamma_{xy}}{\partial x\, \partial y} = \frac{\partial^2}{\partial x\, \partial y}\left(\frac{\partial u}{\partial y} + \frac{\partial v}{\partial x}\right) = \frac{\partial^3 u}{\partial y^2\, \partial x} + \frac{\partial^3 v}{\partial x^2\, \partial y} \tag{4.46}$$

so that, obviously

$$\frac{\partial^2 e_x}{\partial y^2} + \frac{\partial^2 e_y}{\partial x^2} = \frac{\partial^2 \gamma_{xy}}{\partial x\, \partial y} \tag{4.47}$$

This is one of the strain compatibility equations, or conditions. There are five other equations, giving six compatibility equations that the strains must satisfy if the deformations are to be continuous, single-valued and finite in the deformed body. The six equations are obtained as above and are given by†

$$\frac{\partial^2 e_y}{\partial z^2} + \frac{\partial^2 e_z}{\partial y^2} = \frac{\partial^2 \gamma_{yz}}{\partial y\, \partial z} \tag{4.48}$$

$$\frac{\partial^2 e_z}{\partial x^2} + \frac{\partial^2 e_x}{\partial z^2} = \frac{\partial^2 \gamma_{zx}}{\partial z\, \partial x} \tag{4.49}$$

$$\frac{\partial^2 e_x}{\partial y^2} + \frac{\partial^2 e_y}{\partial x^2} = \frac{\partial^2 \gamma_{xy}}{\partial x\, \partial y} \tag{4.50}$$

$$\frac{2\partial^2 e_z}{\partial x\, \partial y} = \frac{\partial}{\partial z}\left(\frac{\partial \gamma_{yz}}{\partial x} + \frac{\partial \gamma_{zx}}{\partial y} - \frac{\partial \gamma_{xy}}{\partial z}\right) \tag{4.51}$$

$$\frac{2\partial^2 e_x}{\partial y\, \partial z} = \frac{\partial}{\partial x}\left(\frac{\partial \gamma_{zx}}{\partial y} + \frac{\partial \gamma_{xy}}{\partial z} - \frac{\partial \gamma_{yz}}{\partial x}\right) \tag{4.52}$$

$$\frac{2\partial^2 e_y}{\partial z\, \partial x} = \frac{\partial}{\partial y}\left(\frac{\partial \gamma_{xy}}{\partial z} + \frac{\partial \gamma_{yz}}{\partial x} - \frac{\partial \gamma_{zx}}{\partial y}\right) \tag{4.53}$$

Because the stresses are related to the strains in the mathematical theory of elasticity, one would expect that a relation given in terms of strains could be transformed into one given in terms of stresses; this, in fact, may be done. Thus, in the case of compatibility of strains, it is possible to give these equations in terms of the stresses. We shall not show them: the interested student will find them, for example, in Ref. 18. However, the important point which we wish to make here is that not only must the strains be compatible, but it is also necessary that the stresses be so. In general, if the strains are compatible, then so also are the stresses, and vice versa. In some cases it is more convenient to consider the strains, and in other cases it is more convenient to consider the stresses. We shall consider both types of compatibility in our later work. In particular we will be concerned with a stress compatibility situation when we discuss the shear stress–bending stress relation for the engineering beam, in a later chapter.

We repeat, that in order to ensure that the deformations satisfy the physical requirement of being finite, single-valued and continuous, it is necessary that the strains satisfy the six strain compatibility equations. If these equations are not satisfied, then the deformations are not single-valued, finite and continuous.‡

†The compatibility conditions of Eqs. 4.48–4.53 are the *linearized* form of these conditions. These also can be given in a more exact non-linear form. See Ref. 4 and 15 for a more complete discussion of this point.

‡It is possible to combine all of the equations of elasticity into three different equations given in terms of the deformations, u, v and w, subject to suitable boundary conditions. If this is done, then separate compatibility conditions are not needed, since a solution to the three equations will themselves indicate whether or not the deformations are finite, single-valued and continuous.

To illustrate, in a physical manner what incompatibility of strains means we shall show two different deformations of a structure — one of which will represent compatible strains and the second incompatible. It would then be found that the strains obtained for the first figure would satisfy the strain compatibility equations whereas the strains obtained for the second would not.

Fig. 4.3 shows a compatible deformation.

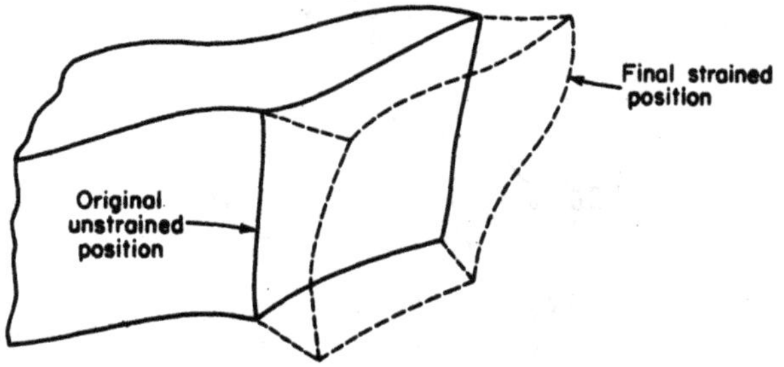

Fig. 4.3

The difference between the original position and final position represents the strains. It is seen that the deformations are continuous and finite, and (we may assume) single-valued.

Now consider the strained body shown in Fig. 4.4. Note the discon-

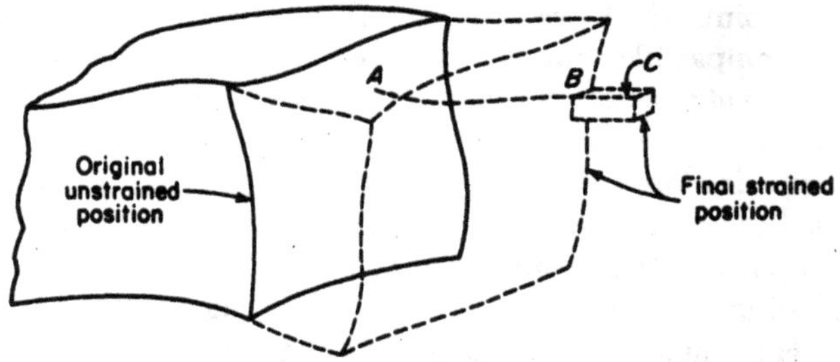

Fig. 4.4

tinuous deformations represented by the projecting portion at the edge. In this case the deformations are obviously not continuous, since a continuous value for the deformations will not describe the deformations of adjacent fibres AB and AC, for example.

4-5 Hooke's Law in the Linearized Theory of Elasticity

In this book we are concerned with materials which are isotropic and homogeneous. This, in effect, means that the materials do not have special directional properties, that is, a particular constant or property of the material in the x direction is assumed to have the same value in the y and z directions.

Also, we are assuming that only small deformations occur. This implies that the strain tensor contains only the linear (first order) terms (see Eq. 3.67).

Under these circumstances, it may be shown that the following relations — which is Hooke's Law — hold between the elements of the stress tensor and the elements of the strain tensor.

$$e_x = \frac{1}{E}[\sigma_x - \nu(\sigma_y + \sigma_z)] \tag{4.54}$$

$$e_y = \frac{1}{E}[\sigma_y - \nu(\sigma_z + \sigma_x)] \tag{4.55}$$

$$e_z = \frac{1}{E}[\sigma_z - \nu(\sigma_x + \sigma_y)] \tag{4.56}$$

$$\gamma_{xy} = \frac{2(1 + \nu)}{E}\tau_{xy} = \frac{\tau_{xy}}{G} \tag{4.57}$$

$$\gamma_{yz} = \frac{2(1 + \nu)}{E}\tau_{yz} = \frac{\tau_{yz}}{G} \tag{4.58}$$

$$\gamma_{zx} = 2\left(\frac{1 + \nu}{E}\right)\tau_{zx} = \frac{\tau_{zx}}{G} \tag{4.59}$$

or, in matrix-tensor notation,

$$\begin{pmatrix} e_x & \frac{1}{2}\gamma_{xy} & \frac{1}{2}\gamma_{xz} \\ \frac{1}{2}\gamma_{yx} & e_y & \frac{1}{2}\gamma_{yz} \\ \frac{1}{2}\gamma_{zx} & \frac{1}{2}\gamma_{zy} & e_z \end{pmatrix} = \frac{1 + \nu}{E}\begin{pmatrix} \sigma_x & \tau_{xy} & \tau_{xz} \\ \tau_{yx} & \sigma_y & \tau_{yz} \\ \tau_{zx} & \tau_{zy} & \sigma_z \end{pmatrix} - \frac{\nu}{E}(\sigma_x + \sigma_y + \sigma_z)\begin{pmatrix} 1 & 0 & 0 \\ 0 & 1 & 0 \\ 0 & 0 & 1 \end{pmatrix} \tag{4.60}$$

which is

$$\eta = \frac{1 + \nu}{E}T - \frac{\nu}{E}\mathcal{I}_1\mathcal{E}_3 \tag{4.61}$$

In the above, E is the tension-compression modulus of elasticity, ν is the Poisson's ratio, and G is the shear modulus of elasticity (also called the modulus of rigidity); η is the strain tensor, T is the stress tensor, $\mathcal{I}_1$ is the first invariant of the stress tensor, and $\mathcal{E}_3$ is the unit tensor.

These equations hold for a homogeneous isotropic material. They are also valid for small deformations only, as noted previously, and in view of the linearization of Hooke's Law, they hold only when the stress is a linear function of the strain. They are occasionally referred to as the "exact" equations of elasticity and solutions which satisfy them are sometimes referred to as "exact" solutions. The student should understand, however, that they are not really exact in the true sense of the word. We may, perhaps, properly speak of the equations as the "exact equations of the *linearized* mathematical theory of elasticity" and of the solutions as "solutions to the exact equations of the *linearized* theory of elasticity;" this, in fact, is how we shall in later portions of the book, occasionally refer to these equations and to the solutions which satisfy them.†

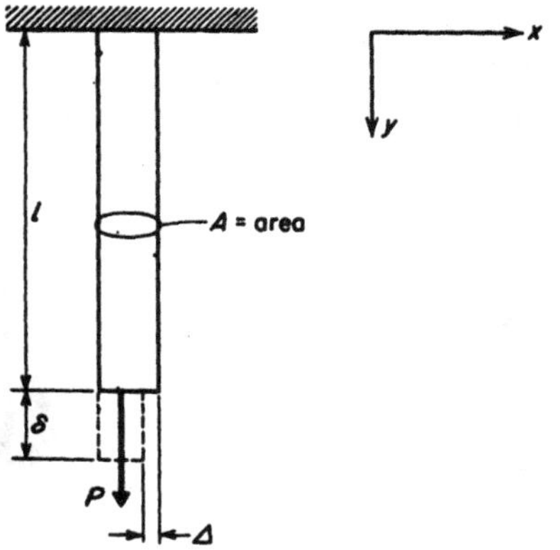

Fig. 4.5

The physical explanation of the three constants which appear in Hooke's Law may be obtained as follows:

As originally stated by Hooke, his law was simply a statement that the deformation is proportional to the load. As an experimental verification of this hypothesis one may proceed as follows (see Fig. 4.5):

Hang a rod of uniform cross-section and length l from a fixed point. Apply a gradually increasing tension load P at the end and at given intervals measure the deflection δ. Continue this until failure of the bar. If this is done and if we plot P/A against δ/l, we find that (depending upon the material of the bar) the curve similar to the one of Fig. 4.6 results.

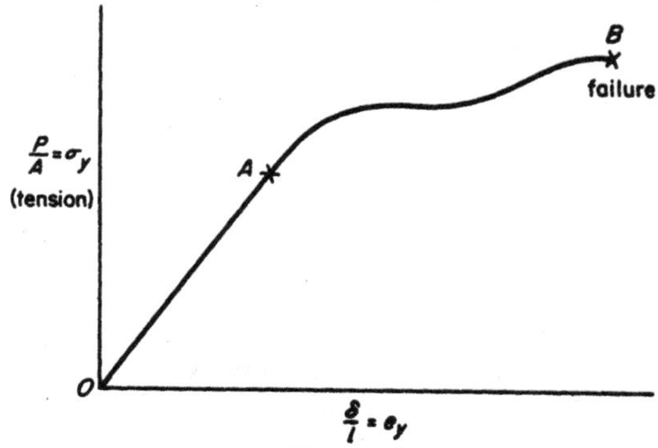

Fig. 4.6

†One non-linear form of Hooke's Law is given in Ref. 4 and 15. See also Prob. 3 at the end of this chapter.

In this curve, the portion OA is essentially a straight line — hence, for this portion of the curve the extension is truly proportional to the load. The slope of the line in this portion is just the tension modulus of elasticity, E. Stresses and strains in this portion of the curve correspond to the "linear theory of elasticity," and they are related by the expression

$$E = \frac{\sigma_y}{e_y} \text{ or } e_y = \frac{\sigma_y}{E} \tag{4.62}$$

Between A and B the strain is no longer proportional to the stress, and we have a plastic (as against elastic) type of action. This is the region of non-linear effects, and the linear theory of elasticity can not predict what will happen in this range.

If the experiment is performed carefully, it will be found that when the bar elongates longitudinally due to the load P it also contracts in the lateral direction. This is the Poisson ratio effect. The amount of contraction (or the lateral strain) is given by

$$e_x = -\frac{\nu\sigma_y}{E} \tag{4.63}$$

If, instead of a tension load, we applied a compression load to a short block of material, we would obtain a curve similar to the one shown in Fig. 4.6. Now, however, instead of the block elongating in the longitudinal direction and contracting in the lateral direction, we would find that it shortens longitudinally and expands laterally. The amount of lateral expansion or the lateral strain is given by

$$e_x = -\frac{\nu(-\sigma_y)}{E} \tag{4.64}$$

$$= +\frac{\nu\sigma_y}{E} \tag{4.65}$$

and the amount of longitudinal contraction is given by

$$e_y = \frac{-\sigma_y}{E} \tag{4.66}$$

Finally, if instead of a pure tension or pure compression, we had applied a pure shear load to a block of material as shown in Fig. 4.7, then we would find that

$$\frac{P}{A} = G\frac{\delta}{l} \tag{4.67}$$

or,

$$\tau_{xy} = G\gamma_{xy} \tag{4.68}$$

The above facts, in the linear range, are included in the statement of Hooke's Law as given above.

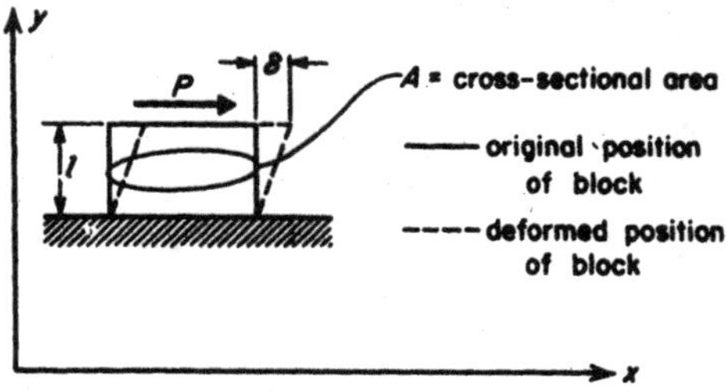

Fig. 4.7

The student should note that although there are three fundamental constants (E, G and ν) in the relations as given above, only two of these are *independent* constants. This is so because of the assumed isotropy of the material. The assumption of isotropy implies that E, G and ν are related by

$$G = \frac{E}{2(1 + \nu)} \tag{4.69}$$

as shown in Eqs. 4.57–4.59. Because of this equation we are finally left with two independent elastic constants — any two of the three, E, G or ν.

4-6 The Two-Dimensional Form of the Equations of the Linearized Theory of Elasticity

The two-dimensional forms of these equations will now be given, in both matrix-tensor and expanded forms.

Equilibrium

$$\frac{\partial \sigma_x}{\partial x} + \frac{\partial \tau_{yx}}{\partial y} + \rho \mathfrak{F}_x = 0 \tag{4.70}$$

$$\frac{\partial \tau_{xy}}{\partial x} + \frac{\partial \sigma_y}{\partial y} + \rho \mathfrak{F}_y = 0 \tag{4.71}$$

or

$$\begin{pmatrix} \frac{\partial}{\partial x} & \frac{\partial}{\partial y} \end{pmatrix} \begin{pmatrix} \sigma_x & \tau_{xy} \\ \tau_{yx} & \sigma_y \end{pmatrix} + \rho(\mathfrak{F}_x \quad \mathfrak{F}_y) = 0 \tag{4.72}$$

or

$$\text{div } T + \rho \mathfrak{F} = 0 \tag{4.73}$$

Boundary Conditions on Stresses

$$\bar{x} = l\sigma_x + m\tau_{yx} \tag{4.74}$$

$$\bar{y} = l\tau_{xy} + m\sigma_y \tag{4.75}$$

or

$$(\bar{x} \; \bar{y}) = (l \; m) \begin{pmatrix} \sigma_x & \tau_{xy} \\ \tau_{yx} & \sigma_y \end{pmatrix} \tag{4.76}$$

Compatibility

$$\frac{\partial^2 e_x}{\partial y^2} + \frac{\partial^2 e_y}{\partial x^2} = \frac{\partial^2 \gamma_{xy}}{\partial x\, \partial y} \tag{4.77}$$

Hooke's Law

$$e_x = \frac{1}{E}[\sigma_x - \nu\sigma_y] \tag{4.78}$$

$$e_y = \frac{1}{E}[\sigma_y - \nu\sigma_x] \tag{4.79}$$

$$\gamma_{xy} = \frac{2(1+\nu)}{E}\tau_{xy} = \frac{\tau_{xy}}{G} \tag{4.80}$$

or

$$\begin{pmatrix} e_x & \frac{1}{2}\gamma_{xy} \\ \frac{1}{2}\gamma_{yx} & e_y \end{pmatrix} = \frac{1+\nu}{E}\begin{pmatrix} \sigma_x & \tau_{xy} \\ \tau_{yx} & \sigma_y \end{pmatrix} - \frac{\nu}{E}(\sigma_x + \sigma_y)\begin{pmatrix} 1 & 0 \\ 0 & 1 \end{pmatrix} \tag{4.81}$$

which is

$$\eta = \frac{1+\nu}{E}T - \frac{\nu}{E}\mathcal{I}_1 \mathcal{E}_2 \tag{4.82}$$

4-7 The Engineering Form of the Equations of the Theory of Elasticity

The Equilibrium Equations

In engineering applications, the equilibrium equations of Eq. 4.21, which represent a set of differential equations that hold at all points (differential elements) in the body (see Fig. 4.1), are replaced by a set of *gross* equilibrium requirements (a sort of integrated differential relation), which hold for large, finite, so-called free-body portions of the structure. In particular, since the three-dimensional body is in static equilibrium, we apply to the finite free body the six rigid-body relations (Eqs. 4.1–4.3 and 4.4–4.6) which follow from Newton's Laws. These six equations, three in terms of forces and three in terms of moments, are (for static equilibrium)

$$\Sigma F_x = 0 \tag{4.83}$$

which says, in words, "the summation of all forces in the x direction acting on the body must be zero;" similarly,

$$\Sigma F_y = 0 \tag{4.84}$$

$$\Sigma F_z = 0 \tag{4.85}$$

In addition, we have the moment equilibrium relation,

$$\Sigma M_x = 0 \tag{4.86}$$

which says, in words, "the summation of the moment of all forces about the x-axis must equal zero;" similarly,

$$\Sigma M_y = 0 \tag{4.87}$$

$$\Sigma M_z = 0 \tag{4.88}$$

The above six equations represent the static equilibrium equations for a body in equilibrium in a three-dimensional space. In two dimensions these become

$$\Sigma F_x = 0 \tag{4.89}$$

$$\Sigma F_y = 0 \tag{4.90}$$

and

$$\Sigma M_{\text{about any point}} = 0 \tag{4.91}$$

or,

$$\Sigma M_z = 0 \tag{4.92}$$

The Two-Dimensional Engineering Boundary Conditions in Terms of Forces and Moments

In engineering applications, the *boundary conditions* are replaced by a set of *support conditions*. Just as the stresses are replaced by forces and moments in the equilibrium equations, so also are the boundary conditions of engineering elasticity given in terms of forces and moments.

In general, there are four different boundary or support conditions possible in the engineering structure:

(a) a built-in end, (b) a hinged end, (c) a roller support, and (d) a free end.

The support (a), for the two-dimensional structure, is shown in Fig. 4.8.

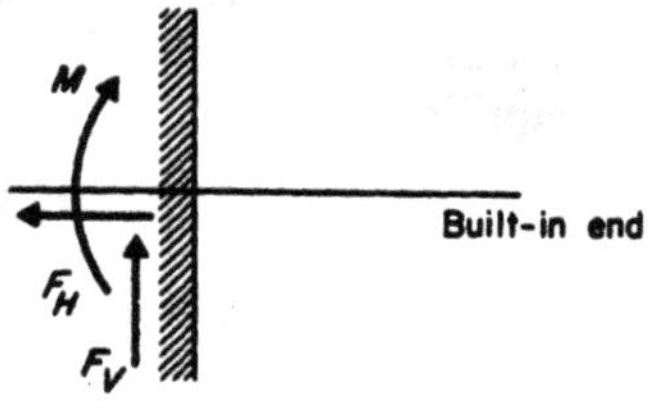

Fig. 4.8

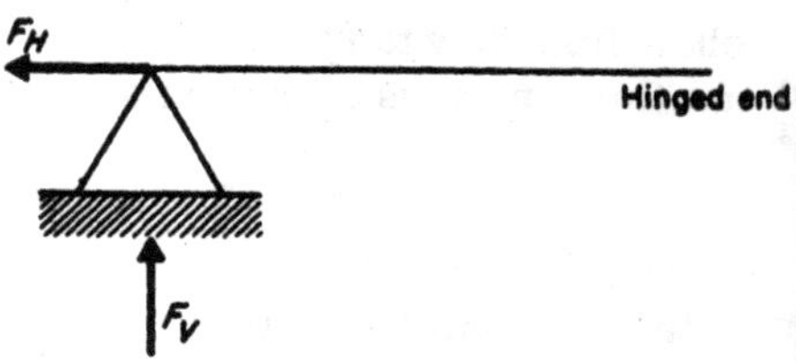

Fig. 4.9

Note that for this support there are, in general, three reactions induced, these being a moment, M, a horizontal force, F_H, and a vertical force, F_V.

Support (b) is shown in Fig. 4.9. Note that for this support, we have in general two reactions: a horizontal force, F_H, and a vertical force, F_V.

The roller support is shown in Fig. 4.10. In this case we have a single reaction force, a vertical force, F_V, which may act either up or down.

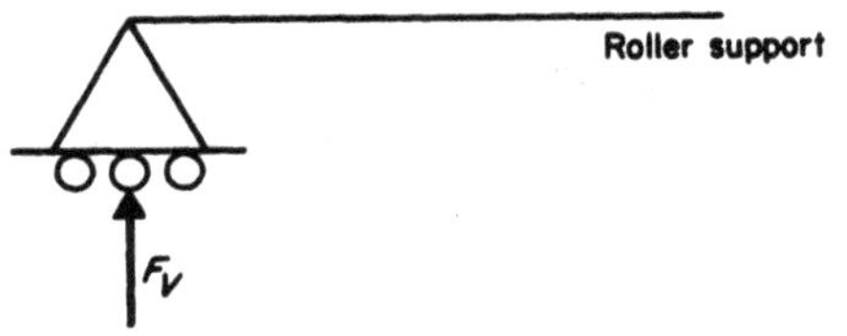

Fig. 4.10

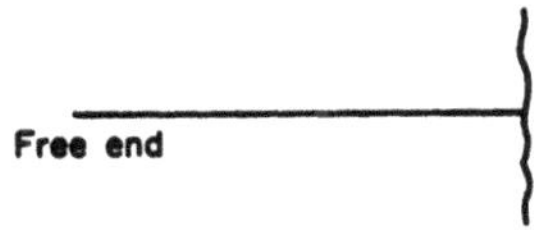

Fig. 4.11

Finally, for the free end, shown in Fig. 4.11, we have, at the free end, neither moment nor force reactions. However, we may, of course, have externally applied moments or forces at this end.

The Engineering Boundary Conditions in Terms of Deformations

It was pointed out in Art. 4-6 that the boundary conditions in the theory of elasticity can be given in terms of strains as well as in terms of stresses. In the same way, in the engineering form of the elasticity relations it is necessary that the boundary deformations be consistent with the type of restraint offered by the boundary. Thus, for the built-in end (see Fig. 4.12), the boundary condition requires that the slope and deflection of the beam at A do not change as the structure deforms.

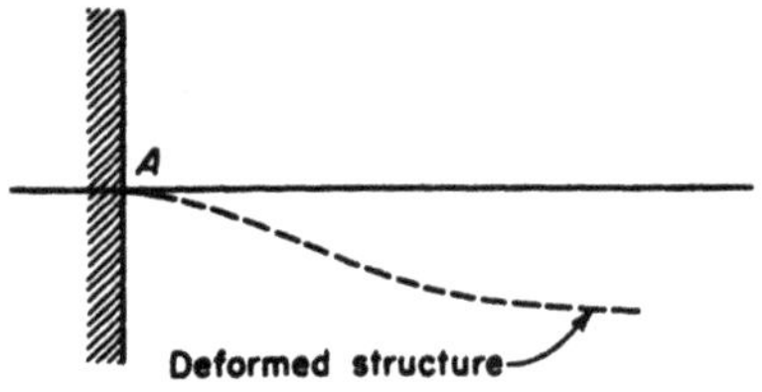

Fig. 4.12

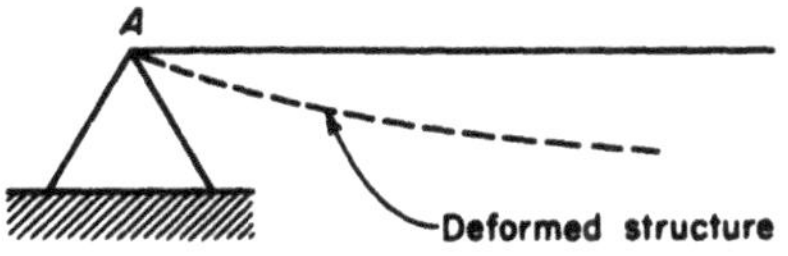

Fig. 4.13

For a hinged end (see Fig. 4.13), the boundary condition requires that the hinged end A does not move horizontally or vertically as the structure deforms.

For the roller support (see Fig. 4.14), the boundary condition requires that the roller end A does not move vertically as the structure deforms.

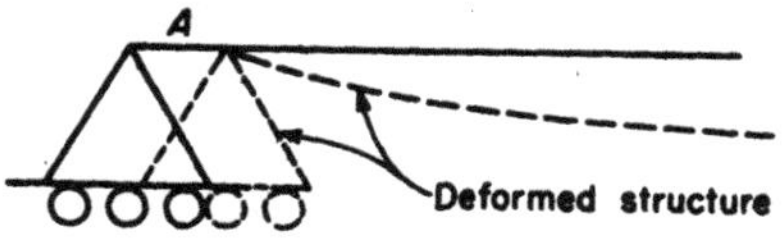

Fig. 4.14

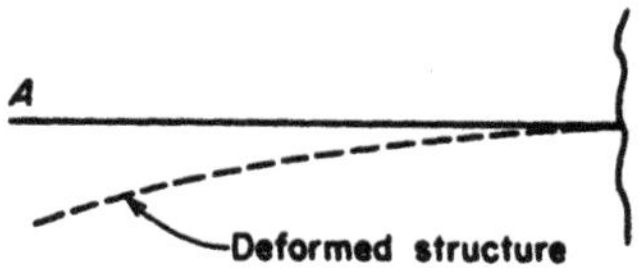

Fig. 4.15

Also, for the free end (see Fig. 4.15), the boundary condition requires that, in general, the end A both rotate and move vertically as the structure deforms.

THE COMPATIBILITY CONDITIONS FOR THE ENGINEERING STRUCTURE

The compatibility conditions of the theory of elasticity are obtained by virtue of the requirement that the deformations of the structure be finite, continuous and single-valued.

In the engineering elasticity we replace these by an essentially equivalent requirement which is, however, also given in a "gross" or "integrated" form. It is, simply stated, the requirement that the deformation of the structure be consistent with the assumed boundary conditions and also with the assumed type of behaviour of the structure and that the deformations be continuous.

The consistency of deformation requirement will be illustrated in Art. 5-5 in connection with the engineering analysis of trusses.

The engineering requirement of continuity of deformation for the beam or beam combination structure (which includes plates, shells and similar structures) may be stated as follows (see Fig. 4.16): the deflection and the

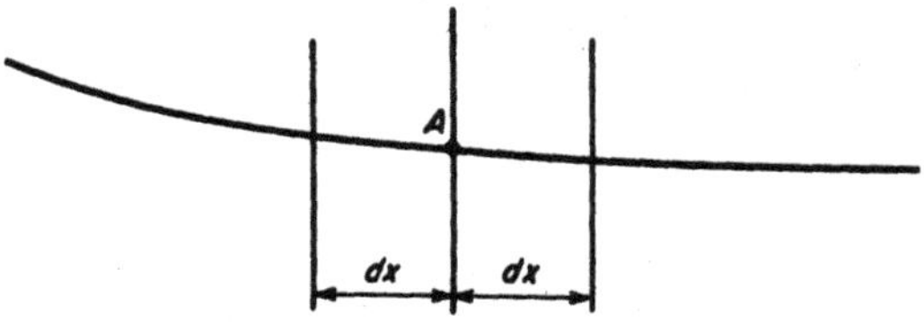

Fig. 4.16

change in slope of a deformed structure a differential distance to one side of any point, such as A, is the same as the deflection and the change in slope a differential distance to the other side of the same point.

An exception to this occurs when A is a hinge point — but this is essentially an artificial discontinuity in slope built into the structure.

Hooke's Law for the Engineering Structure

In engineering applications we frequently use a degenerate form of the linearized Hooke's Law, in which ν, Poisson's ratio, is neglected. Thus, we say

$$e_x = \frac{\sigma_x}{E} \tag{4.93}$$

$$e_y = \frac{\sigma_y}{E} \tag{4.94}$$

$$\gamma_{xy} = \frac{\tau_{xy}}{G} \tag{4.95}$$

However, it should be noted that very frequently the more accurate form of Hooke's Law, as expressed by Eq. 4.60, is used in the engineering elasticity.

4-8 The St. Venant Principle

Some remarks may be in order in connection with the engineering boundary conditions as shown in Figs. 4.8–4.10 of Art. 4-7.

In engineering applications of elasticity we generally assume that the reactions are concentrated, i.e., applied at a *point*. Mathematically, this is the equivalent of a "singularity" which furthermore generally implies an infinity — in this case an infinity in the stress at the point of the reactions.

Actually, of course, the reaction can not be applied at a single point; it must be spread over a region. But, in a sense, conditions at the reactions still correspond to singular ones in that the stress distribution at the reaction is generally fairly concentrated.

One might expect that these singular support conditions could lead to peculiar stress conditions in the overall structure — if this were true, then, depending upon the manner in which a structure actually were supported, one would obtain different results for the stress distribution throughout the structure. In other words, it would be practically impossible to obtain unique results for any particular solution to a given problem.

Fortunately, this is not so, and the mathematical statement which explains why it is not so is given in the very famous, extremely important St. Venant Principle. We may state this principle as follows (refer to the single-reaction case of Fig. 4.10):

"At distances further away from the support than the depth of the beam, the stress distribution does not depend on how the force F_y is applied at the support. It is necessary only that the *resultant force* be F_y." In the neighborhood of the support, the stress distribution certainly depends upon how the reaction is applied. For example, in Fig. 4.17a we have stresses near A which are different from the stresses near A in Fig. 4.17b. But near points B, the stresses are essentially the same in both cases.

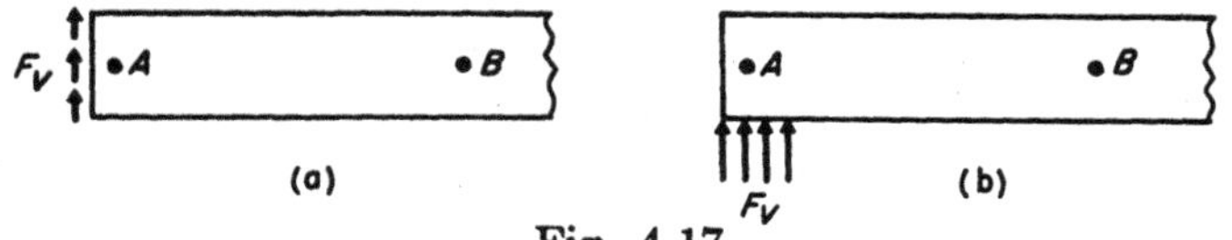

Fig. 4.17

4-9 Summary

Summarizing — in this chapter we first obtained the fundamental equations of the linearized mathematical theory of elasticity, these being (a) the equilibrium equations, (b) the boundary conditions, (c) the compatibility conditions, and (d) Hooke's Law, the relation between the stress tensor and the strain tensor.

Following this, we showed how all of these relations and equations are modified for use in the engineering form of elasticity.

Finally we discussed the St. Venant Principle, which assures uniqueness of solutions for the many actual support and loading conditions, — which generally are different from the idealized conditions assumed in the various solutions, and may be different from each other.

Problems

1. Given a two-dimensional problem with

$$\sigma_x = x^2$$
$$\sigma_y = y^2$$
$$\tau_{xy} = -2xy$$

(a) Are the equilibrium equations satisfied everywhere in the body?
(b) Show that the strain compatibility equations are not satisfied. Hint: use Hooke's Law to obtain the strain components.
(c) By integrating the strains ($\partial u/\partial x$, $\partial v/\partial y$, etc.) directly show that the above stresses cannot represent a strain-compatible solution to the problem.

2. Hooke's Law for stresses in terms of strains is given (in tensor notation) by

$$T = \lambda \mathcal{I}_1 \mathcal{E}_s + 2G\eta$$

in which

$$\mathcal{I}_1 = \text{the first strain invariant}$$

$$\lambda = \frac{\nu E}{(1+\nu)(1-2\nu)}$$

Write out the six components of this form of Hooke's Law.

3. One form of the nonlinear Hooke's Law (see Ref. 15) is given by

$$T = \lambda \mathcal{J}_1 \mathcal{E}_3 + 2G\eta + (l\mathcal{J}_1^2 - 2m\mathcal{J}_2)\mathcal{E}_3 + 2m\mathcal{J}_1\eta + \eta\text{co}\eta$$

in which l and m are new elastic constants, $\mathcal{J}_2$ is the second strain invariant, and coη is the cofactor of the strain tensor. Write out the six terms of this nonlinear form of Hooke's Law.

4. Given a nonlinear strain tensor (see Ref. 15)

$$\eta = \frac{1}{2}\begin{pmatrix} 0 & k & 0 \\ k & k^2 & 0 \\ 0 & 0 & 0 \end{pmatrix}$$

determine Hooke's Law in its nonlinear form. Write out the six terms in their expanded form.

5. Determine the equations for τ_{zx}, τ_{zy}, and σ_z in order that the equilibrium equations be satisfied if

$$T = \begin{pmatrix} x^2 + 2yz & 3yz + xy & \tau_{xz} \\ ? & 2z^2y & \tau_{yz} \\ \tau_{zx} & \tau_{zy} & \sigma_z \end{pmatrix}$$

6. A series of equations that occur in flat plate theory (see Ref. 4) are

$$\sigma_x = \frac{Ez}{1-\nu^2}\left(\frac{1}{\rho_{xx}} + \nu\frac{1}{\rho_{yy}}\right)$$

$$\sigma_y = \frac{Ez}{1-\nu^2}\left(\frac{1}{\rho_{yy}} + \nu\frac{1}{\rho_{xx}}\right)$$

$$\tau_{xy} = -\frac{Ez}{1+\nu}\frac{1}{\rho_{xy}} = -\frac{Ez}{1+\nu}\frac{1}{\rho_{yx}} = \tau_{yx}$$

Show the matrix-tensor form of this equation in terms of the stress and curvature tensors given by

$$T = \begin{pmatrix} \sigma_x & \tau_{xy} \\ \tau_{yx} & \sigma_y \end{pmatrix}, \quad \mathcal{R} = \begin{pmatrix} -\dfrac{1}{\rho_{xx}} & \dfrac{1}{\rho_{xy}} \\ \dfrac{1}{\rho_{yx}} & -\dfrac{1}{\rho_{yy}} \end{pmatrix}$$

7. (a) The three-dimensional stress equilibrium equations (body forces neglected) are given by

$$\text{div } T = 0$$

or

$$\begin{pmatrix} \dfrac{\partial}{\partial x} & \dfrac{\partial}{\partial y} & \dfrac{\partial}{\partial z} \end{pmatrix}\begin{pmatrix} \sigma_x & \tau_{xy} & \tau_{xz} \\ \tau_{yx} & \sigma_y & \tau_{yz} \\ \tau_{zx} & \tau_{zy} & \sigma_z \end{pmatrix} = 0$$

Show the expanded form of these three equations.

(b) In thin plate theory (see Ref. 4), the equilibrium equations are

$$\frac{\partial M_{xx}}{\partial x} + \frac{\partial M_{yx}}{\partial y} - Q_x = 0$$

$$-\frac{\partial M_{xy}}{\partial x} + \frac{\partial M_{yy}}{\partial y} - Q_y = 0$$

$$\frac{\partial Q_x}{\partial x} + \frac{\partial Q_y}{\partial y} + q = 0$$

in which M_{xx}, $M_{yx} = -M_{xy}$, M_{yy} are moments per unit length of side of plate, Q_x and Q_y are shear forces per unit length of side of plate, and q is the transverse load per unit area. All are functions of (x, y) only. These three equations may be given in the form

$$\text{div}\ \mathcal{M} = 0$$

in which $\mathcal{M}$ is a *symmetrical* moment tensor. Determine $\mathcal{M}$.

8. A generalized Hooke's Law relation for thin plates can be given in the form

$$\mathcal{R} = -\frac{1+\nu}{D(1-\nu^2)}\mathcal{M} + \frac{\nu}{D(1-\nu^2)}\mathcal{I}_1\mathcal{E}_3$$

in which $\mathcal{R}$, a generalized curvature tensor, is equal to

$$\begin{pmatrix} -\dfrac{1}{\rho_{xx}} & \dfrac{1}{\rho_{xy}} & \dfrac{1}{\rho_{xz}} \\ \dfrac{1}{\rho_{yx}} & -\dfrac{1}{\rho_{yy}} & \dfrac{1}{\rho_{yz}} \\ \dfrac{1}{\rho_{zx}} & \dfrac{1}{\rho_{zy}} & -\dfrac{1}{\rho_{zz}} \end{pmatrix}$$

$\mathcal{M}$ = moment tensor of Prob. 7,
ν = a scalar, Poisson's ratio,
D = a scalar, the plate stiffness,
$\mathcal{I}_1$ = the first invariant of the moment tensor,
$\mathcal{E}_3$ = the 3×3 unit tensor.

Obtain the six equations which are the expanded form of this equation.

9. (a) Show the tensor, $\mathcal{R}$, Prob. 8, in a form analogous to the strain tensor of Eq. 3.65 if w is the deflection of the thin plate and if we define the curvature elements of tensor $\mathcal{R}$ as

$$-\frac{1}{\rho_{xx}} = \frac{\partial^2 w}{\partial x^2}$$

$$\frac{1}{\rho_{xy}} = \frac{\partial^2 w}{\partial x \partial y}$$

etc.

(b) Show that the elements of this tensor also satisfy six *curvature* compatibility conditions, identical to the six strain compatibility conditions, Eq. 4.48–4.53

Chapter 5

THE SIMPLE TENSION-COMPRESSION STRUCTURE — THE TRUSS

5-1 Introduction

In this chapter we shall discuss the very simplest form of the stress tensor which satisfies the equations of the linearized mathematical theory of elasticity and which also has physical significance: the condition corresponding to tension-compression acting on a straight bar. We shall then discuss the engineering generalization of this type of action — the simple pin-ended truss — after pointing out the necessary simplifications and assumptions in order that the simple tension-compression action may be valid for the truss members.

Following this, we discuss briefly the engineering analysis of this structure.

5-2 The Tension-Compression Bar in Elasticity

The equations of the linearized mathematical theory of elasticity are as follows:

1. Equilibrium (see Eq. 4.21), body forces neglected:

$$\frac{\partial \sigma_x}{\partial x} + \frac{\partial \tau_{yx}}{\partial y} + \frac{\partial \tau_{zx}}{\partial z} = 0 \tag{5.1}$$

$$\frac{\partial \tau_{xy}}{\partial x} + \frac{\partial \sigma_y}{\partial y} + \frac{\partial \tau_{zy}}{\partial z} = 0 \tag{5.2}$$

$$\frac{\partial \tau_{xz}}{\partial x} + \frac{\partial \tau_{yz}}{\partial y} + \frac{\partial \sigma_z}{\partial z} = 0 \tag{5.3}$$

2. Boundary Conditions (see Eq. 4.42):

$$\bar{x} = l\sigma_x + m\tau_{yx} + n\tau_{zx} \tag{5.4}$$

$$\bar{y} = l\tau_{xy} + m\sigma_y + n\tau_{zy} \tag{5.5}$$

$$\bar{z} = l\tau_{xz} + m\tau_{yz} + n\sigma_z \tag{5.6}$$

3. Compatibility (see Eqs. 4.48–4.53):

$$\frac{\partial^2 e_y}{\partial z^2} + \frac{\partial^2 e_z}{\partial y^2} = \frac{\partial^2 \gamma_{yz}}{\partial y\, \partial z} \tag{5.7}$$

$$\frac{\partial^2 e_z}{\partial x^2} + \frac{\partial^2 e_x}{\partial z^2} = \frac{\partial^2 \gamma_{zx}}{\partial z\, \partial x} \tag{5.8}$$

$$\frac{\partial^2 e_x}{\partial y^2} + \frac{\partial^2 e_y}{\partial x^2} = \frac{\partial^2 \gamma_{xy}}{\partial x\, \partial y} \tag{5.9}$$

$$\frac{2\partial^2 e_z}{\partial x\, \partial y} = \frac{\partial}{\partial z}\left(\frac{\partial \gamma_{yz}}{\partial x} + \frac{\partial \gamma_{zx}}{\partial y} - \frac{\partial \gamma_{xy}}{\partial z}\right) \tag{5.10}$$

$$\frac{2\partial^2 e_x}{\partial y\, \partial z} = \frac{\partial}{\partial x}\left(\frac{\partial \gamma_{zx}}{\partial y} + \frac{\partial \gamma_{xy}}{\partial z} - \frac{\partial \gamma_{yz}}{\partial x}\right) \tag{5.11}$$

$$\frac{2\partial^2 e_y}{\partial z\, \partial x} = \frac{\partial}{\partial y}\left(\frac{\partial \gamma_{xy}}{\partial z} + \frac{\partial \gamma_{yz}}{\partial x} - \frac{\partial \gamma_{zx}}{\partial y}\right) \tag{5.12}$$

4. Hooke's Law (see Eq. 4.60):

$$e_x = \frac{1}{E}[\sigma_x - \nu(\sigma_y + \sigma_z)] \tag{5.13}$$

$$e_y = \frac{1}{E}[\sigma_y - \nu(\sigma_z + \sigma_x)] \tag{5.14}$$

$$e_z = \frac{1}{E}[\sigma_z - \nu(\sigma_x + \sigma_y)] \tag{5.15}$$

$$\gamma_{xy} = \frac{\tau_{xy}}{G} \tag{5.16}$$

$$\gamma_{yz} = \frac{\tau_{yz}}{G} \tag{5.17}$$

$$\gamma_{zx} = \frac{\tau_{zx}}{G} \tag{5.18}$$

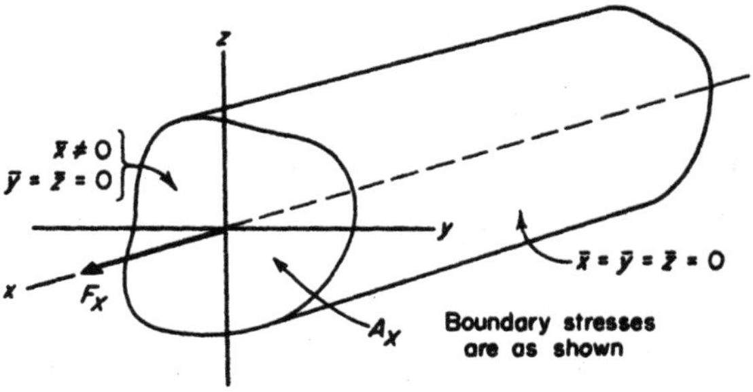

Fig. 5.1

Let us assume that we are interested in investigating the behaviour of a cylindrical (not necessarily circular) cross-section bar, as shown in Fig. 5.1. Let us assume further that we are interested in the behaviour of this bar when it is subjected to a stress in the x direction only.

If we assume the stress tensor is given by

$$T = \begin{pmatrix} C & 0 & 0 \\ 0 & 0 & 0 \\ 0 & 0. & 0 \end{pmatrix} = \begin{pmatrix} \sigma_x & \tau_{xy} & \tau_{xz} \\ \tau_{yx} & \sigma_y & \tau_{yz} \\ \tau_{zx} & \tau_{zy} & \sigma_z \end{pmatrix} \tag{5.19}$$

in which $\sigma_x = C$, a constant, then (the student should verify this and the following statements) obviously the equilibrium equations are satisfied, and we have from Hooke's Law

$$e_x = \frac{\sigma_x}{E} = \frac{C}{E} \tag{5.20}$$

$$e_y = -\frac{\nu C}{E} \tag{5.21}$$

$$e_z = -\frac{\nu C}{E} \tag{5.22}$$

$$\gamma_{xy} = \gamma_{yz} = \gamma_{zx} = 0 \tag{5.23}$$

so that the strain compatibility conditions are also satisfied. For a cylindrical body, shown in Fig. 5.1, on the end faces which are perpendicular to the x-axis (so that $l = 1$, $m = n = 0$), the boundary conditions give

$$\bar{x} = l\sigma_x = \sigma_x \tag{5.24}$$

and the other equations which hold for the end faces and also for the longitudinal outside surface (for which $l = 0$, m and $n \neq 0$), are identically zero. This means that on the end planes normal to the x-axis the surface force per unit area must equal the constant stress, C. On all other boundary surfaces the stress is zero. This is in agreement with the conditions on the body which we wish to investigate.

In other words, the assumed form of the stress tensor, Eq. 5.19, corresponds to simple tension or compression in a cylindrical bar — tension if C is positive, compression if C is negative.

Let us now determine the deformations of the structure.

In order to do this we must use Hooke's Law. We have, from Eqs. 5.20–5.23,

$$e_x = \frac{C}{E} = \frac{F_x}{A_x E} = \frac{\partial u}{\partial x} \tag{5.25}$$

$$e_y = -\frac{\nu F_x}{A_x E} = \frac{\partial v}{\partial y} \tag{5.26}$$

$$e_z = -\frac{\nu F_x}{A_x E} = \frac{\partial w}{\partial z} \tag{5.27}$$

$$\gamma_{xy} = \frac{\partial u}{\partial y} + \frac{\partial v}{\partial x} = 0 \tag{5.28}$$

$$\gamma_{yz} = \frac{\partial v}{\partial z} + \frac{\partial w}{\partial y} = 0 \tag{5.29}$$

$$\gamma_{zx} = \frac{\partial w}{\partial x} + \frac{\partial u}{\partial z} = 0 \tag{5.30}$$

in which F_x/A_x is the constant stress, caused by a force F_x (applied at the centroid) in the x direction, acting on a plane, A_x normal to the x-axis.†

If we integrate the first of the above, using as the datum point or point of zero elongation the origin, $x = 0$, we get

$$u = \frac{F_x x}{A_x E} \tag{5.31}$$

Similarly, from the second and third equations,

$$v = -\frac{\nu F_x y}{A_x E} \tag{5.32}$$

$$w = -\frac{\nu F_x z}{A_x E} \tag{5.33}$$

It may easily be verified by substitution that these values for u, v, and w satisfy the partial differential strain expressions given above.

The expression for u, the deformation in the x direction, will be recognized as the elementary relation for the elongation (if F_x is tension) predicted by Hooke's Law. (See Eq. 4.62 of the last chapter.)

The equations for v and w (the deformations in the y and z directions, respectively) represent the Poisson ratio effects corresponding to lateral

†We have in this example an illustration of St. Venant Principle, described in Art. 4–8. Although the exact requirement of the solution is that the end face be subjected to a *unit stress* of value σ_x, for engineering purposes this is equivalent to an applied *force* on the end face, of such value that the force divided by area is equal to σ_x. Then, at distances greater than the depth of the bar away from the end face, the stress distribution in the bar is essentially the same in both cases. The stress σ_x is, of course, applied at *both* end faces so that the bar is in equilibrium.

contractions (or elongations) caused by longitudinal elongation (or contraction); see Eq. 4.63.

The above represents an exact solution of a problem in the linearized theory of elasticity. It represents the solution to the problem previously discussed in Fig. 4.5, in which we described the results of a simple experimental investigation. We wish to emphasize that although the mathematical solution obtained above is a simple one it is by no means a trivial one. Indeed, it represents an exact solution to a very important physical problem and as such is of considerable practical interest. And because it is an exact solution, it represents a possible point of departure for simplifications and approximations.

We consider one of these approximations in the next article, in which is discussed the engineering generalization of the above exact solution — the simple truss.

5-3 The Truss in the Theory of Elasticity

The structural truss, in ordinary usage, is a structure made up of straight bars and subjected to some external loading; see Fig. 5.2.

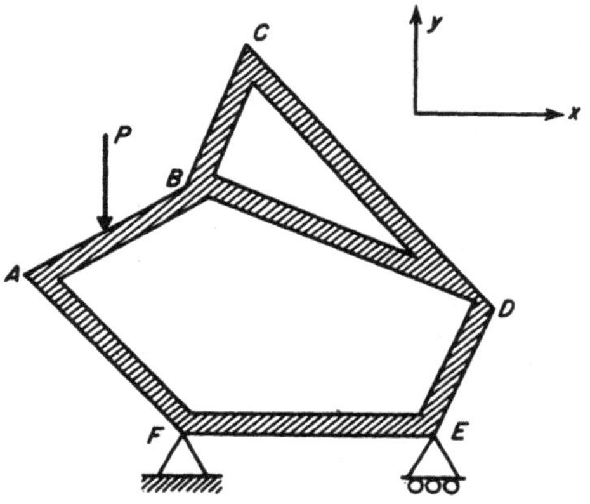

Fig. 5.2

The structure $ABCDEF$ is the truss, the supports are at E and F, the external loading is P, and AB is a typical member of this truss. The truss shown is a plane truss, since all members and the loading are in the x-y plane. A more general, and more complicated, structure would be the space truss, in which members and loads may lie in any direction.†

†As noted, we consider in this discussion *plane trusses* only, that is, trusses whose members may all be assumed to lie in the x-y plane. Although in the general case the truss structure is three-dimensional, the two-dimensional truss is used more frequently than the latter, and in any event, the solution of the three-dimensional truss simply utilizes the space generalizations of the planar relations developed here.

For the truss shown, the linearized theory of elasticity requires that we find stresses and strains which (neglecting body forces) satisfy the following relations (shown in tensor form):

Equilibrium —

$$\operatorname{div} T = 0 \quad \text{(See Eq. 4.72)} \tag{5.34}$$

Boundary conditions —

$$\bar{x} = \bar{N}\, T \quad \text{(See Eq. 4.76)} \tag{5.35}$$

Compatibility condition and Hooke's Law (assuming stresses are proportional to strains) —

$$\frac{\partial^2 e_x}{\partial y^2} + \frac{\partial^2 e_y}{\partial x^2} = \frac{\partial^2 \gamma_{xy}}{\partial x\, \partial y} \tag{5.36}$$

(see Eq. 4.77), and

$$\eta = \frac{1+\nu}{E} T - \frac{\nu}{E} \mathcal{J}_1 \mathcal{E}_2 \tag{5.37}$$

(see Eq. 4.82).

No solution to the above equations for the general truss structure has ever been obtained.

5-4 The Engineering Solution of the Truss

In order to obtain solutions, the engineering approximations of the more exact linearized theory of elasticity must be used. The particular approximations utilized are the following:

1. We assume all bars are pin-connected or hinged at their ends; in other words, the support condition at the end of each bar is such that forces only and not moments can be induced at the ends.‡

A typical hinged or pinned end for a truss is shown in Fig. 5.3. Note

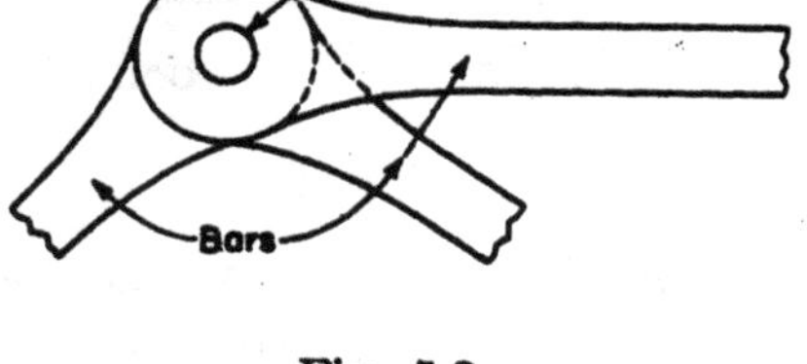

Fig. 5.3

‡Although most trusses being built today have riveted or bolted joints at the ends of all members (so that the true pin-ended condition is rarely met with in practice), experience and many tests indicate that we may still analyze the lighter building and bridge trusses using the assumption of pin-endedness. It is only for the very heavy trusses, having massive members with very rigid end gusset plates, that this assumption is not permissible; in these cases it is necessary to consider the so-called "secondary stresses" introduced by the rigid end connections. See Ref. 26 for a more complete discussion of this.

that the bars have holes or eyelets through which a pin or bolt is inserted. The bearing of the pins on the bars is assumed to be frictionless.

2. We assume that all loads are applied only at the joints.

This means, for example, that a truss load — even if not applied at the joint — must be transferred to the joints in some reasonable manner. One way of doing this is shown for the concentrated load of Fig. 5.4a.

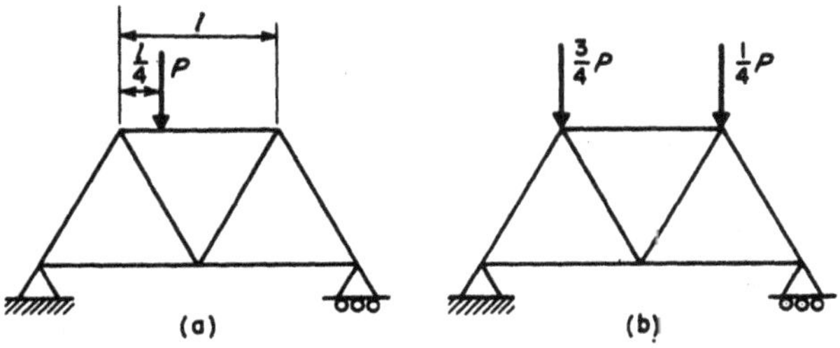

Fig. 5.4

Load P, acting at the quarter point, is transferred to the joints on the assumption of simple statics, i.e., three-quarters of the load is transferred to the near joint and one-quarter to the far joint.

Assumptions (1) and (2) together mean that the resultant loads in each bar must be applied along the center line of the bar — for, if the end loads had a resultant at one end (see Fig. 5.5) that was not along the line of the

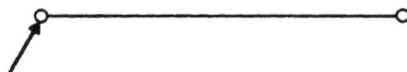

Fig. 5.5

bar, then (as a free body analysis of the bar indicates), the bar would tend to turn about the other support and it could not be in static equilibrium.

Thus, we are able to analyze the truss as a composite structure, each member or bar of which is a simple tension-compression member corresponding to the exact solution obtained in the previous section.

With the assumptions made above, it follows that at each joint of the truss we can apply the equilibrium relations

$$\Sigma F_H = 0 \tag{5.38}$$

$$\Sigma F_V = 0 \tag{5.39}$$

Therefore, if j is equal to the number of joints in the truss, there are $2j$ independent equations of statics which can be utilized. Also, if r is equal to the number of reactions and m is equal to the number of members, it follows that we can completely solve the truss (i.e., determine the reactions and loads in the members) if†

$$2j = m + r \tag{5.40}$$

Under these circumstances, the truss is said to be "statically determinate," i.e., we can completely solve it by utilizing the equations of statics only.

To illustrate all of the above, let us consider the truss and loading shown in Fig. 5.6.

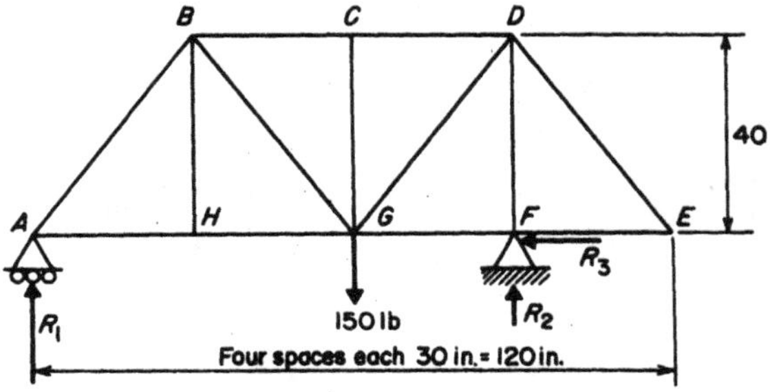

Fig. 5.6

First, we check:

$$2j \stackrel{?}{=} m + r \tag{5.41}$$

We have

$$\left.\begin{aligned} j &= 8 \\ r &= 3 \\ m &= 13 \end{aligned}\right\} \tag{5.42}$$

so that

$$2(8) = 3 + 13 \tag{5.43}$$

and the structure is statically determinate.

In order to solve the structure we may utilize the general engineering form of the equilibrium equations, *applied to any part of the truss as a free body*. Thus, although we noted that at each joint, we have

$$\Sigma F_V = 0 \tag{5.44}$$

$$\Sigma F_H = 0 \tag{5.45}$$

†The criterion of Eq. 5.40 is a *necessary* condition that the truss be statically determinate, but it is not a *sufficient* one. In addition to determinateness, in truss analysis we must consider the question of *stability*, the condition which keeps a truss from collapsing under any possible loading. The complete requirement on statical determinateness, therefore, implies a truss (a) satisfying Eq. 5.40 and (b) *being stable*.

and that we therefore have $2j$ independent equations of statics, still we are at liberty to use other permissible equations of equilibrium for this structure, if we find it convenient or expedient to do so. We generally find it is convenient to use these. In the present case, for example, we could go from joint to joint, applying at each the equations

$$\Sigma F_H = 0 \tag{5.46}$$

$$\Sigma F_V = 0 \tag{5.47}$$

This technique is frequently called *the method of joints.* We can, however, save considerable time and labor by utilizing, in addition to the method of joints, two other very commonly used techniques, *the method of sections* and *the method of moments.*

We shall illustrate all three for the truss in question.

First, we determine R_1 by taking moments about joint F.

$$R_1(90) = 150(30) \tag{5.48}$$

$$R_1 = 50 \text{ lb} \tag{5.49}$$

To get R_2 we take, for the entire structure,

$$\Sigma F_V = 0 \tag{5.50}$$

or,

$$R_1 + R_2 = 150 \tag{5.51}$$

$$R_2 = 150 - 50 = 100 \text{ lb} \tag{5.52}$$

R_3 is obtained by noting, that since $\Sigma F_H = 0$,

$$R_3 = 0 \tag{5.53}$$

Let us apply the method of joints to joint A; see Fig. 5.7.

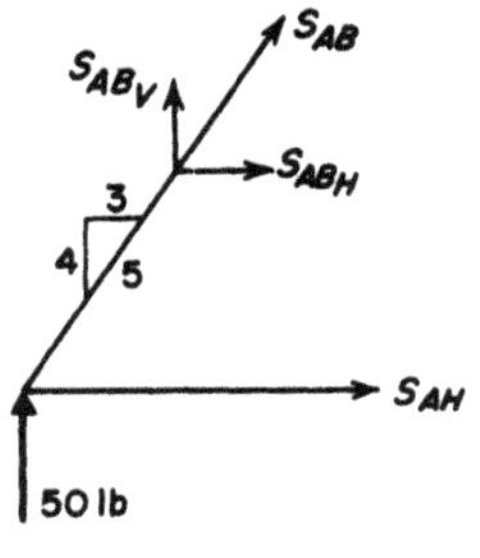

Fig. 5.7

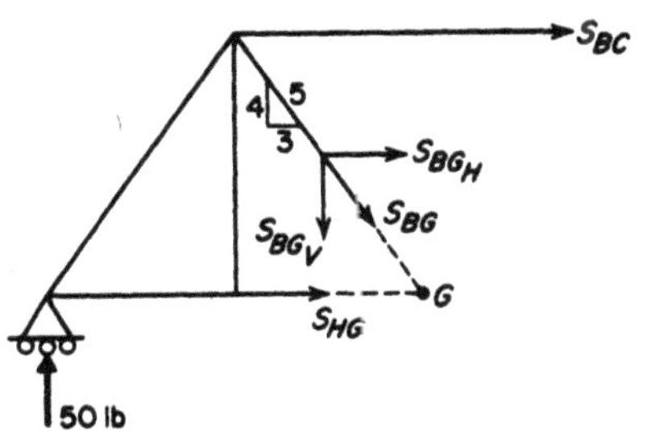

Fig. 5.8

Since the force in each member must act along the member, we assume the unknown loads S_{AB} and S_{AH} to be pure tension loads as shown. Applying to the joint the equation

$$\Sigma F_V = 0 \tag{5.54}$$

we have

$$50 + S_{AB_V} = 0 \tag{5.55}$$

and since

$$S_{AB_V} = \tfrac{4}{5} S_{AB} \tag{5.56}$$

we get

$$S_{AB} = -\frac{250}{4} \text{ lb} \tag{5.57}$$

and we see that the force in the member is a compression force. Now applying

$$\Sigma F_H = 0 \tag{5.58}$$

we have

$$S_{AB_H} + S_{AH} = 0 \tag{5.59}$$

and since

$$S_{AB_H} = \tfrac{3}{5} S_{AB} \tag{5.60}$$

we get

$$S_{AH} = \frac{3}{5} \times \frac{250}{4} \tag{5.61}$$

$$= +\frac{150}{4} \text{ lb} \tag{5.62}$$

The load in member AH is tension, as assumed.

The above procedure is typical of joint analysis. In general, it can be carried out for each joint of the truss. However, as stated above, great savings in time and effort can be made by using, in addition to the method of joints, the other commonly used procedures. For example, to determine the loads in BG, take a section through the second panel of the truss, and consider the left-hand portion as a free body. This is the method of sections, and it gives the free body structure shown in Fig. 5.8.

Note again that the unknown bar loads are assumed to be tension loads. Now, applying

$$\Sigma F_V = 0 \tag{5.63}$$

to the entire free body shown, we have

$$50 - S_{BG_V} = 0 \tag{5.64}$$

or

$$S_{BG_V} = +50 \tag{5.65}$$

and since

$$S_{BG} = \tfrac{5}{4} S_{BG_V} \tag{5.66}$$

we have

$$S_{BG} = +\frac{250}{4} \text{ lb} \tag{5.67}$$

which indicates that the load in member *BG* is a tension load, as assumed. This technique is typical of the method of sections.

Finally, we illustrate the method of moments by referring once more to Fig. 5.8. Let us take moments for the free body shown about point *G*, which is the intersection of cut members *BG* and *HG* extended. Because point *G* is along the lines of action of *BG* and *HG*, the forces in these members have no moment about an axis through *G* normal to the plane of the truss. We have (see Eq. 4.91)

$$\Sigma M_G = 0 \tag{5.68}$$

$$50(60) + S_{BC}(40) = 0 \tag{5.69}$$

$$S_{BC} = -75 \text{ lb} \tag{5.70}$$

and the minus sign indicates that the load in member *BC* is a compression load.

In this manner, using the method of joints, the method of sections and the method of moments wherever each one can best be employed, one may solve for the loads in all members of the truss.

5-5 The Deformation of the Engineering Truss

The elongation of each bar is given by Eq. 5.31, which is the elementary Hooke's Law relation. Compatibility of deformation requires that the structure deform in such a manner that all members meet at the joints corresponding to their end points in the undeformed position.

A simple example of a deformation analysis for a truss which involves the compatibility of deformation concept is given for the truss of Fig. 5.9.

We wish to determine δ_B, the deflection of the joint *B* under the load shown. The area of each bar is shown on the figure. The material of each bar is steel, having a modulus of elasticity of 30×10^6 psi.

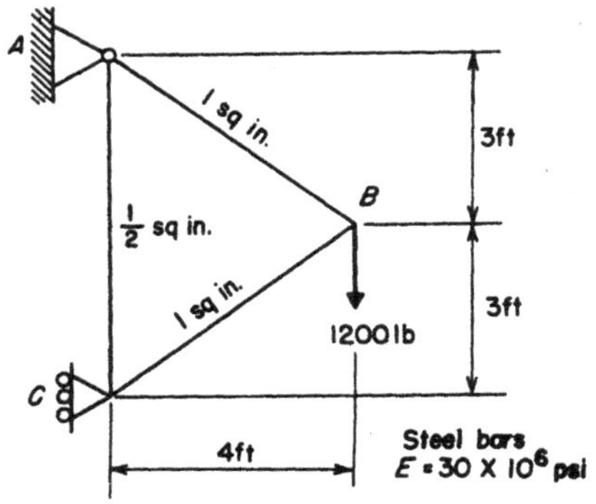

Fig. 5.9

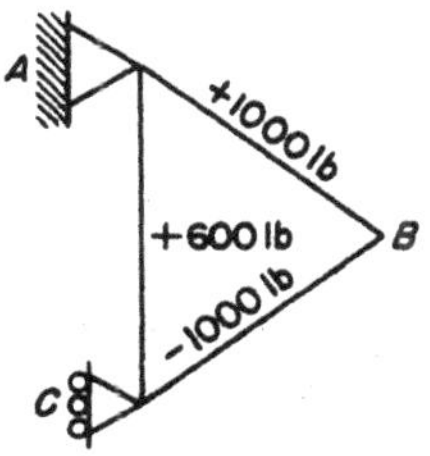

Fig. 5.10

The first step in the solution is to determine the load in each bar. This is done using any of the methods — joints, sections, or moments — whichever is most suitable. The loads in the members determined in this way are shown on Fig. 5.10. Note that tension loads are shown +, compression loads are shown −.

Now the elongation or shortening of each bar is determined, using Eq. 5.31 or

$$\epsilon = \frac{Fl}{AE} \tag{5.71}$$

so that

$$\epsilon_{AB} = +\frac{1000(60)}{1(30 \times 10^6)} = 2 \times 10^{-3\prime\prime} \text{ elongation} \tag{5.72}$$

and

$$\epsilon_{BC} = 2 \times 10^{-3\prime\prime} \text{ shortening} \tag{5.73}$$

and

$$\epsilon_{AC} = +\frac{(600)(72)}{(\frac{1}{2})(30 \times 10^6)} = 2.88 \times 10^{-3\prime\prime} \text{ elongation} \tag{5.74}$$

Next, starting at point A, which is a fixed point, and proceeding in the direction AC, which is a fixed direction, we can say that joint C moves to point C' corresponding to ϵ_{AC}, as shown in Fig. 5.11.

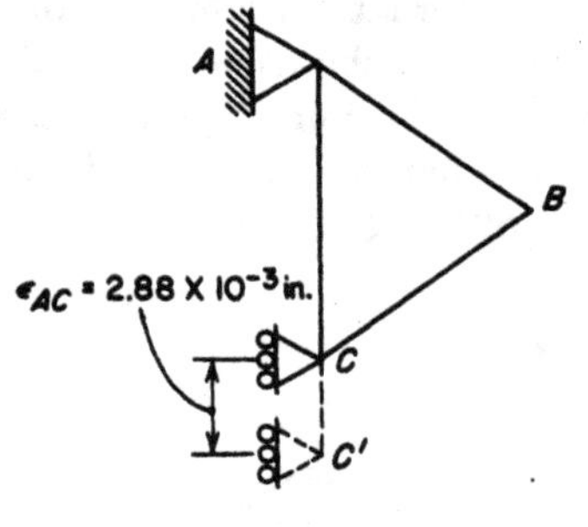

Fig. 5.11

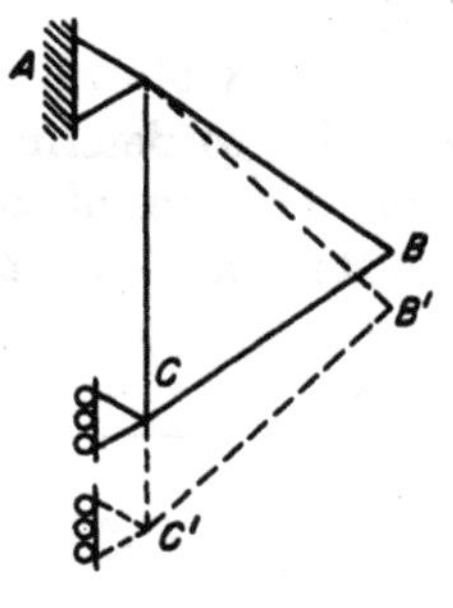

Fig. 5.12

To find the location of B', the point to which B moves, we swing arcs from A and from C', of lengths $(AB + \epsilon_{AB})$ and $(CB + \epsilon_{CB})$, respectively. The intersection of these arcs will locate the true point, B', and the deformed truss will be as shown in Fig. 5.12. The dotted lines represent the truss with compatible deformations.

It should be noted that although the deformation procedure given above is one which in theory may be used to determine the deformations of a truss, in practice it is inconvenient to swing arcs of the necessary lengths. Hence simpler methods must be used — and these are available. See in this connection Refs. 25 and 26.

5-6 Summary

In this chapter we discussed first the very simplest form which the stress tensor of the theory of elasticity can assume and showed that it corresponded to simple tension-compression of a bar.

The generalized engineering form of this structure, the truss, was then introduced. After pointing out the necessary assumptions required to permit one to solve this structure, the method of solution was presented and a typical simple truss was solved. Following this, a typical truss deformation problem was solved in order to illustrate the significance and meaning of compatibility of deformations when applied to a truss.

Problems

1. If $m + r > 2j$, then the structure is *statically indeterminate*, and the difference between $m + r$ and $2j$ represents the *degree of redundance* or number of redundants. In the following trusses,

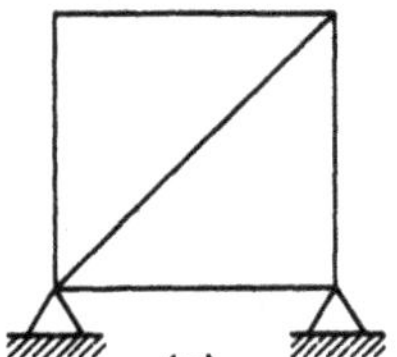

(a)

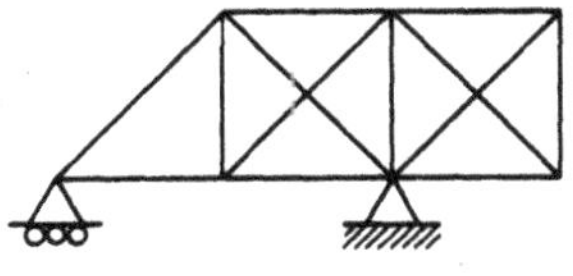

(d)

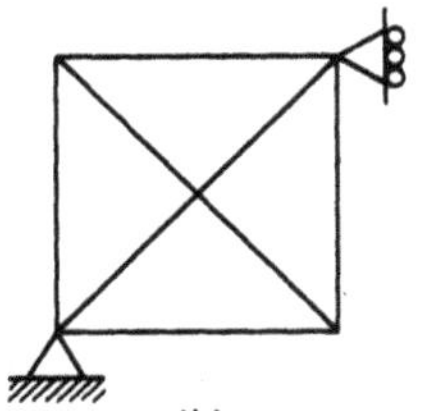

(b)

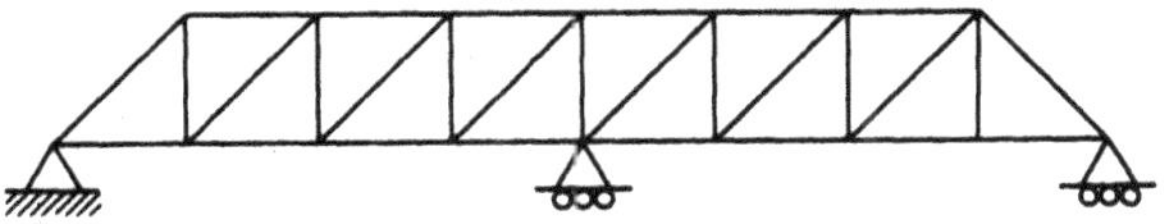

(e)

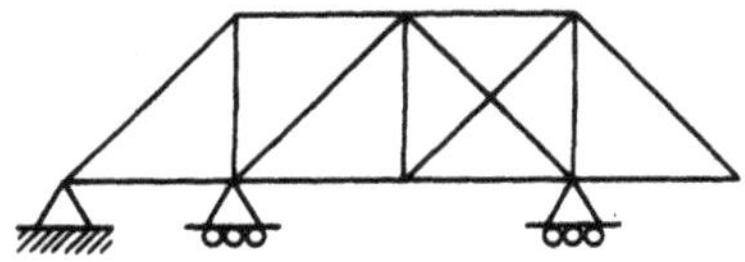

(c)

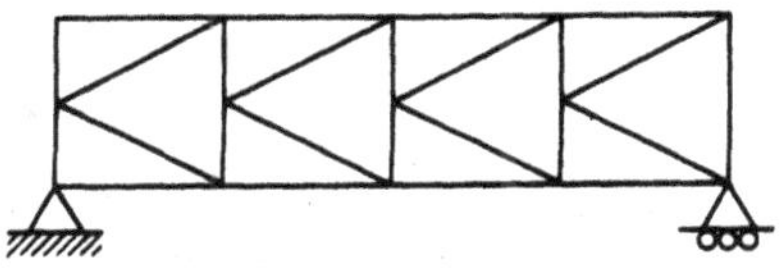

(f)

(a) determine whether the trusses are statically determinate;
(b) if they are not determinate, find the number of redundants, or extra members, and/or reactions.

2. Solve each of the following trusses for all reactions and member loads.

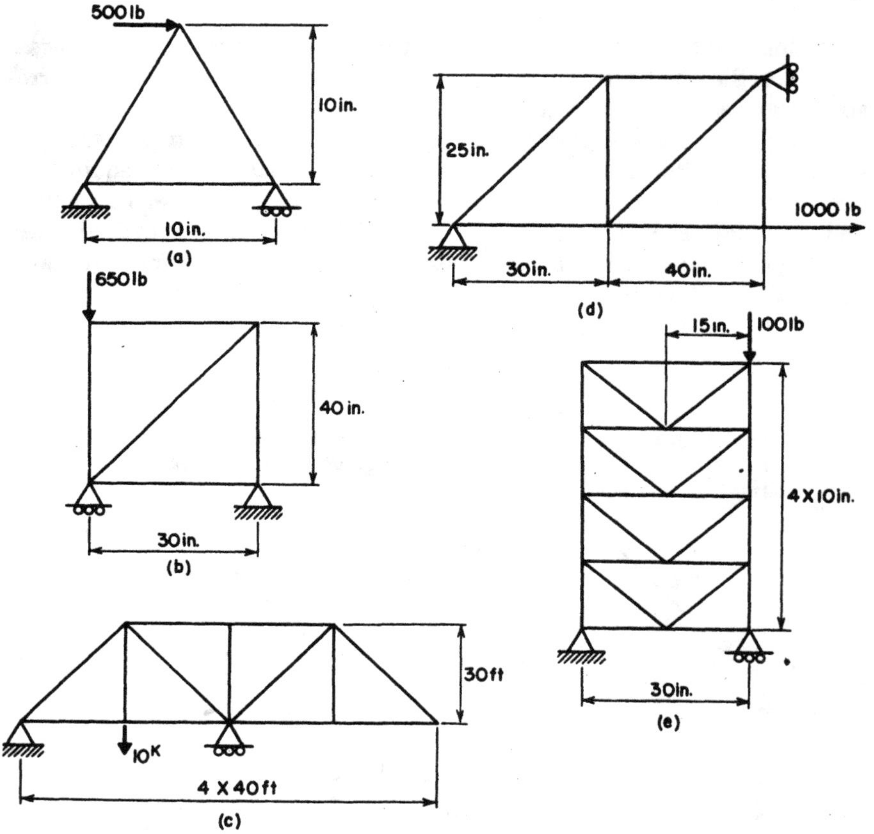

3. Show that the resultant force, F_x, of the uniform stress, σ_x, acting on a plane normal to the x-axis, is applied at the *centroid* of the plane.

4. By adding equal and opposite colinear pairs of forces (and therefore by adding, essentially, nothing) show that the force, F, in the figure is equivalent to the sum of
(a) a force, F, acting at the centroid, O;
(b) a couple, Fa, about the x-axis;
(c) a couple, Fb, about the y-axis.

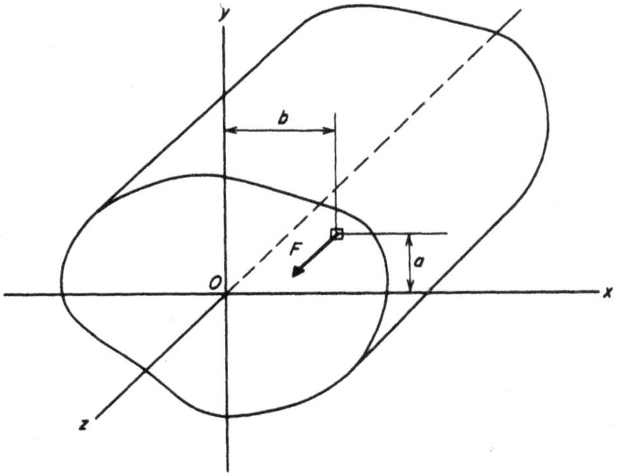

5. (a) By considering the small deformation of a unit cube, show that the *change in volume* of an elementary volume due to a pure dilatation (i.e., normal strains only) is given by

$$\Delta V = e_x + e_y + e_z$$

(b) Determine what the value of Poisson's ratio, ν, must be in order that the deformation due to $\sigma_x = c$ may be an *incompressible* one, i.e., in order that the volume may not change.

(c) Can you suggest, from (b), why the value, $\nu = \frac{1}{2}$, should be the maximum possible value of Poisson's ratio?

(d) If

$$\sigma_x = C$$
$$\sigma_y = D$$
$$\sigma_z = F$$

where C, D and F are constants, determine what ν must be in order that the deformation may be an incompressible one.

(e) If all deformations are assumed to be small, show that (with only first order effects being considered) a pure shear strain does not change the volume of the elementary volume.

Chapter 6

THE SIMPLEST SHEAR-TENSION-COMPRESSION STRUCTURE — THE BEAM.

6-1 Introduction

In the last chapter we considered the simplest form that the stress tensor may assume and found that this corresponded to the tension-compression bar, which generalizes into the engineering structure, the truss.

In this chapter we consider the stress tensor one degree more complex, that is, instead of the single tensor element being a constant, we assume that it varies linearly. It will be seen that in the exact linearized theory of elasticity this corresponds to the bending of a bar, i.e., a beam subjected to a bending moment.

Next we discuss the engineering approximation of this structure: the Bernoulli-Euler beam solution. We shall indicate the connection between this form of solution and the more exact solution, and shall derive the various differential relations between deflection, slope, moment and shear which follow from the Bernoulli-Euler formulation.

Following this we discuss the necessary variation that the shear stress must have across the cross section of the engineering beam, in order that stress compatibility be satisfied.

Finally, we present a development which gives a measure of the amount of approximation involved in the engineering elasticity beam analysis as compared to the more exact solution given by the linearized mathematical theory of elasticity.

6-2 The Pure Moment Acting on a Bar — the Theory of Elasticity Solution

Let us consider the linearized theory of elasticity solution for the stress tensor given by

$$T = \begin{pmatrix} \sigma_x & 0 & 0 \\ 0 & 0 & 0 \\ 0 & 0 & 0 \end{pmatrix} = \begin{pmatrix} Ky & 0 & 0 \\ 0 & 0 & 0 \\ 0 & 0 & 0 \end{pmatrix} \qquad (6.1)$$

for a cylindrical body as shown in Fig. 6.1. It is clear that K is the stress at a unit distance, y, from the origin. We shall neglect body forces.

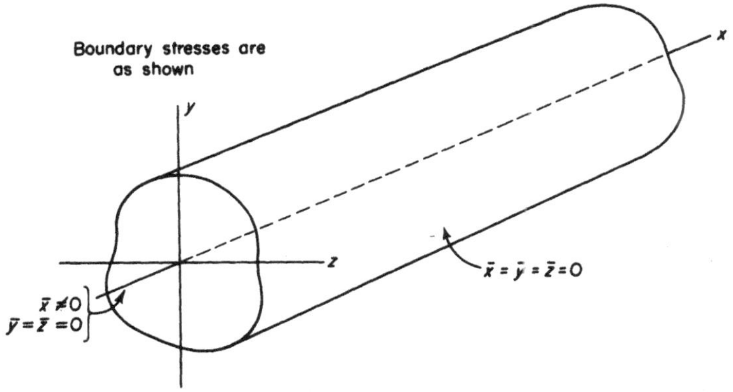

Fig. 6.1

The equilibrium equations (see Eq. 4.22),

$$\text{div } T = 0 \tag{6.2}$$

are obviously indentically satisfied by the assumed stress tensor (see Eq. 6.1).

Since on the end faces $l = \pm 1$, $m = 0$, $n = 0$, and on the side surface $l = 0$, $-1 \leq m \leq 1$, $-1 \leq n \leq 1$, the boundary conditions on the end faces become

$$\bar{x} = \sigma_x \tag{6.3}$$

and all other boundary conditions are identically satisfied, $0 = 0$.

Hooke's Law gives

$$e_x = \frac{1}{E}\sigma_x = \frac{Ky}{E} \tag{6.4}$$

$$e_y = -\frac{\nu\sigma_x}{E} = -\frac{\nu Ky}{E} \tag{6.5}$$

$$e_z = -\frac{\nu\sigma_x}{E} = -\frac{\nu Ky}{E} \tag{6.6}$$

$$\gamma_{xy} = \gamma_{yz} = \gamma_{zx} = 0 \tag{6.7}$$

so that the compatibility equations are also satisfied identically, as the student may verify.

Thus, all formal requirements of the linearized mathematical theory of elasticity are satisfied by the assumed stress tensor.

We shall insist on the additional requirement that the *net* force be zero

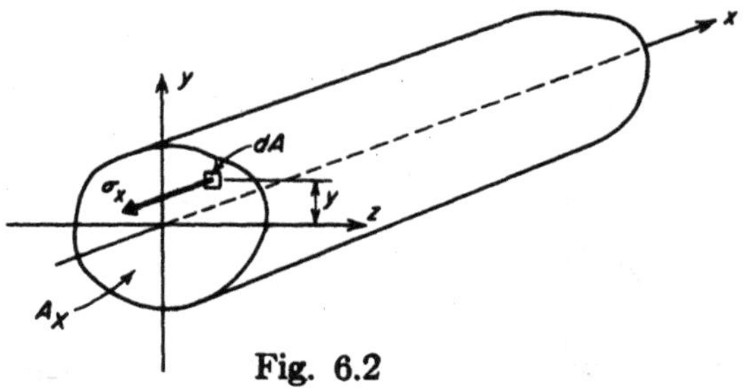

Fig. 6.2

on the end faces (and hence on all planes) which are perpendicular to the x-axis; see Fig. 6.2. This means that (at all cross-sections, A_x)

$$\int_{A_x} \sigma_x dA = 0 \tag{6.8}$$

or

$$K \int_{A_x} y dA = 0 \tag{6.9}$$

Hence the z-axis is a centroidal axis.† The z-axis, which is the line of zero stress, is called the neutral axis, and is usually designated as "N.A."

Let us determine the moment of σ_x about the z-axis. This is given by (at all cross-sections, A_x)

$$\int_{A_x} \sigma_x y dA \tag{6.10}$$

which gives

$$K \int_{A_x} y^2 dA \tag{6.11}$$

We call this quantity M_z, the bending moment about the z-axis. Thus,

$$M_z = KI_z \tag{6.12}$$

in which I_z is the area moment of inertia of the cross section about the z-axis; see Sec. 3-2. Note: Eq. 6.12 holds at all cross-sections, A_x.

Hence, K, the stress per unit distance from the centroidal axis, is given by

$$K = \frac{M_z}{I_z} \tag{6.13}$$

and the stress σ_x is given by

$$\sigma_x = Ky = \frac{M_z y}{I_z} \tag{6.14}$$

†In fact, the requirement that $M_y = 0$ (i.e., that we have a pure moment about the z axis), leads to $I_{yz} = 0$, or the z axis is also a *principal* axis. See Art. 3-2.

The above equation shows that at any section the stress varies linearly across the cross section from the neutral axis. This is as shown in Fig. 6.3.

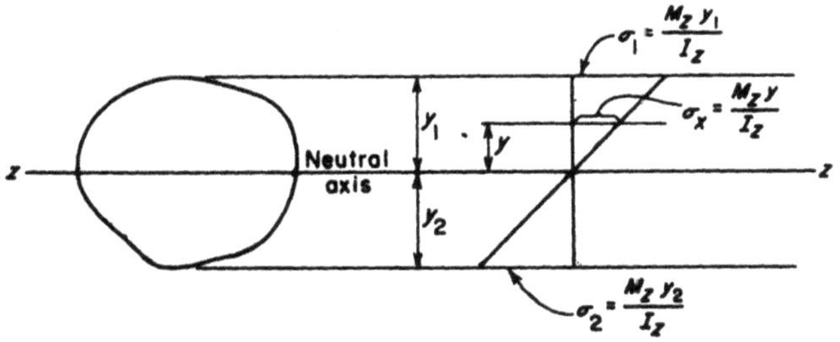

Fig. 6.3

It is clear that the maximum stress on the cross section occurs at the point furthest away from the neutral axis. This is of fundamental importance in the actual design of engineering beams, as will be described in Chapter 8.

To determine the deformations, we have from Hooke's Law, given in Eq. 4.60, and from Eqs. 6.4–6.7,

$$\frac{\partial u}{\partial x} = \frac{Ky}{E} \tag{6.15}$$

$$\frac{\partial v}{\partial y} = -\frac{\nu Ky}{E} \tag{6.16}$$

$$\frac{\partial w}{\partial z} = -\frac{\nu Ky}{E} \tag{6.17}$$

$$\frac{\partial u}{\partial y} + \frac{\partial v}{\partial x} = \frac{\partial v}{\partial z} + \frac{\partial w}{\partial y} = \frac{\partial w}{\partial x} + \frac{\partial u}{\partial z} = 0 \tag{6.18}$$

and if these are solved for the deformations u, v and w, we find (see Ref. 18, for example) that

$$u = \frac{Kxy}{E} \tag{6.19}$$

$$v = -\frac{\nu Ky^2}{2E} + \frac{\nu Kz^2}{2E} - \frac{Kx^2}{2E} \tag{6.20}$$

$$w = -\frac{\nu Kyz}{E} \tag{6.21}$$

These deformations, for a rectangular cross section, represent the anticlastic, or saddle, surface, shown in Fig. 6.4. The student can verify this configuration by grasping a soft rubber eraser at its ends and bending it into an arc.

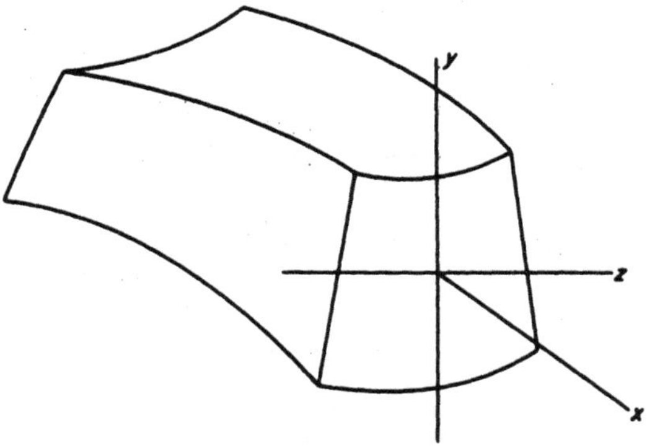

Fig. 6.4

Also the line x = variable ($y = z = 0$) deflects, as given by

$$u = 0 \tag{6.22}$$

$$v = -\frac{Kx^2}{2E} \tag{6.23}$$

$$w = 0 \tag{6.24}$$

In particular, we may note that, since for this line,

$$v = -\frac{Kx^2}{2E} \tag{6.25}$$

it follows that, for this line,

$$\frac{d^2v}{dx^2} = -\frac{K}{E} = -\frac{M_z}{EI_z} \tag{6.26}$$

The right-hand side of Eq. 6.26 is a sort of *invariant* form in that it depends only on the applied moment, the cross section of the bar and the material of the bar — but not on the coordinates. We shall make use of this very important relation in our engineering solution, which will be discussed in the next section.

Summarized here is the exact solution obtained fer the assumed stress tensor:

1. The solution obtained is for a pure moment, M_z, constant along the length of the bar.†
2. The stress varies linearly from the neutral axis (which is also a principal centroidal axis of the cross section) and is given by

$$\sigma_x = \frac{M_z y}{I_z} \tag{6.27}$$

3. The cross section deformation is given by

$$u = \frac{Kxy}{E} \tag{6.28}$$

$$v = -\frac{\nu K y^2}{2E} + \frac{\nu K z^2}{2E} - \frac{K x^2}{2E} \tag{6.29}$$

$$w = -\frac{\nu K y z}{E} \tag{6.30}$$

4. The line x = variable ($y = z = 0$) deflects in accordance with

$$v = -\frac{K x^2}{2E} \tag{6.31}$$

and

$$\frac{d^2 v}{dx^2} = -\frac{M_z}{E I_z} \tag{6.32}$$

6-3 The Engineering Elasticity Solution for the Beam — the Bernoulli-Euler Solution

In the previous article it was shown that the exact solution of the linearized equations of elasticity corresponding to a pure moment, M_z, constant along the axis, x, of a cylindrical bar, gives

$$\sigma_x = \frac{M_z y}{I_z} \tag{6.33}$$

and

$$\frac{d^2 v}{dx^2} = -\frac{M_z}{E I_z} \tag{6.34}$$

In these equations σ_x is the stress at a distance y from the z-axis, I_z is the moment of inertia of the cross section about the z-axis, and E is the tension-compression modulus of elasticity; v is the deflection of the x-axis in the y direction.

†Although the solution obtained assumed that on the end faces the stress distribution is given by Eqs. 6.1 and 6.3, which were shown to correspond to a pure moment, in view of St. Venant Principle we can expect that the solution given will hold for sections away from the end no matter *how* the pure moment is applied on the end face. The only requirement is that the *resultant* of the stresses on the end face be equal to the moment of Eq. 6.12.

In the engineering application of the exact bending moment solution, the so-called Bernoulli-Euler solution, we take as a starting point the above two equations, which hold for a *constant* moment on the bar.

However, we make the assumption that *Eqs. 6.33 and 6.34 hold at all points on the bar even if the moment varies from point to point on the beam.* In other words, the equations

$$\sigma_x = \frac{M_z y}{I_z} \tag{6.35}$$

and

$$\frac{d^2v}{dx^2} = -\frac{M_z}{EI_z} \tag{6.36}$$

are assumed to be valid, locally, at all points on the beam.

In the engineering analysis, we say nothing about the deformations u and w: indeed, as a consequence of the above two relations, it follows (for stresses σ_x proportional to strain e_x), that a plane section normal to the x-axis before bending is also a plane section after the bending deformation (which is a pure rotation of the cross section about the neutral axis). Furthermore, the deformation of this cross section (due to bending only) is limited to this rotation only — no distortion of the cross section in the y and z directions is assumed to occur.† This means that the deflection of the line $y = 0$, $z = 0$, which is given by Eq. 6.23, is also, in the engineering beam, the deflection of the plane $y = 0$, the neutral plane.‡

Finally, since the curvature of the deflected line $y = 0$, $z = 0$, is given by

$$\frac{(d^2v)/(dx^2)}{\left[1 + \left(\frac{dv}{dx}\right)^2\right]^{3/2}} = \frac{1}{\rho_z} \tag{6.37}$$

it follows that, if we limit the beam to small deflections, so that

$$\frac{dv}{dx} \ll 1 \tag{6.38}$$

we can neglect $\left(\frac{dv}{dx}\right)^2$ in comparison to unity and we have

$$\frac{1}{\rho_z} = \frac{d^2v}{dx^2} = -\frac{M_z}{EI_z} \tag{6.39}$$

In other words, the local curvature of the beam at any point, x, is, in the engineering approximation, given by the right-hand side of Eq. 6.39. This expression is made use of later (see Art. 11-5) in the solution of the elastica.

†We are, in effect, using the approximate form of Hooke's Law as given in Eq. 4.93–4.94.

‡In fact, in all engineering solutions of beams we generally indicate the beam by means of a line. This is consistent with the above, and in deflection analyses the line may be assumed to represent the neutral axis.

Eq. 6.35 and 6.39 are the fundamental equations for the deflected beam in the Bernoulli-Euler theory of beams. They can also be obtained, starting with the assumption that plane sections of the beam before bending are plane after bending; however, for our purposes it was deemed preferable to arrive at them as a logical consequence of the exact solution of the linearized equations of elasticity, subject to the assumption noted above.

6-4 The Shear, Moment, Slope, Deflection Relations

In Art. 6-3 it was pointed out that the engineering beam may be analyzed by assuming that the exact solution of the linearized theory of elasticity (for a pure moment acting on a beam) applies also to the engineering beam, even though the moment varies in this latter structure.

Now, in the engineering beam, the moment varies from point to point on the beam because, in general, there are transverse loads acting on the beam. These transverse loads introduce *lateral shear forces* and stresses. In this section we shall consider these lateral forces. In particular, we shall derive the basic differential relations between the load, shear, moment, slope and deflection for the engineering structure. To do this, we consider a typical, laterally loaded beam, Fig. 6.5.

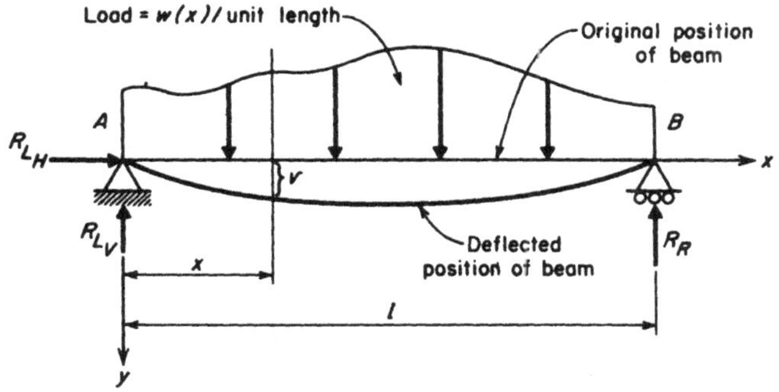

Fig. 6.5

Note that the beam is shown by a single line, which essentially represents the neutral plane, $y = 0$, of the beam. This is a permissible representation in view of the deformation approximations made in the engineering beam analysis; see Art. 6-3.

Also shown are the original and final positions of the beam. The student should note particularly that the assumption is made that the deformed horizontal projection length is the same as the undeformed horizontal length of the beam. In the linearized theory, that assumption is consistent with the assumption of small deformations. It may be shown that the horizontal movement of the roller, due to the lateral deformation, (see Art. 13-3) is given by

$$\Delta = \frac{1}{2}\int_0^l \left(\frac{dv}{dx}\right)^2 dx \tag{6.40}$$

This square of a small quantity, the slope dv/dx, can be neglected in the linear theory.

Note also that the left support is a hinged support and the right support is a roller support. That is, we have the reactions shown in the figure. However, the equation

$$\Sigma F_H = 0 \tag{6.41}$$

shows at once that

$$R_{LH} = 0 \tag{6.42}$$

We determine the other two reactions, R_{LV} and R_R by utilizing the engineering equilibrium equations:

$$\Sigma F_V = 0 \tag{6.43}$$

$$\Sigma M_{\text{about any point}} = 0 \tag{6.44}$$

Having the reactions and the lateral load (which is given), we can now determine the shear and moment at any point x on the beam. Before doing this, however, we define the shear and bending moment at any point x, and state the sign convention which will be used in this book.

Consider the beam and loading of Fig. 6.5. The shear, V, at any section, x, is the net value of all transverse loads to the left (or right) of the section. This shear is assumed positive if the net value of these transverse forces to the left of the section is up. This is shown in Fig. 6.6, which represents a

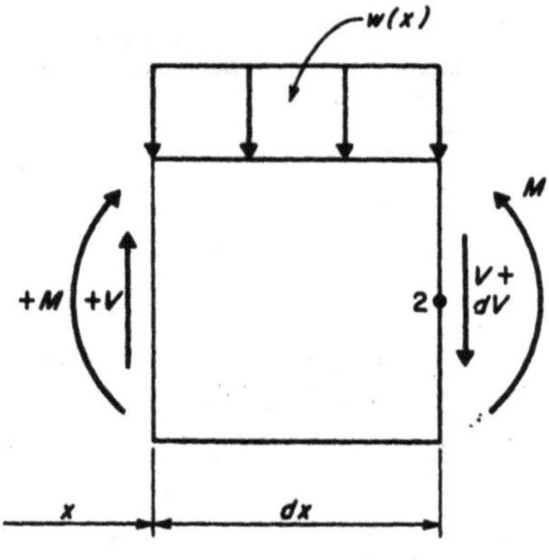

Fig. 6.6

differential portion of the beam at x. Also shown on this figure is the moment, positive as shown (see the next paragraph), and the lateral loading at x.

The bending moment at any station, x, on the beam is the moment at that section of all loads to the left (or to the right) of the section. We assume a moment is positive if it tends to bend the beam so that it is concave upwards. Otherwise, it is a negative moment. This is occasionally shown schematically as in Fig. 6.7.

Fig. 6.7

It is now possible to derive the four fundamental differential relations for the Bernoulli-Euler beam. We assume a beam and loading as shown in Fig. 6.8, with positive directions of axes as shown, and with positive shear and positive moment as defined above.

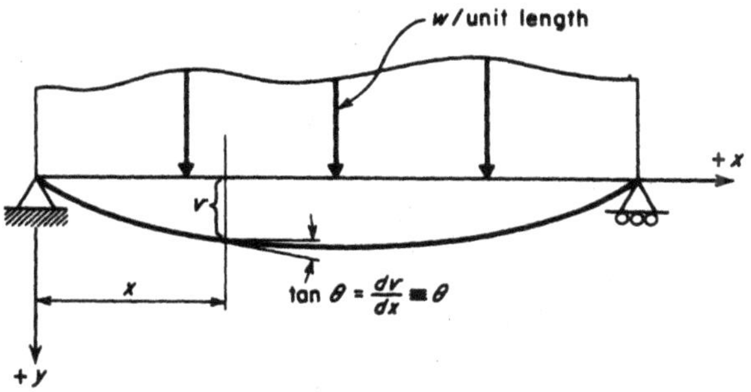

Fig. 6.8

We have, see the figure,

$$\frac{dv}{dx} = +\tan\theta \cong +\theta \tag{6.45}$$

Also, we have

$$\frac{d^2v}{dx^2} = -\frac{M}{EI_z} \tag{6.46}$$

and the sign before the M/EI_z is obtained as follows:

The curvature of the beam at x is obviously that due to a positive moment. The slope at x is positive, but it is getting smaller as x increases. Hence,

$$\frac{d\left(\frac{dv}{dx}\right)}{dx} < 0 \tag{6.47}$$

and, since the moment is positive, the equality of Eq. 6.46 will only hold if we have a negative sign before the M/EI_z term.

Return to Fig. 6.6. Take moments about axis 2, normal to the plane of the page:

$$M + V dx - w(x)\frac{(dx)^2}{2} - (M + dM) = 0 \tag{6.48}$$

The third term in this expression is a higher order differential and hence may be neglected. Therefore,

$$\frac{dM}{dx} = V = -\frac{d}{dx}\left(EI_z \frac{d^2v}{dx^2}\right) \tag{6.49}$$

and, if EI_z is constant along the beam, this becomes

$$-\frac{V}{EI_z} = \frac{d^3v}{dx^3} \tag{6.50}$$

Similarly, summing vertical forces of Fig. 6.6,

$$+V - w(x)dx - (V + dV) = 0 \tag{6.51}$$

or

$$w(x) = -\frac{dV}{dx} \tag{6.52}$$

and if EI_z is constant

$$\frac{w(x)}{EI_z} = +\frac{d^4v}{dx^4} \tag{6.53}$$

We repeat, next, for convenience, the four differential relations just obtained, for constant EI_z:

$$\frac{dv}{dx} = +\theta \tag{6.54}$$

$$\frac{d^2v}{dx^2} = -\frac{M}{EI_z} \tag{6.55}$$

$$\frac{d^3v}{dx^3} = -\frac{V}{EI_z} \tag{6.56}$$

$$\frac{d^4v}{dx^4} = +\frac{w(x)}{EI_z} \tag{6.57}$$

We emphasize this very important fact: the signs given on the right-hand sides of the above equations depend upon

1. the assumed positive directions for the coordinate axes, x and y;
2. the assumed sign convention for M, V and $w\ (x)$

The convention we had chosen was entirely arbitrary. It is possible to have sign conventions for M and V which are just opposite the ones we chose. Also the y-axis could have been taken as positive upward. In each case, the signs which would result in the four differential relations would have to be determined by analysing the particular beam subject to the conventions chosen. See, in this connection, Prob. 4 at the end of this chapter.

6-5 The Loads on an Engineering Beam

We may now consider in greater detail the ways in which actual beams are loaded. This will give us additional understanding of the manner in which shears and moments are involved in beam behaviour.

It will be remembered that the bending moment stress and deflection relations for the exact solution hold for a constant moment along the beam. However, the engineering beam is generally subjected to a transverse loading quite different from the constant moment. For example, a typical

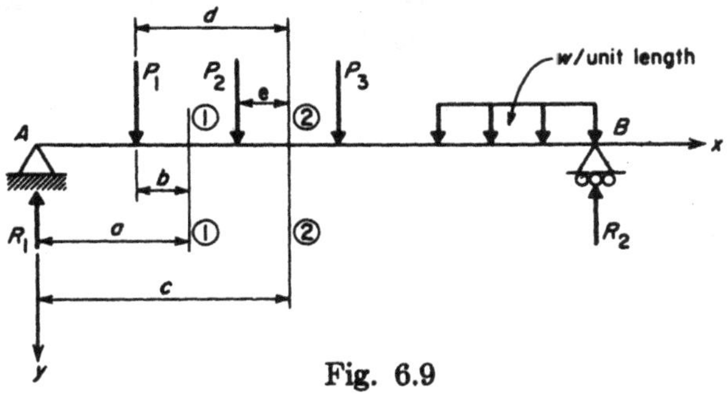

Fig. 6.9

beam and loading is shown in Fig. 6.9. In this beam the bending moment varies from point to point on the beam. Thus, at 1 - 1 it is given by

$$M_{1-1} = R_1 a - P_1 b \tag{6.58}$$

and at 2 - 2 we have

$$M_{2-2} = R_1 c - P_1 d - P_2 e \tag{6.59}$$

and so on for other stations on the beam. Note how the signs agree with the convention stated in Art. 6-4.

In the engineering elasticity, as pointed out before, we assume that the exact moment stress (given by Eq. 6.14) and moment deflection relations (given by Eq. 6.26) *hold locally at all points on the beam*, even though the moment varies along the beam. Thus, we assume for example that, at Sec. 1 - 1,

$$\sigma_x = \frac{M_{1-1} y}{I_z} \tag{6.60}$$

and

$$\frac{d^2 v}{dx^2} = -\frac{M_{1-1}}{EI_z} \tag{6.61}$$

Similarly, at Sec. 2 - 2 ,

$$\sigma_x = \frac{M_{2-2}y}{I_z} \tag{6.62}$$

and

$$\frac{d^2v}{dx^2} = -\frac{M_{2-2}}{EI_z} \tag{6.63}$$

By virtue of this assumption, it follows that for any portion of the beam, say between A and Sec. 1 - 1 ,

$$\Sigma F_x = 0 \tag{6.64}$$

and

$$\Sigma M_{\text{about any point}} = 0 \tag{6.65}$$

However, in the engineering structure we also have loads in the y direction, these being the transverse loads on the beam, the reactions, and in the case of a cut section such as 1 - 1 , also the transverse force acting on the section, the *shear force*.

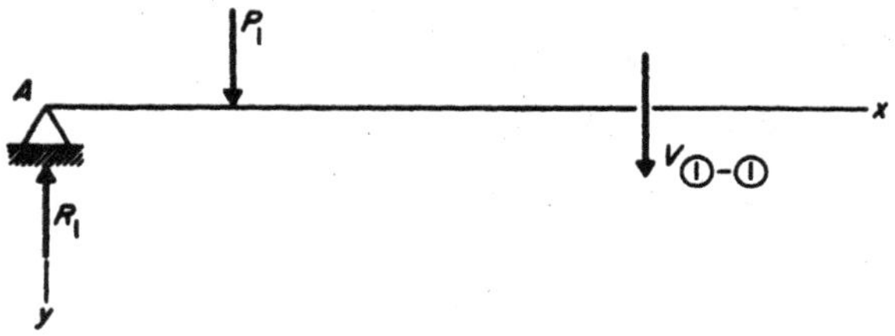

Fig. 6.10

It follows, from equilibrium of a typical free body, portion A to Sec. 1 - 1 (see Fig. 6.10), that at Sec. 1 - 1 we have acting a shear force V determined by

$$\Sigma F_y = 0 \tag{6.66}$$

or,

$$R_1 - P_1 - V_{1-1} = 0 \tag{6.67}$$

so that

$$V_{1-1} = R_1 - P_1 \tag{6.68}$$

In the same way, at Sec. 2 - 2 we have

$$\Sigma F_y = 0 \tag{6.69}$$

$$R_1 - P_1 - P_2 - V_{2-2} = 0 \tag{6.70}$$

or

$$V_{2-2} = R_1 - P_1 - P_2 \tag{6.71}$$

and so on at all other sections of the beam. Note how the signs chosen agree with our convention of the previous section.

Equations 6.64 through 6.71 are just the equations of equilibrium, Eqs. 4.89 through 4.91, which for emphasis we list once more below:

$$\Sigma F_x = 0 \tag{6.72}$$

$$\Sigma F_y = 0 \tag{6.73}$$

$$\Sigma M_{\text{about any point}} = 0 \tag{6.74}$$

6-6 The Shear Stress in the Engineering Beam

Eq. 6.14 gives the bending or normal stress distribution across any cross section subjected to a moment M_z. Because, in the engineering beam, we also have a net shear force acting on each cross section, it follows that there are shear stresses distributed in some fashion across each cross section.

However, as pointed out in the last chapter, it is necessary that the stresses be compatible in any structure. This requirement in the present case means that the shear *stress* distribution across the cross section can not be taken arbitrarily (as, for example, a uniformly distributed stress, or any other assumed distribution), but must be compatible with the bending stress distribution given by Eq. 6.14. To determine what this compatible shear stress distribution must be, we analyze a portion of the beam of length dx, at a distance x from the support; see Fig. 6.11a.

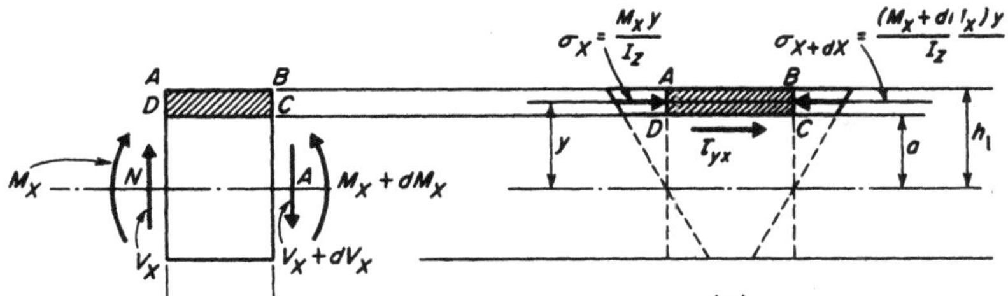

Fig. 6.11

On the face at x we have a moment M_x (we have changed the notation from M_z to M_x because we wish to emphasize here the variation of the moment with the coordinate x) and on the face at $x + dx$ we have a moment $M_x + dM_x$. We also have shears V_x and $V_x + dV_x$ acting as shown.

Due to the moments M_x and $M_x + dM_x$, we have bending stresses σ_x and σ_{x+dx}, as shown in Fig. 6.11b.

Let us take as a free body the portion of the beam bounded by $ABCD$, as shown in the figure. The width of the beam at plane DC is equal to b.

Now, because $M_z + dM_z$ is different from M_z, it follows that σ_{x+dx} will be different from σ_x and therefore the total force, due to σ_x, along AD will be different from the total force, due to σ_{x+dx}, along face BC. Hence, there *must* be a balancing force acting along the face DC. We show the stress corresponding to this force as τ_{yx}. Note τ_{yx} acts on the face DC. But in accordance with the relations given in Eq. 3.74 it follows that along face AD (at point D) we have a stress τ_{xy} which is just equal to τ_{yx}. In fact, it is this stress which we wish to determine, since it is the shear stress caused by the shear force V_x. The stress τ_{xy} is the compatible shear stress along the cross section due to the shear force V_x. Thus, conditions are as shown on the enlarged portion $ABCD$ shown in Fig. 6.12.

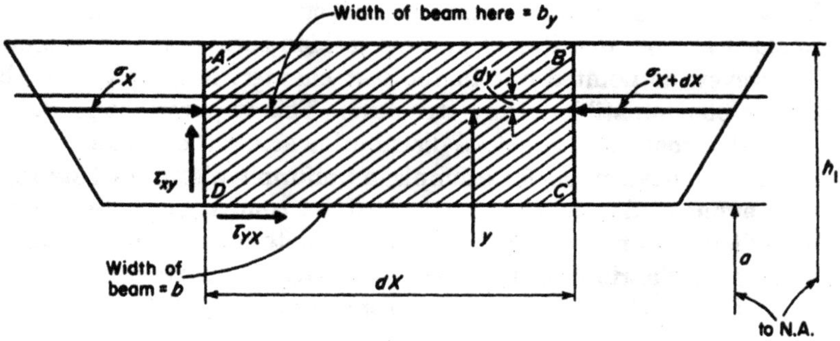

Fig. 6.12

Then, for the free body of Fig. 6.12, taking

$$\Sigma F_x = 0 \tag{6.75}$$

we have

$$\int_a^{h_1} \sigma_x b_y \, dy + \tau_{yx} b \, dx - \int_a^{h_1} \sigma_{x+dx} \, b_y dy = 0 \tag{6.76}$$

or since

$$\sigma = \frac{My}{I_z} \tag{6.77}$$

This becomes,

$$\frac{1}{I_z}\int_a^{h_1} M_x b_y y \, dy + b\tau_{yx} dx - \frac{1}{I_z}\int_a^{h_1} (M_x + dM_x) b_y y \, dy = 0 \tag{6.78}$$

or,

$$b\tau_{yx} dx = \frac{1}{I_z}\int_a^{h_1} dM_x b_y y \, dy \tag{6.79}$$

or,

$$b\tau_{yx} = \frac{1}{I_z}\int_a^{h_1} \frac{dM_x}{dx} b_y y \, dy \tag{6.80}$$

Now, as shown in Eq. 6.49,

$$\frac{dM_z}{dx} = V_z, \quad \text{the shear force} \tag{6.81}$$

and, since this is constant on the cross section and independent of y, it may be taken from under the integral sign, and we have finally

$$\tau_{yz} = \frac{V_z}{bI_z}\int_a^{h_1} b_y y\, dy \tag{6.82}$$

which is frequently shown as

$$\tau_{yz} = \frac{V_z Q}{bI_z} \tag{6.83}$$

In the above,

τ_{yz} is the shear stress, due to V_z at point D on the cross section;
V_z is the total shear force on the cross-section;
b is the width of the cross section at point D;
I_z is the moment of inertia *of the entire cross section* about the N.A.;
Q is the moment, about the N.A., of the cross section between the point at which the shear stress τ_{yz} is given and the top of the beam.

We repeat, for emphasis: τ_{yz} is the shear stress on the cross section, which is compatible with the bending stress given by $\sigma = My/I_z$. Eq. 6.83 is the formula which, in the engineering elasticity, shows how the shear stress is distributed across the cross section of a beam.

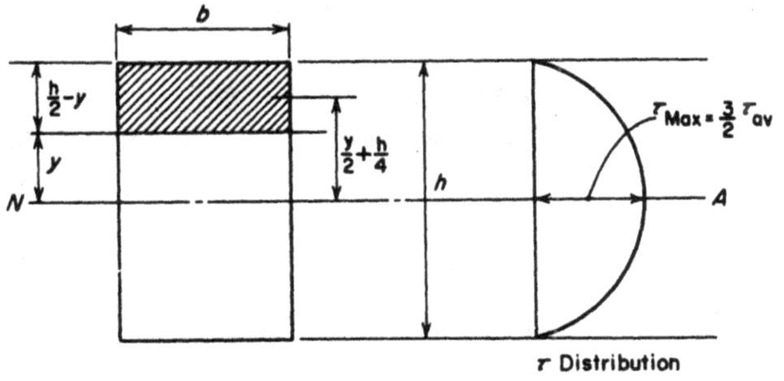

Fig. 6.13

For a rectangular cross section, see Fig. 6.13, at any point y from the N.A. we have

$$\tau_{yz} = \frac{V_z b\left(\frac{h}{2} - y\right)\left(\frac{y}{2} + \frac{h}{4}\right)}{bI_z} \tag{6.84}$$

$$= \frac{12V_z}{bh^3}\left(\frac{h^2}{8} - \frac{y^2}{2}\right) \tag{6.85}$$

This indicates that the shear stress distribution across the cross section is parabolic, with maximum value at the N.A. and minimum (zero) value at the outside fibers. At the N.A., since $y = 0$, we have

$$\tau_{\max} = \frac{3}{2}\frac{V_z}{bh} = \frac{3}{2}\frac{V_z}{A_z} \tag{6.86}$$

and, if we define the *average* shear τ_{AV} as V_z/A_z, this gives

$$\tau_{\max} = \tfrac{3}{2}\,\tau_{\mathrm{av}} \tag{6.87}$$

In other words, in a beam of rectangular cross section, the actual compatible shear stress distribution gives a maximum shear stress which is 50% greater than the *average* shear stress.

For the cross section of the usual rolled steel beams, the so-called

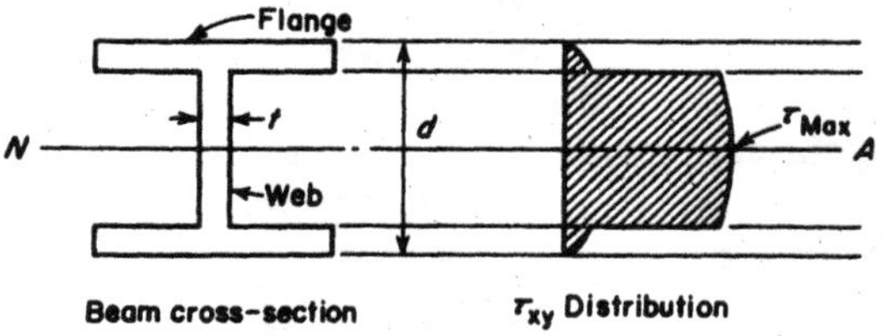

Fig. 6.14

H-section (see Fig. 6.14), WF-section (wide-flange) or I-section, we have a shear stress distribution as shown. In this case it is found that, very nearly,

$$\tau_{\max} = \frac{V_z}{A_{\mathrm{web}}} = \frac{V_z}{(d)(t)} \tag{6.88}$$

and this relation is frequently used in shear analyses for the rolled sections.

6-7 The Shear and Moment Curves and the Design of Beams

We have shown up to this point that the engineering beam, under ordinary loadings, is subjected to bending stresses and to shear stresses. The bending stresses are given (see Eq. 6.14) by

$$\sigma_x = \frac{M_z y}{I_z} \tag{6.89}$$

and $(\sigma_x)_{\max}$ occurs at the fiber furthest from the neutral axis.

The shear stress is given (see Eq. 6.83) by

$$\tau = \frac{V_x Q}{I_z b} \tag{6.90}$$

and in most engineering cross sections, τ_{max} occurs at the neutral axis.†

The engineer is concerned with the design of the beam, that is, he must choose a beam of such material and cross section that it will adequately withstand the loads to which it is subjected. In other words, he must choose a beam such that the induced stresses $(\sigma_x)_{max}$ and τ_{max} are equal to or less than the *allowable* values of these stresses, $\sigma_{allowable}$ and $\tau_{allowable}$. This, in essence, is what is involved in engineering design, and we tabulate the technique or procedure in order to bring out its essential simplicity:‡

1. Determine the loads on the beam. (Note: the beam must support itself; hence, the weight of the beam must be included in the load applied to it.)
2. Knowing the material and shape of the beam, determine the allowable stresses.
3. Knowing the load on the beam, determine the bending moment and shear at all points.
4. Having the bending moment and shear at all points, determine the maximum or critical stress conditions. For bending stress this will generally occur at the point of maximum moment. For shear stress this will generally occur at the point of maximum shear. However, if the shape of the beam varies in the span, it may be that the critical station is other than the point of maximum moment or maximum shear.
5. Compare the maximum stresses with the allowable. Make certain that

$$\text{Maximum actual stress} \leq \text{Allowable stress}$$

In the procedure outlined above, it can be seen that a key step is (3), the determination of the shear and moment at all points on the beam. We discussed the *methods* for computing these in the previous sections. In the next chapter we describe the manner in which we *show* the variation of the shear and moment along the beam. This is done graphically by means of the *shear* and *moment curves*. However, before discussing this we shall describe a theoretical treatment which gives a measure of the approximation involved in the Bernoulli-Euler theory of beams.

†Since the shear stress distribution is dependent on Q and b, which vary depending upon the shape of the cross section, it is possible to have combinations of these which will give a maximum stress at some point other than the neutral axis.

‡The design of a beam occasionally depends upon other factors than allowable stress. For example, sometimes it is necessary to limit the allowable deflection of the beam in order that undesirable cracking of plaster materials be avoided. In these cases the design of the beam is governed by the deflection requirement and not by the allowable stress, although in every case the actual stresses in the beam must be equal to or less than the allowable stresses.

6-8 The Approximation Involved in the Bernoulli-Euler Solution for the Beam

The development which follows is due to the English mathematician, Pearson. He established an amount of approximation of the Bernoulli-Euler solution as follows:

If Eqs. 4.54–4.59 of the linearized theory of elasticity are solved for the stresses as functions of the strains, we would obtain

$$\sigma_x = A(e_x + e_y + e_z) + 2Ge_x \tag{6.91}$$

in which

A is a constant,
G is the shear modulus of elasticity,
$e_x + e_y + e_z = \mathcal{J}_1$, the invariant of the strain tensor.

In the Bernoulli-Euler solution we assume that $e_y = e_z = 0$ and hence

$$\sigma_x = Ee_x = E\frac{\partial u}{\partial x} \tag{6.92}$$

E being the modulus of elasticity.

From the two equations given above we find

$$\mathcal{J}_1 = \frac{E - 2G}{A}\frac{\partial u}{\partial x} \tag{6.93}$$

and therefore,

$$\frac{\partial \mathcal{J}_1}{\partial x} = \frac{E - 2G}{A}\frac{\partial^2 u}{\partial x^2} \tag{6.94}$$

Now the equilibrium equation of the theory of elasticity (see Eq. 4.18) is

$$\frac{\partial \sigma_x}{\partial x} + \frac{\partial \tau_{yx}}{\partial y} + \frac{\partial \tau_{zx}}{\partial z} = 0 \tag{6.95}$$

and since

$$\tau_{yx} = G\gamma_{yx} = G\left(\frac{\partial u}{\partial y} + \frac{\partial v}{\partial x}\right) \tag{6.96}$$

$$\tau_{zx} = G\gamma_{zx} = G\left(\frac{\partial u}{\partial z} + \frac{\partial w}{\partial x}\right) \tag{6.97}$$

$$\sigma_x = A\mathcal{J}_1 + 2G\frac{\partial u}{\partial x} \tag{6.98}$$

this gives

$$(A + G)\frac{\partial \mathcal{J}_1}{\partial x} + G\left(\frac{\partial^2 u}{\partial x^2} + \frac{\partial^2 u}{\partial y^2} + \frac{\partial^2 u}{\partial z^2}\right) = 0 \tag{6.99}$$

Using the above expression for $\partial \mathcal{J}_1/\partial x$ gives

$$K\frac{\partial^2 u}{\partial x^2} + K_1\left(\frac{\partial^2 u}{\partial y^2} + \frac{\partial^2 u}{\partial z^2}\right) = 0 \tag{6.100}$$

in which K and K_1 are constants.

Now the Bernoulli-Euler expression is also given by

$$\sigma_x = \frac{M(x)y}{I_z} \tag{6.101}$$

$$= E\frac{\partial u}{\partial x} \tag{6.102}$$

in which $M(x)$ is a function of the x only.
Hence, differentiating and integrating this we get,

$$E\frac{\partial^2 u}{\partial x^2} = \frac{y}{I_z}\frac{dM(x)}{dx} \tag{6.103}$$

and

$$Eu = \frac{y}{I_z}\int M(x)dx + \psi(y, z) \tag{6.104}$$

in which $\psi(y, z)$ is a function of y and z only.

Putting this last expression for u in Eq. 6.100, above, we get

$$K_2 y\frac{dM(x)}{dx} + K_3\left(\frac{\partial^2\psi(y, z)}{\partial y^2} + \frac{\partial^2\psi(y, z)}{\partial z^2}\right) = 0 \tag{6.105}$$

and since the last bracket of this equation is independent of x, then so also is the first term.

Hence,

$$\frac{dM(x)}{dx} = \text{constant} \tag{6.106}$$

and

$$M(x) = K_5 x + K_6 \tag{6.107}$$

which means $M(x)$ can only be due to a concentrated load or a concentrated moment. In other words, for beams subjected to loadings other than these, there is a conflict between the Bernoulli-Euler theory and the linearized theory of elasticity, indicating that the Bernoulli-Euler theory is in error. However, tests indicate that the approximate Bernoulli-Euler theory gives results which are sufficiently accurate for engineering purposes.†

†The interpretation to be placed on the Pearson result given above is as follows: The Bernoulli-Euler assumptions — namely (1) $\sigma_x = \frac{M(x)y}{I_z}$ everywhere in the beam, and (2) $e_y = e_z = 0$ everywhere in the beam — are consistent with the exact equation of equilibrium and with the exact linearized form of Hooke's Law only if the moment is either a constant or varies linearly along the length of the beam. This result is, therefore, *one possible* measure of the inaccuracy of the Bernoulli-Euler theory within the framework of the more exact relations of the mathematical theory of elasticity. However, there are *other* possible measures of this inaccuracy. For example, one may test when the failure of the equivalent Bernoulli-Euler assumption — that plane sections before bending remain plane after bending — occurs. Or one may consider the compatibility of strain equation and determine the inaccuracy introduced by neglecting this requirement.

6-9 Summary

We discussed the exact linearized mathematical theory of elasticity solution for pure bending of a beam, and pointed out the connection between this and the engineering solution — the so-called Bernoulli-Euler theory of beams.

We discussed also the necessary introduction of transverse shear in the cases where we have variable moments in the beam, as is commonly the case in the engineering beam. This led to a discussion of the differential relations for the transversely loaded engineering beam and also to a treatment for the transverse shear stress. In connection with this last, it was pointed out that compatibility of stresses requires a particular distribution of shear stress across the cross section.

Finally, we described Pearson's theoretical treatment, which gives a measure of the approximation involved in the engineering treatment of beams.

Problems

In the problems which follow, use the beam section properties given in Tables 4 and 5, on pages 131 and 139.

1. An 8 WF 24 has a maximum moment of 350,000 in.-lb and a maximum shear of 18,000 lb. Determine the maximum bending stress and maximum shear stress.

2. If the allowable bending stress for a steel beam is 20,000 psi and the allowable shear stress is 13,000 psi, what are the allowable moments and shear forces for the following beams?

   36 WF 160
   27 WF 102
   14 WF 78
   12 WF 40
   18 I 54.7 (American Standard beam)
   10 [15.3 (channel)

3. If the allowable bending stress and shear stress for a timber beam are 1200 psi and 100 psi, respectively, what are the allowable moment and shear for the following beams?

2 × 4	3 × 6
2 × 10	3 × 12
2 × 16	3 × 18

4. Given positive axes assumed as shown and positive moments and shears as indicated, determine the four differential relations for the cases shown. Assume $w(x)$ to be positive in the direction of the positive y-axis.

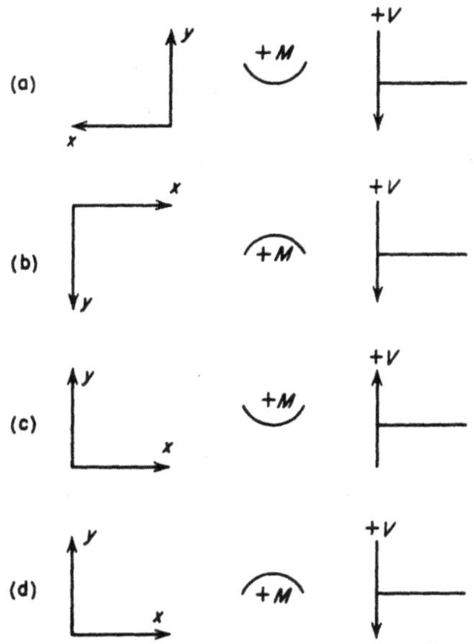

5. Determine the shear stresses at the planes 1, 2 and 3 due to a 1000-lb shear force for the cross sections shown.

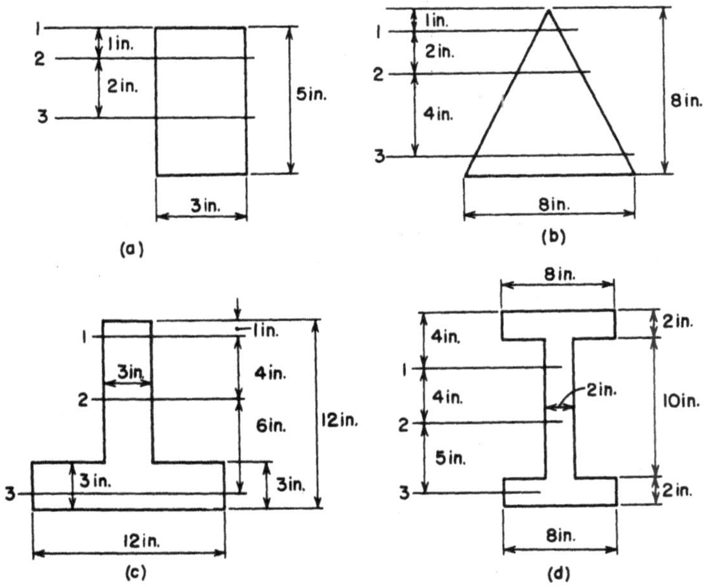

6. See Prob. 4 at the end of Chapter 5. Note that the normal force applied off the centroid of the cross section is, in general, equivalent to a force at the centroid

plus moments about the pair of axes through the centroid. The stress at any point due to the original force, therefore, is equal to the *sum* of the stresses due to the centroidal force plus the couples. In the cross sections which follow, assume that a 1000-lb load is applied at the point, o, indicated. Determine the four normal corner stresses due to this load.

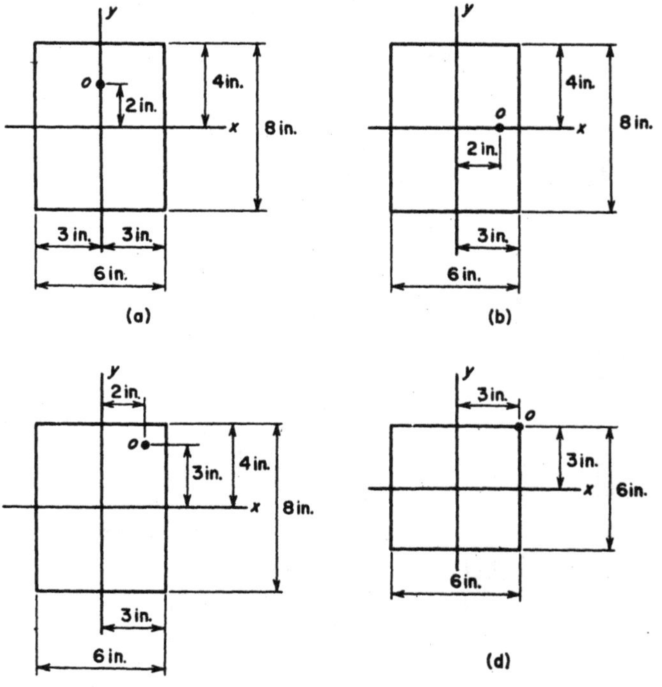

(c)

7. In many cases it is desirable that the net stress on a cross section not be tensile, since the material may be particularly weak when subjected to this type of

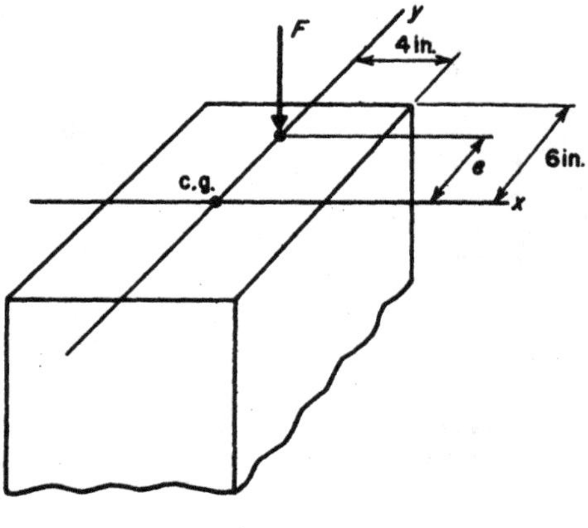

stress. For the cross section shown, what is the maximum value of e in order that the stress due to the compressive load, F, may not be tensile anywhere in the cross section.

8. For the cross section of Prob. 7, show the region within which F must be applied in order that no tensile stress may be developed on the cross section.

9. Determine the shears and moments at points A and B for the beams and loadings shown.

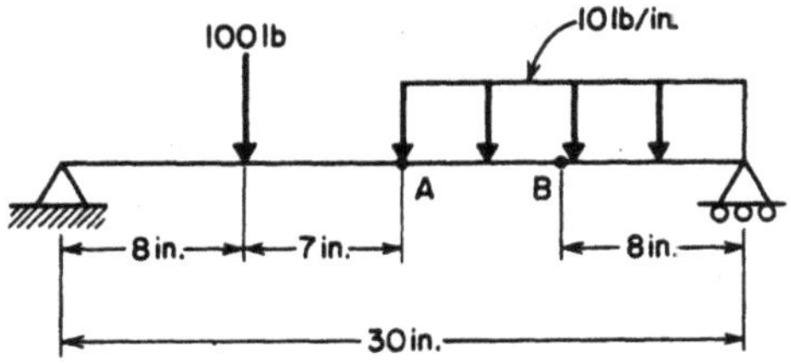

(a)

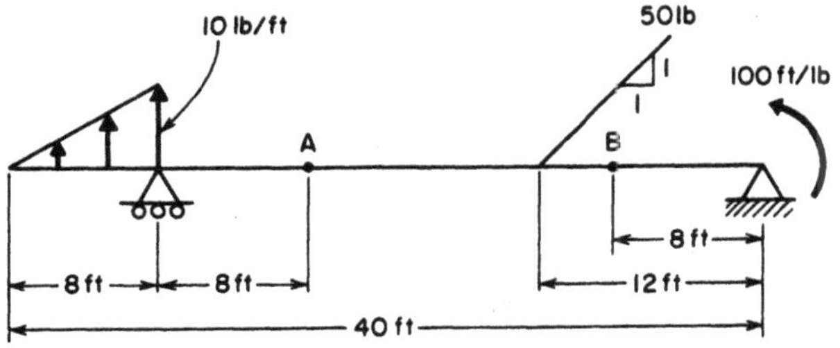

(b)

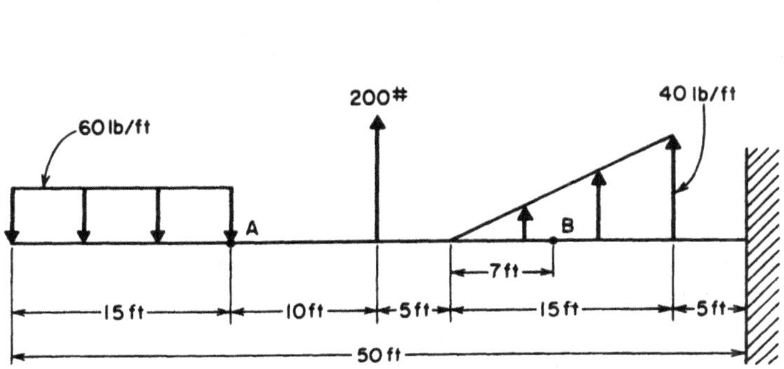

(c)

Chapter 7

SHEAR AND MOMENT DIAGRAMS

7-1 Introduction

In Chapter 6 it was pointed out that the design of an engineering beam requires a knowledge of the shears and moments at all points on the beam. In this chapter we discuss the method that is used by engineers to represent this variation, the so-called *shear curve* and *moment curve.*

We present first the shear and moment curves for certain basic loadings. This then leads naturally to a discussion of the *parts moment curve*, which will be found most useful in later sections of the text.

Following this, we point out that in many cases a *combined* form of

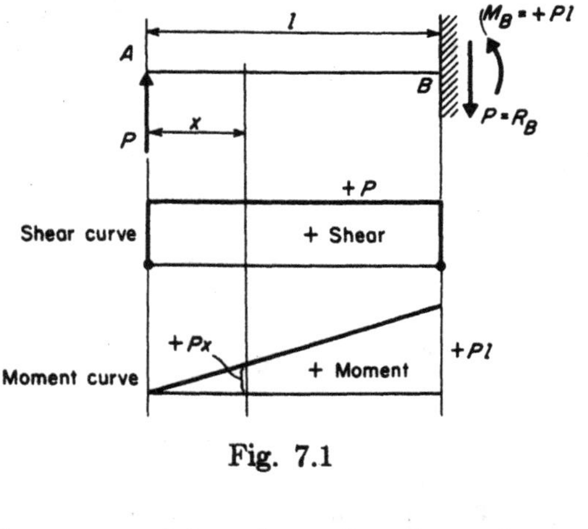

Fig. 7.1

moment curve is more useful, and we describe a simple procedure for preparing this type of moment curve.

7-2 The Basic Shear and Moment Curves

The shear curve is simply a curve whose ordinate at any point on the beam represents, in value and sign, the shear at that point on the beam. In like manner, the moment curve is one whose ordinate at any point on the beam represents, in value and sign, the moment at that point on the beam. For example, consider the beam and loading of Fig. 7.1.

Note that at all sections x, to the right of the load P, the shear (the net force to the left of the section) is due to an upward force of value P. Hence the shear has zero value at A, jumps discontinuously to $+P$ at A, and is constant along the beam to point B, at which place it returns discontinuously to zero. The student should note this point particularly: *the shear curve always starts at zero and ends at zero.*

In the same way, referring to the figure, the moment at any point x is given by $+Px$; hence the moment curve is a straight line from zero to value $+Pl$ at the point B.

As another example, consider the beam and loading of Fig. 7.2. In this

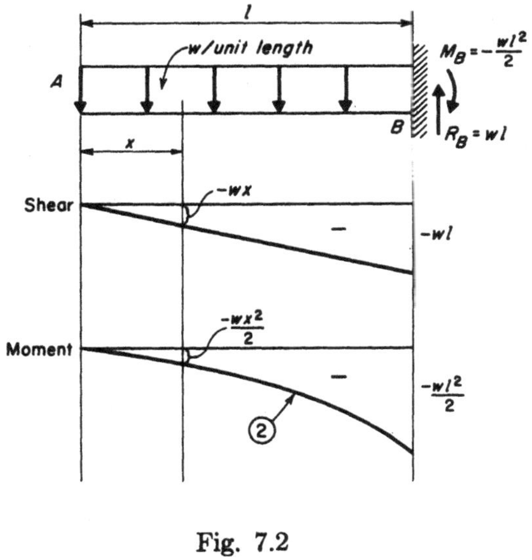

Fig. 7.2

case, at the station x, the net force to the left is given by wx and it is down. Hence, $V_x = -wx$ and the shear curve is the sloped line shown, returning to zero at B.

At station x, the net moment of all forces to the left is $-wx^2/2$, so that the moment curve is a second order parabola (shown by the symbol 2 on the curve), and the moment at the wall is $-wl^2/2$ as shown.

Consider the beam and loading of Fig. 7.3. In this case the shear is zero everywhere along the beam, and is so shown on the shear curve. The

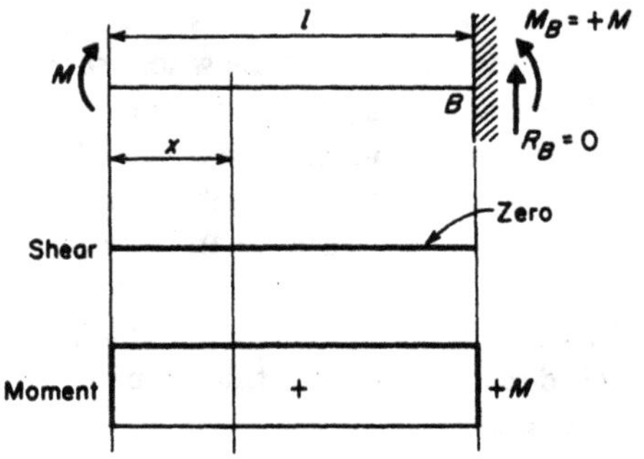

Fig. 7.3

moment is positive and constant at all points on the beam. Hence the curve is as indicated.

One final, simple example will suffice: the case of the triangular load on a cantilever, Fig. 7.4. Here the loading rate at B is w/unit length. At A it is

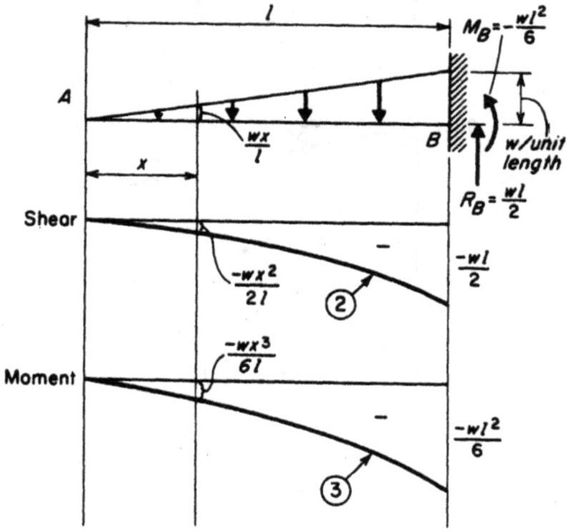

Fig. 7.4

zero, varying linearly between these two points. It is clear therefore that the total load on this beam is $wl/2$. At any point, x, the ordinate of the loading is wx/l.

It will be found, therefore, that the shear curve is a second order parabola, and this is shown by attaching the symbol 2 to the curve.

In a similar manner we find the moment at any point, noting now that the centroid of a triangular load is two-thirds the distance from the vertex to the base; hence,

$$M_x = -\left(\frac{wx}{l}\right)\left(\frac{x}{2}\right)\left(\frac{x}{3}\right) \tag{7.1}$$

$$= -\frac{wx^3}{6l} \tag{7.2}$$

so that the moment curve is a third order parabola. This is shown by placing the symbol 3 on the moment curve.

The shapes which the shear curves have for the basic loadings as represented by the previous cases are summarized in Table 1.

TABLE 1

Loading	Representation of Load and Shear Curve Form
Constant moment	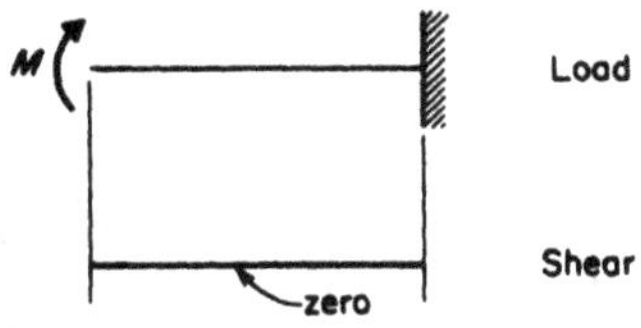
Concentrated load	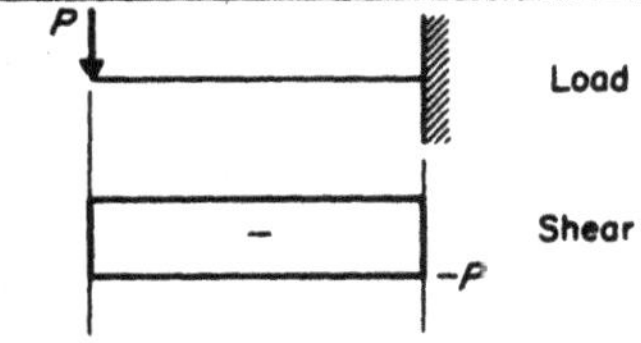
Uniform load	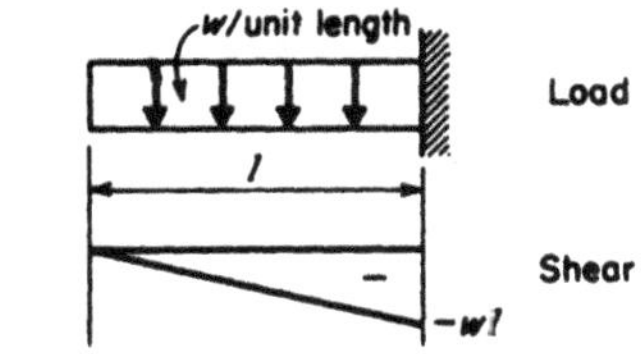
Triangular load	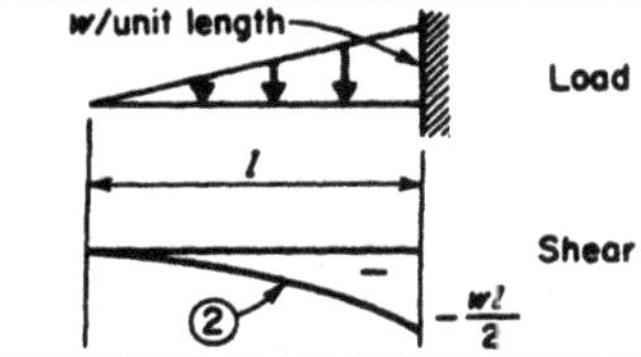

Table 2 shows the moment curves corresponding to the same basic loadings.

TABLE 2

Loading	Representation of Load and Moment Curve Form
Constant moment	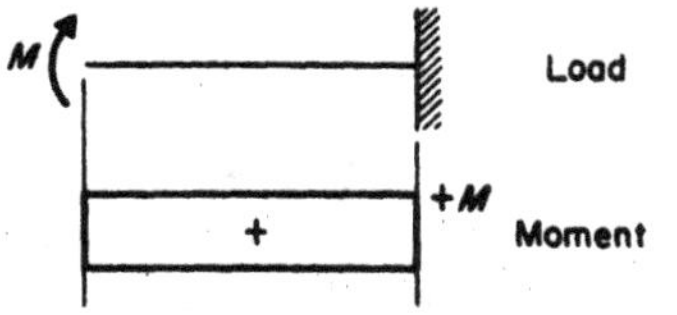
Concentrated load	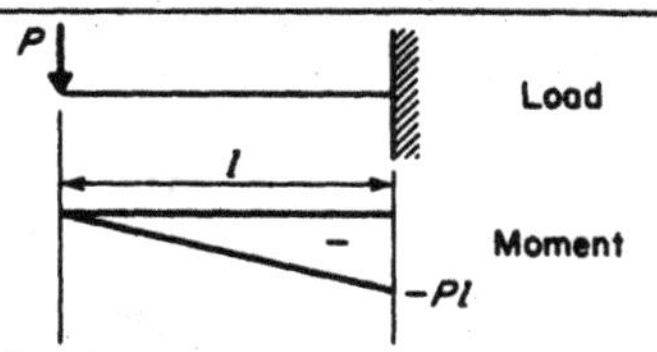
Uniform load	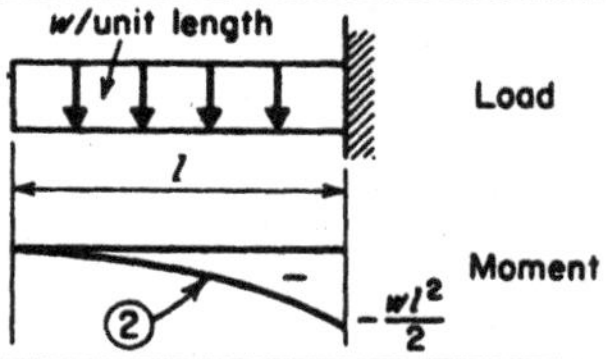
Triangular load	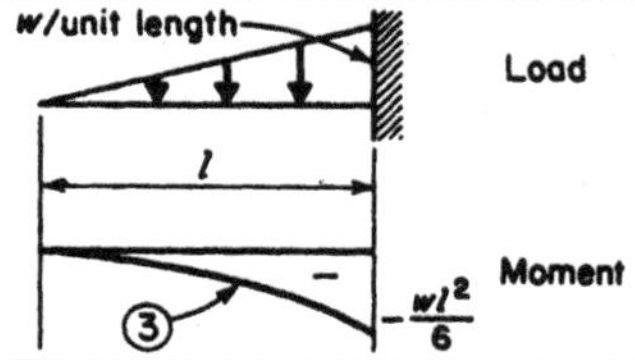

The area and center-of-gravity data for the simple moment curves given in Table 2 will be extremely useful in later portions of the text. The data are summarized in Table 3.

7-3 The Shear Curve for Several Loads on the Beam

If the beam has a combination of the basic loads discussed in Art. 7-2, then the shear curve is drawn by simply using the shape of curve corresponding to the given load, noting that

1. The shear curve always starts at the left, where it is zero, and ends up at zero at the right end.
2. Over portions of the beam having no load, the shear curve is horizontal.

TABLE 3

<table>
<tr><th>Shape and
Center of Gravity Location</th><th>Area</th></tr>
<tr><td>Rectangle
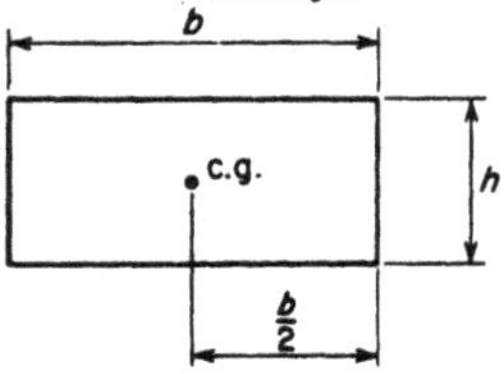
</td><td>bh</td></tr>
<tr><td>Triangle
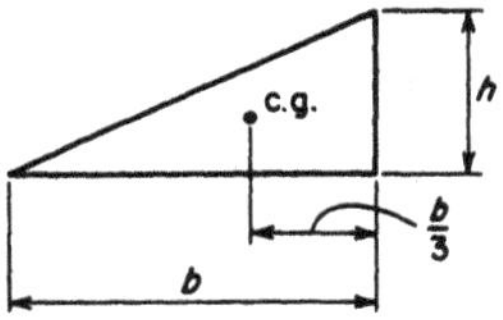
</td><td>$\frac{bh}{2}$</td></tr>
<tr><td>Second order parabola
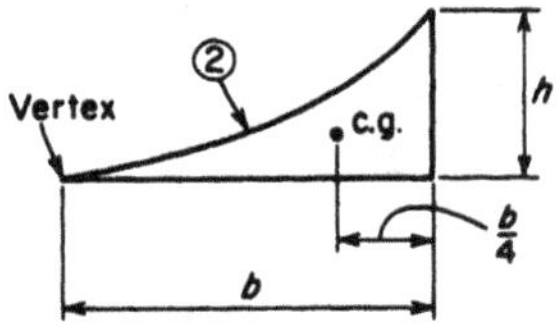
</td><td>$\frac{bh}{3}$</td></tr>
<tr><td>Third order parabola
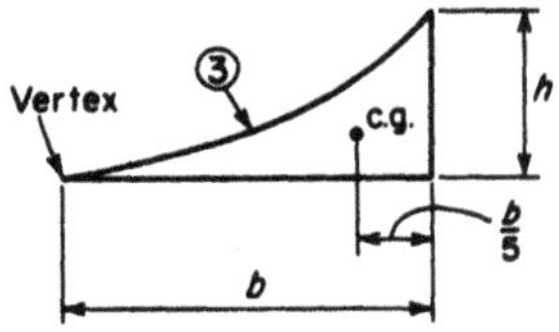
</td><td>$\frac{bh}{4}$</td></tr>
</table>

3. At a concentrated load there is an abrupt change in the shear.
4. Under a uniform load there is a linear variation in shear.
5. Under a triangular load there is a parabolic variation in shear.
6. The signs and directions of shear curves will be opposite those shown for loads in the opposite directions.

To illustrate, we shall draw shear curves, qualitatively, for several typical loadings. The student should study these carefully, noting how the basic shapes given in the table and the hints given above are utilized in each case.

Example 1 In connection with the curve of Fig. 7.5, note that

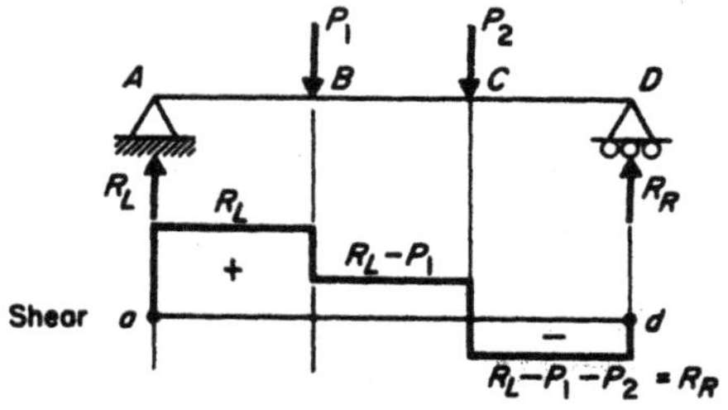

Fig. 7.5

1. The shear is zero at a, and jumps, discontinuously at this point, corresponding to the concentrated force (reaction, R_L) here.
2. Between A and B the shear is positive, constant, and just equal to the value of the left reaction, R_L.
3. At B, the shear decreases discontinuously, by amount P_1, the down force at this point.
4. Between B and C the shear is constant, and equal to R_L-P_1.
5. At C, the shear decreases discontinuously, by amount P_2, the down force at this point.
6. Between C and D the shear is constant, and equal to R_L-P_1-P_2. It is negative, as shown in the figure.
7. At point D, the shear jumps discontinuously to zero, as shown at d at the shear curve. The amount of discontinuous shear at D is just equal to R_R, the right reaction.

Example 2 In connection with the curve of Fig. 7.6, we may note the following:

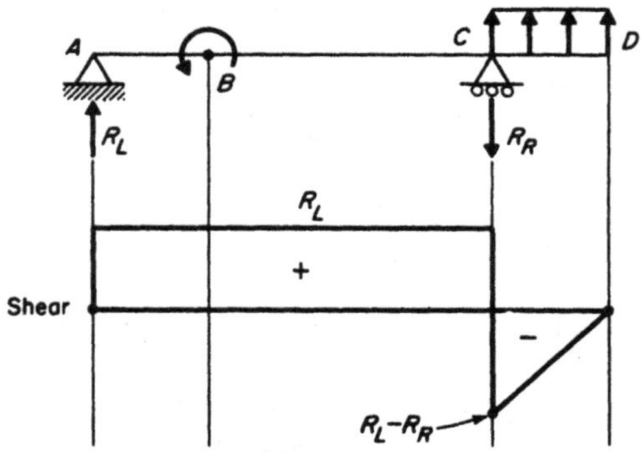

Fig. 7.6

1. The concentrated moment applied at B does *not* cause a change in shear at this point. We will see shortly what effect it *does* have.
2. The uniform load between C and D causes a linear variation in shear, as shown on the figure.

EXAMPLE 3 We note the following facts in connection with the shear curve shown in Fig. 7.7:

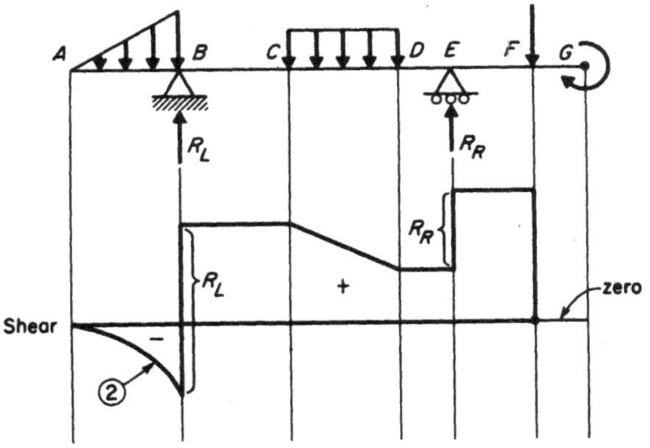

Fig. 7.7

1. Between A and B the shear varies as a second order parabola, starting at zero, with zero slope. We show on the curve the numeral 2 to indicate that it is a second order parabola.
2. Between F and G there is zero shear. The moment applied at the end does not introduce shear in this portion of the beam. It *does* introduce a moment; and we discuss this shortly.

EXAMPLE 4

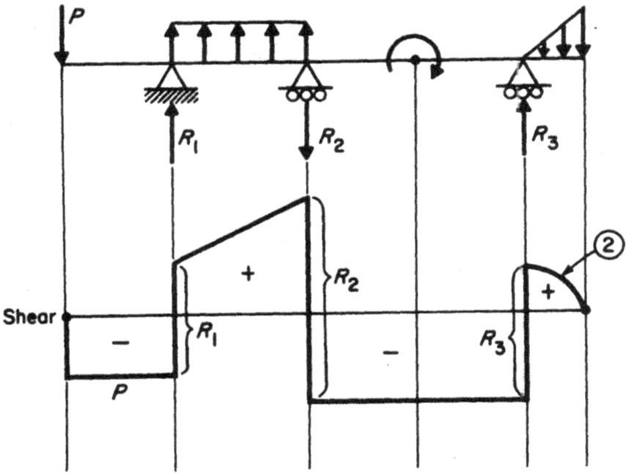

Fig. 7.8

7-4 The Moment Diagram by Parts

It was pointed out earlier that the moment diagram is a means for showing graphically the variation of the bending moment along the beam.

Now, if the beam has a combination of the basic loads of Table 2, we draw a "parts" moment diagram by simply drawing a curve for each load separately, noting that

1. We may start at the left end (or the right end) and proceed across the

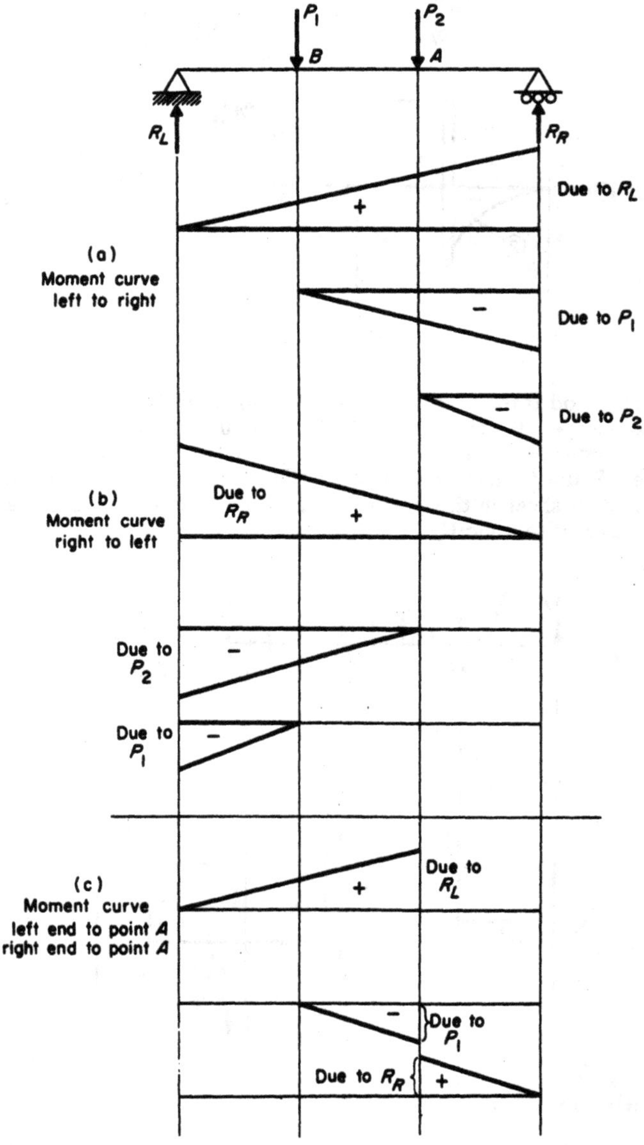

Fig. 7.9

beam considering each load in turn until the right end (or the left end) is reached. If desired, one can start at both ends, proceeding to a common point within the beam.

2. The curve due to each load is as shown in the table.
3. The signs and directions of the curves will be opposite those shown for loads in the opposite directions.

To illustrate the procedure, we shall draw the parts diagrams for the same beams and loadings considered in Art. 7-3. We shall use the various forms given in (1) above. Note that the curves shown are qualitative only — numerical values were not considered. In any given numerical problem we would simply add the proper numbers to the curves.

EXAMPLE 1 Each of the sets in Fig. 7.9 represents a complete moment curve for the beam and loading shown. The moment at any point, such as B, is given by the *net* ordinate at that point in each set. All three sets are equivalent in that, at any point, all three will give the same value for the moment.

In set (a) we start at the left end and draw a separate curve for *each load, as it is reached*. As shown on the curve, each triangular moment diagram is due to a separate load.

Set (b) is similar, except that here we are moving from right to left.

Set (c) is an example in which we move from *both ends* to a common interior point within the beam. The general procedure, however, is similar to that used in sets (a) and (b).

It will be found that the *net* ordinate at the right end of set (a) will be zero, as required for zero moment here. The same is true for the net ordinate at the left end of set (b). Finally, the net ordinate at A for the left curves and the right curve will be found equal in set (c).

EXAMPLE 2 In the solution of Fig. 7.10 we may note the following:

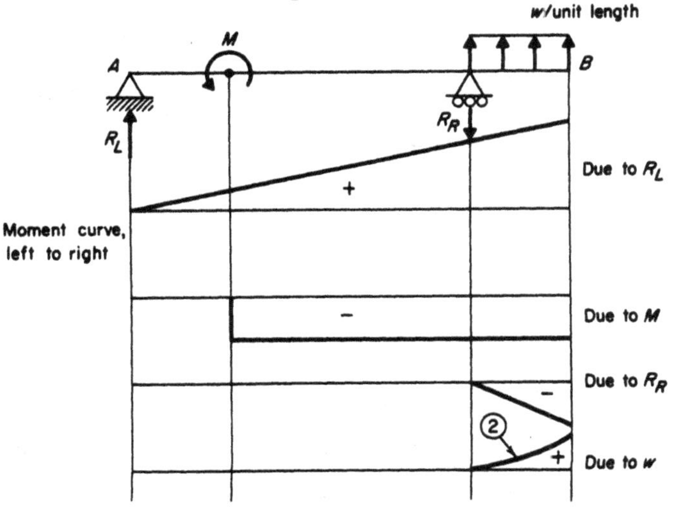

Fig. 7.10

1. The reaction R_L gives a triangular moment curve, positive as shown.
2. The concentrated moment gives the constant negative curve, since at any point to its right, it has a moment of value M.
3. R_R gives a negative triangular moment curve as shown.
4. The uniform load causes a second order parabolic variation in moment, as shown by the 2 connected to the curve.
5. The moment at A is zero, as required.
6. The moment at B will be found to be zero, as required, that is, the *net* sum of all the ordinates at B will be zero.

Example 3 In Fig. 7.11, the student should note the following:

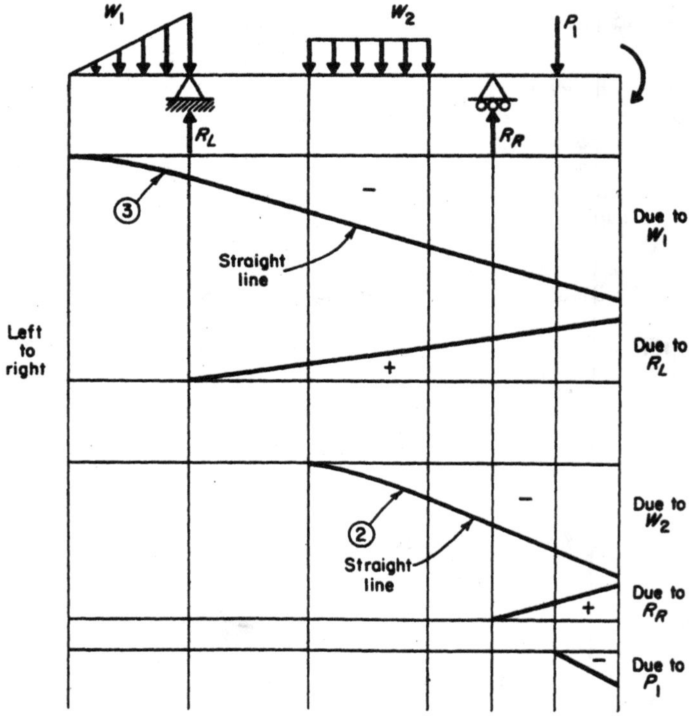

Fig. 7.11

1. For the triangular load, the curve is a third order parabola up to point A. After point A, the curve is a straight line, sloped at the value of the slope of the third order curve at A.
2. The above is also the case for the uniform load. For this load, the common slope is the value of the slope of the 2nd order parabola at B.
3. The end moment (at the right end) does not appear directly in the moment curve, since we are proceeding from left to right. However, the net value of the moment at the right end will be found to be just equal to this end moment.

EXAMPLE 4 The *seven* curves shown in Fig. 7.12 represent the "parts" moment

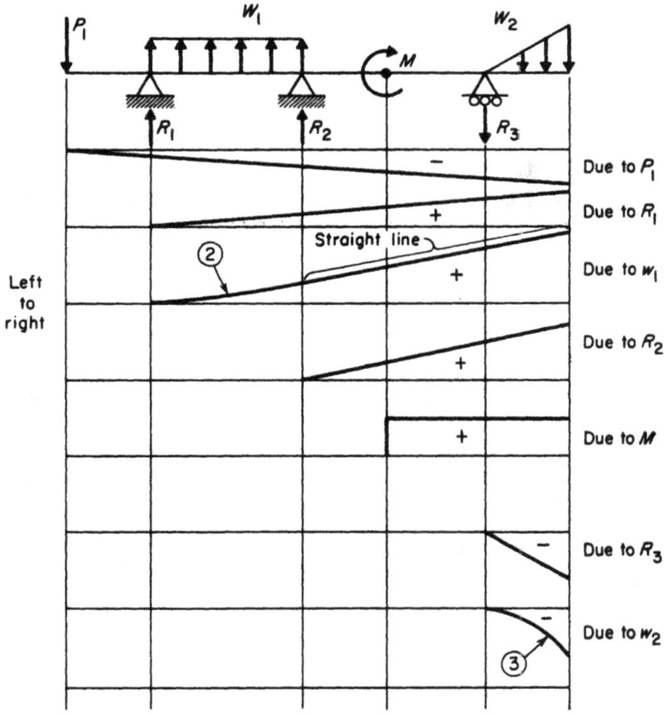

Fig. 7.12

curve for the beam and loading. At any point on the beam, the moment is given by the net ordinate of all the curves at the point.

The parts diagram representation of moment is extremely useful in connection with the "moment-area" and "conjugate beam" deflection analysis of beams; we shall use this form of moment curve when we discuss these methods in a later chapter. However, for the *design* of beams, it is much more convenient to have a single combined curve, and we discuss this form of the moment curve next.

7-5 The Combined Moment Curve

It was pointed out in Chapter 6 that an important step in the design of the engineering beam is the determination of the maximum shear and bending moment acting on the beam. The locations of the points of maximum shear and moment can, in general, be determined from the shear and moment curves. However, if the curve is drawn by "parts," then it is difficult to determine, by inspection, where the maximum moment occurs, because the net moment at any point is given by the sum of all the parts curve ordinates at the point in question. For this reason, it is very convenient, in designing beams for bending stresses, to have a "combined" moment diagram — this being a single curve which shows the variation of moment over the entire beam.

The preparation of this curve is simplified to a considerable extent if the student will utilize the single relation proven in Chapter 6 (see Eq. 6.49).

$$\frac{dM}{dx} = V \tag{7.3}$$

which states, in words, "The slope of the moment curve at any point is given, in value and sign, by the shear at that point."

Thus, in drawing the moment curve by utilizing this relation, a necessary first step is to prepare the shear curve. This is done as described in Art. 7.3. Having the shear curve, recalling the shape of moment curve due to the various loads (see Table 2), and utilizing the simple requirement given above relating the shear to the slope of the moment curve, we can then

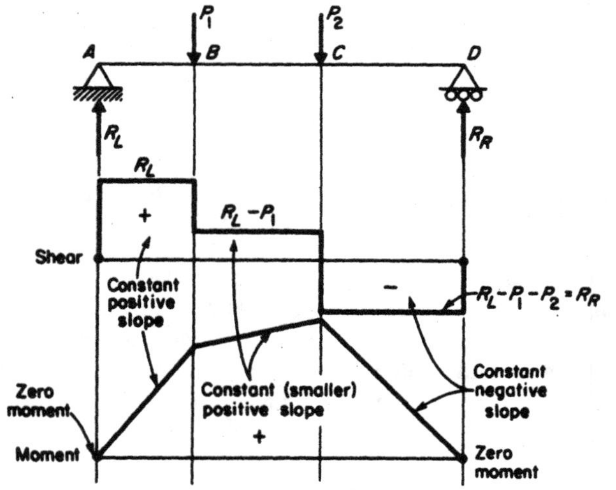

Fig. 7.13

draw the combined moment curve without difficulty. Note that the moment will be a maximum or minimum (a) where the shear is zero (or where the shear goes *through* zero) as at point C of Fig. 7.13, or (b) at a point of concentrated moment as at point B of Fig. 7.14.

To illustrate all of the above, we draw the combined moment curves for the same beams and loadings considered before. We repeat the shear curves as given before and emphasize that the student should note how, in all cases, the following simple relation will give the shape of the moment curve:

$$\frac{dM}{dx} = V \tag{7.4}$$

EXAMPLE 1 In Fig. 7.13, we note first that at the ends of the beam the moments are zero. These are key values and are shown on the moment curve. Then, starting at the left end, from zero moment, we draw a line with a positive, constant slope (corresponding to the positive, constant shear) between A and B.

Between B and C the shear is still positive and constant — but its value is smaller than between A and B. Hence the straight line of the moment curve between B and C has a smaller slope.

Between C and D the shear is constant and negative, hence the moment curve has a constant, negative slope, to zero moment at the right end.

EXAMPLE 2 For the beam of Fig. 7.14, again we have zero moment at A and D.

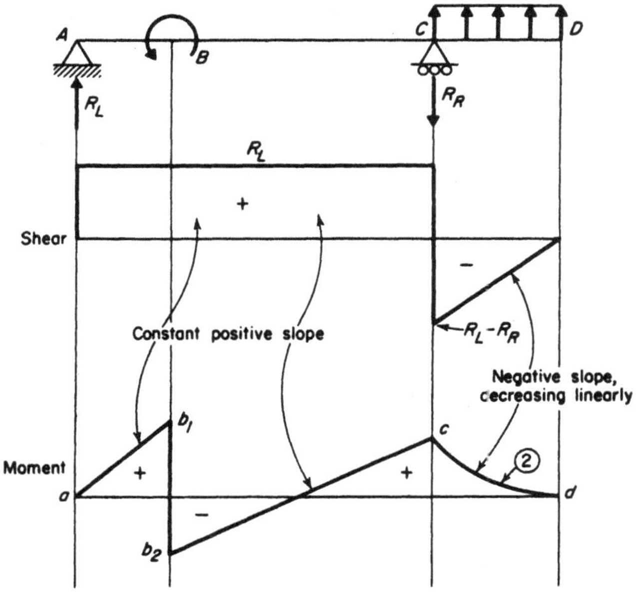

Fig. 7.14

Between A and B we have a positive constant shear, corresponding to a positive constant slope of the curve ab_1. At B we have an abrupt *decrease* in the moment, shown by the vertical line $b_1 b_2$ and then between b_2 and c we have the same positive constant slope as for the line $a_1 b_1$ — corresponding to the same positive constant shear. At C on the beam we have an abrupt change in shear. Just to the right of C the shear is negative and it decreases negatively to zero. This means that just to the right of C, the moment curve has a negative slope, and it decreases negatively to zero. This is as shown in portion *cd* of the moment curve.

Example 3 In connection with the curve of Fig. 7.15, note the following facts:

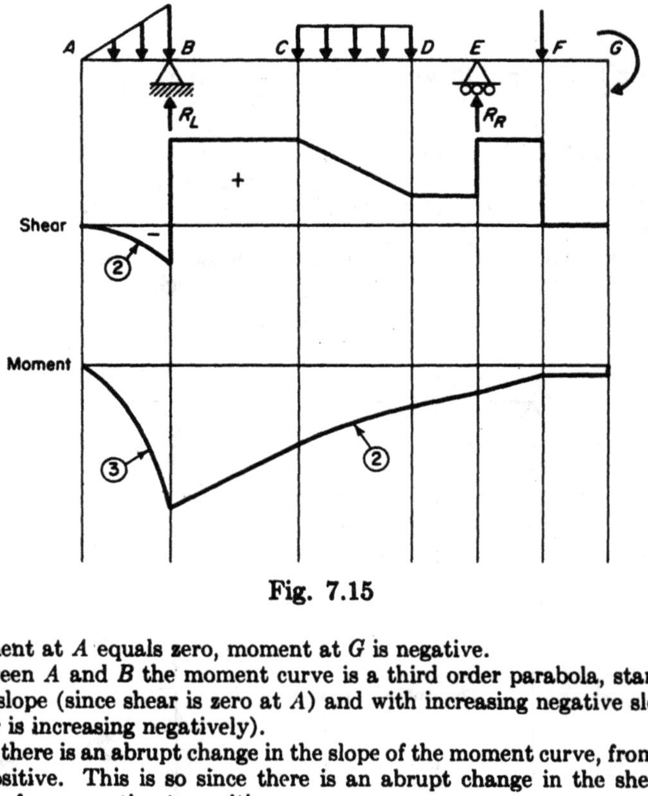

Fig. 7.15

1. Moment at A equals zero, moment at G is negative.
2. Between A and B the moment curve is a third order parabola, starting with zero slope (since shear is zero at A) and with increasing negative slope (since shear is increasing negatively).
3. At B there is an abrupt change in the slope of the moment curve, from negative to positive. This is so since there is an abrupt change in the shear at this point, from negative to positive.
4. Between B and C the moment curve is a straight line with constant positive slope since shear is constant positive.
5. Between C and D the slope of the moment curve decreases positively. Hence it is the second order curve as shown on the figure.
6. Between D and E the moment curve is a straight line with constant positive slope since shear is constant positive — but this slope is less than the slope of the line between B and C, since the shear here is less than between B and C.
7. At E there is an abrupt change in the slope and between E and F the slope of the moment curve is positive constant.
8. Between F and G the moment is constant, of negative value. This is as shown. Note that the zero slope of the moment curve in this region is in agreement with the zero shear here.

EXAMPLE 4 In connection with the curve of Fig. 7.16, we note the following:

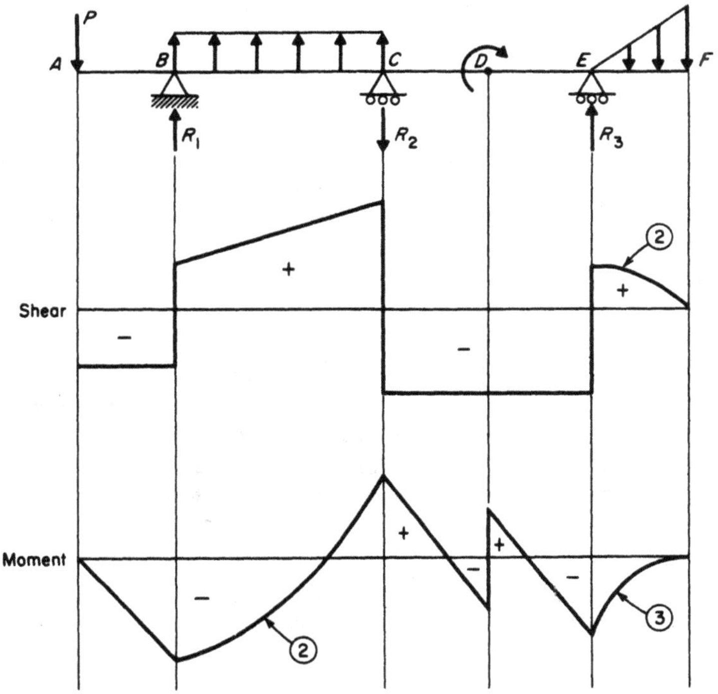

Fig. 7.16

1. Moments at A and at F are zero.
2. Moment curve between A and B has constant negative slope, since shear is constant negative.
3. At B there is an abrupt change in the slope of the moment curve from negative to positive, and between B and C the slope of the moment curve increases positively. This is shown by the curve marked as a second order parabola.
4. At C there is another abrupt change in the slope of the moment curve, from positive to negative.
5. Between C and D the moment curve has a constant negative slope.
6. At D there is an abrupt change in the moment. In going from just to the left of D to just to the right of D, the moment *increases* positively. This is as shown on the moment curve.
7. Between D and E the moment curve has a negative constant slope. This slope has the same value as the slope of the curve between C and D, since the *shear* has the same value in both regions.
8. Between E and F the moment curve has a positive slope, decreasing to a zero slope. This is a third degree parabola.

The examples given above are representative of the types of beams and loadings which might be met in practice. The ability to prepare shear and moment curves is a necessity in the beam design technique and a thorough mastery of the construction of these curves should be obtained. It is strongly recommended that the student solve the problems included at the end of this chapter.

7-6 Summary

The techniques of preparing shear and moment curves were discussed in detail. Two types of moment curves were described: (1) a "parts" diagram, which is essentially a superposition solution in which a moment curve is drawn for each load considered separately, and (2) a combined diagram, in which a single curve is drawn to show the variation of the moment over the beam.

It was pointed out that curves (1) are useful in deflection and related beam problems. Curves (2) on the other hand are more useful in the actual design of beams, since one can determine directly from the curve where the maximum and minimum moments occur. This is necessary because the design of the beam depends upon these maxima and minima.

In the next chapter we shall discuss the technique for designing beams for bending and shear stress.

Problems

1. In the figure which follows, *qualitative* loadings are shown. Draw shear and moment diagrams consistent with these loadings.

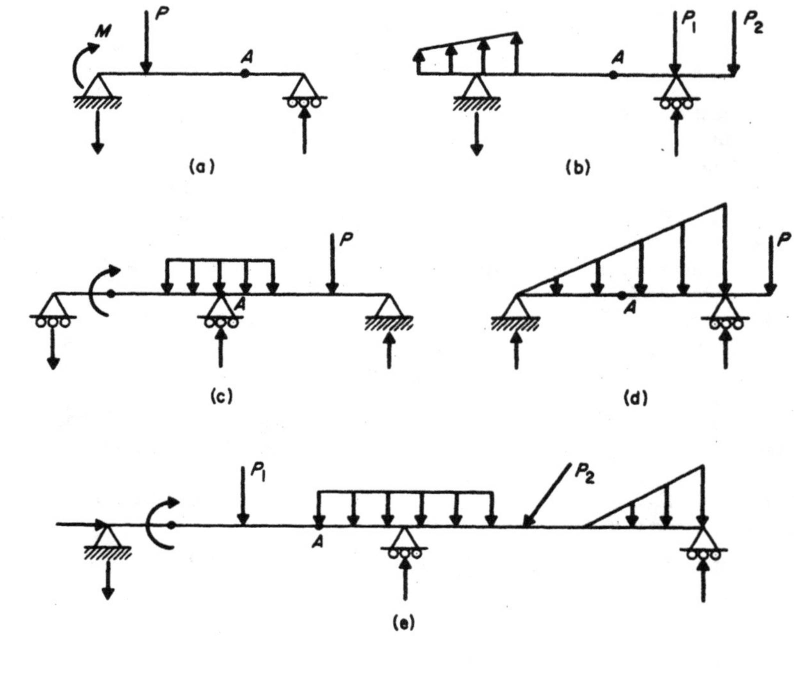

2. For the beam and loadings of Prob. 1, prepare moment diagrams by parts —
   (a) going from the left end to the right end;
   (b) going from the right end to the left end;
   (c) going from *both* ends to point A.

3. *Qualitative* moment diagrams are shown in the figures which follow. Draw qualitative shear and loading diagrams which are consistent with these.

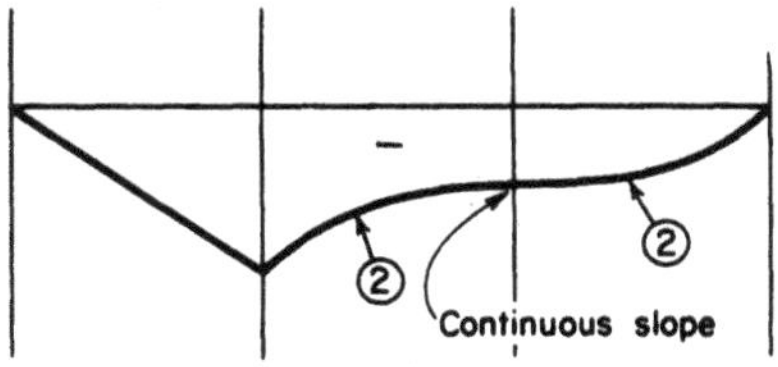

(a)

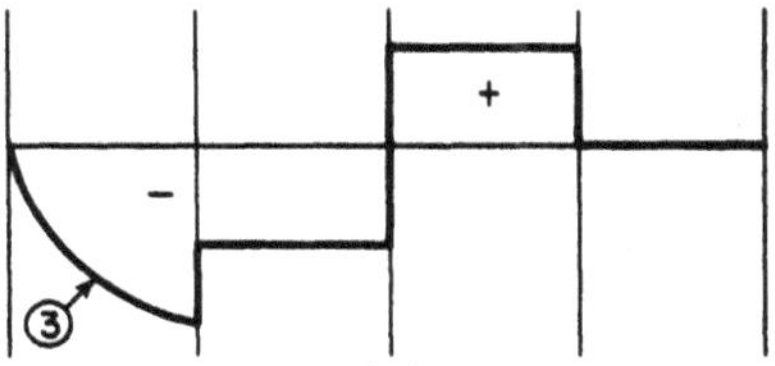

(b)

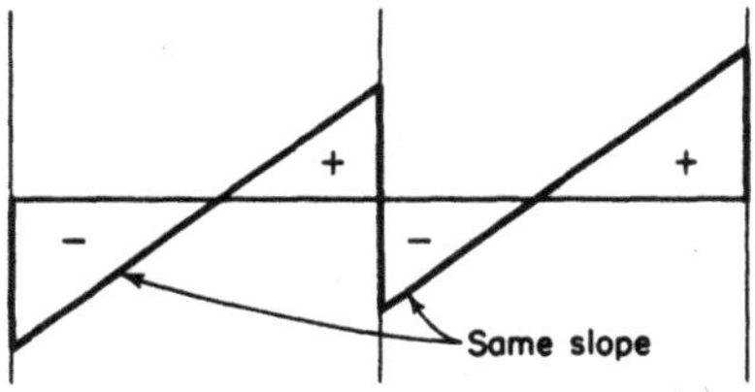

(c)

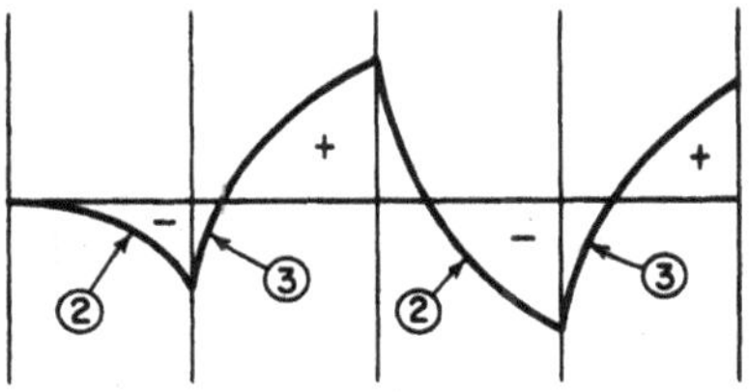

(d)

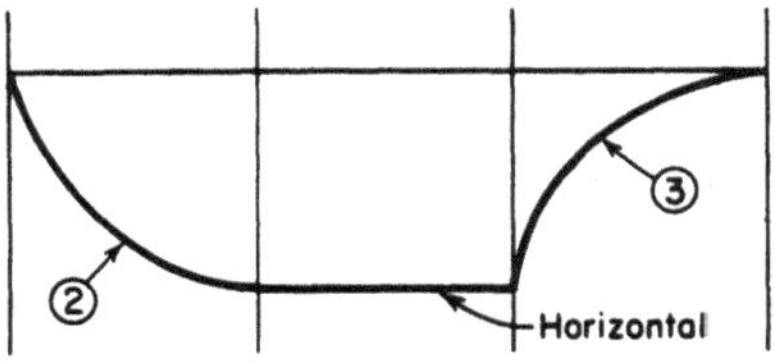

(e)

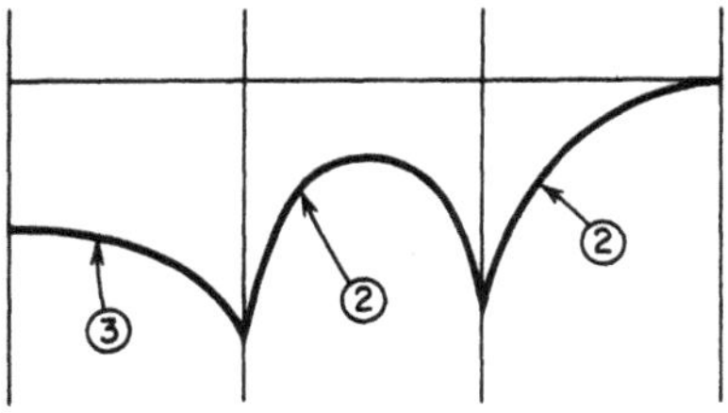

(f)

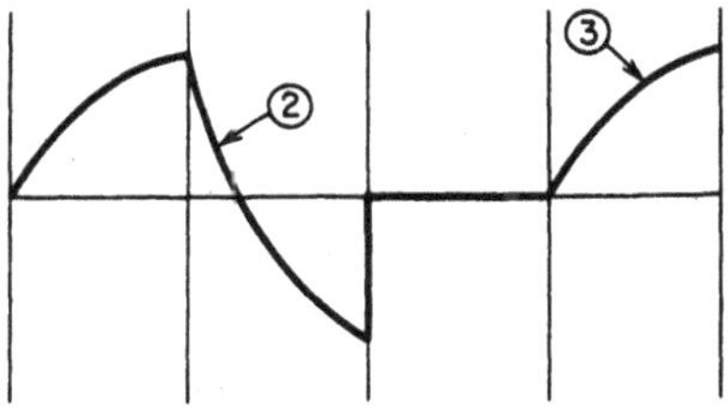

(g)

4. Draw shear and moment curves for the beam and loading shown. Determine all points (and values) of maximum moment and maximum shear.

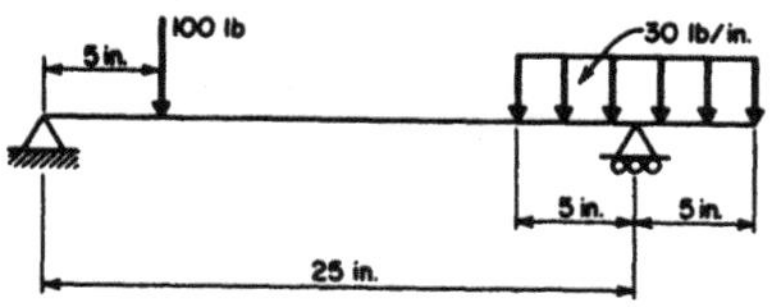

(a)

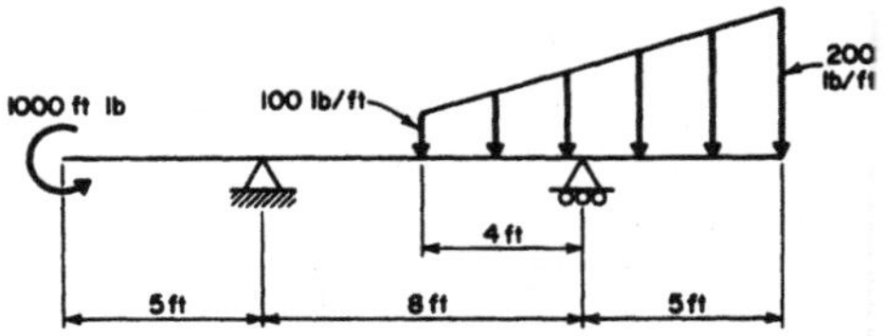

(b)

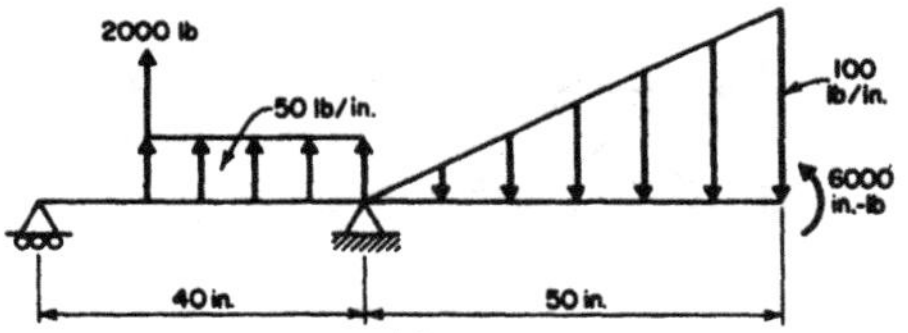

(c)

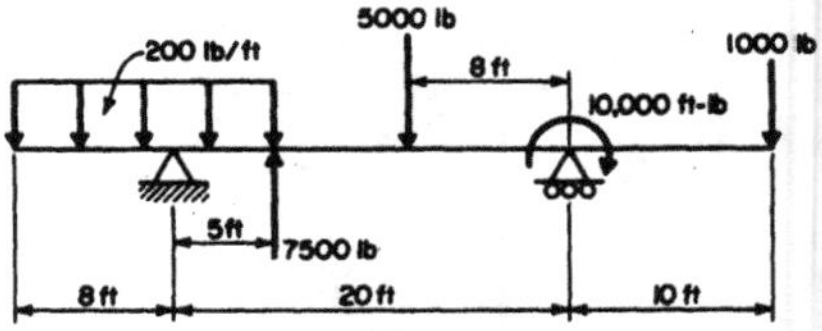

(d)

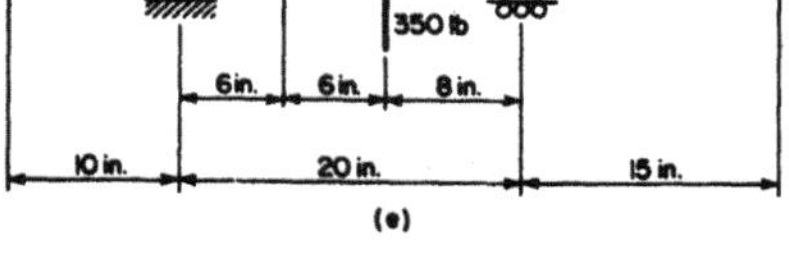

(e)

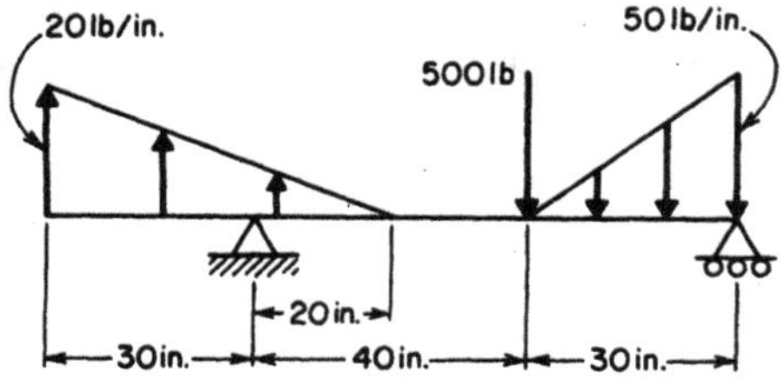

(f)

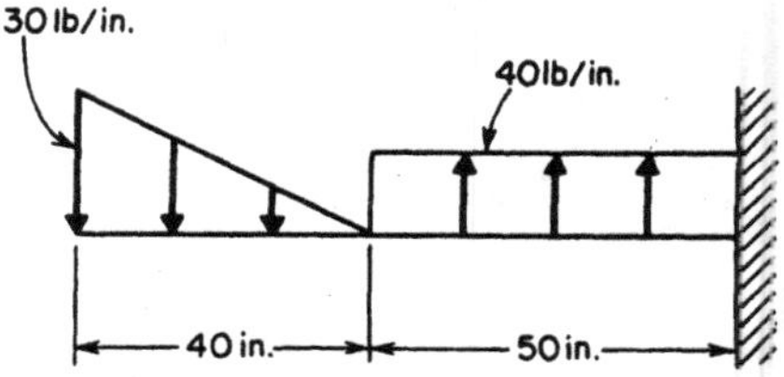

(g)

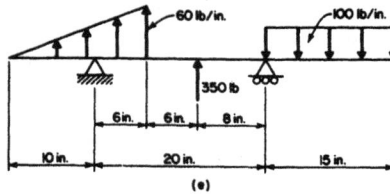

Chapter 8

THE BENDING STRESS AND SHEAR DESIGN OF BEAMS

8-1 Introduction

In this chapter we shall describe the technique used in the engineering design of beams.

Our primary concerns at this point will be the design for bending moments and (when it applies) shear loads. In practice there are other factors which must be considered in the design of beams, but these are usually taken care of in accordance with specific instructions given in the "specifications" which govern the design.†

We shall limit our discussion to steel beams and to wood beams.‡

8-2 Steel and Timber Beam Design

As pointed out in Art. 6-7, a major step in the design of a beam is the choice of member which will be strong enough to withstand the applied bending moment and shear force.

In the last chapter we pointed out how the variation of these two loadings along the beam is represented, namely, by means of shear and moment curves. In Chapter 6, we pointed out that when a beam is subjected to a bending moment, normal tension and compression stresses are induced across the cross-section, and the maximum value of these is given by

$$\sigma_{max} = \frac{Mc}{I} \tag{8.1}$$

in which M is the bending moment on the beam at the point in question, c is the fiber furthest away from the neutral axis, and I is the moment of inertia of the cross section about the neutral axis.

†Among these other factors we may mention
(a) buckling (stability) considerations;
(b) allowable deflection;
(c) headroom and similar considerations which may limit the depths of beams.

‡Although this discussion is limited to beams of steel and wood, the general principles obviously apply to *all* materials. The allowable stresses certainly vary for different materials, and special factors such as ductility, toughness, stiffness, etc. may require consideration for particular materials. Buckling, which may be critical in various forms, is also a major design parameter for many shapes in different materials.

It was shown also that the shear force results in shear stress distributions across the cross section given by

$$\tau = \frac{VQ}{Ib} \tag{8.2}$$

in which τ is the shear *stress* at a point in the beam, V is the shear *force* at the section where τ is determined, b is the beam width at the point where τ is determined, I is the moment of inertia of the cross section about the neutral axis, and Q is the moment, about the neutral axis, of the portion of the cross section between the point where τ is determined and the outside fiber of the beam.

Furthermore, we pointed out that for the wide-flange, or H, section, this formula may be approximated (see Eq. 6.88) by

$$\tau_{\max} = \frac{V}{dt} \tag{8.3}$$

in which t is the web width and d is the full depth of the cross section (see

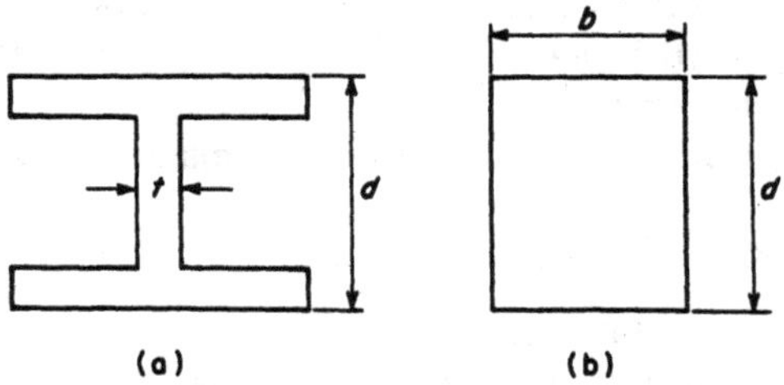

Fig. 8.1

Fig. 8.1a). For the rectangular cross section (see Fig. 8.1b) the formula gives

$$\tau_{\max} = \frac{3}{2}\frac{V}{bd} \tag{8.4}$$

The steel beams used in engineering are very often of the shape shown in Fig. 8.1a. Hence, in designing steel beams for bending stresses and shear stresses we proceed as follows:

1. Determine the load on the beam. Note: this should include the weight of the beam itself. But the beam size (and hence weight) is not known until it is designed; the design, in turn, depends upon the weight of the beam. The procedure in these cases is to *assume* a weight of the beam. This can be done quite accurately after a few beams have been designed.
2. Draw the shear and moment curves for the beam and loading.
3. Determine the maximum moment and shear.

For bending moment design —

4. Knowing the *allowable bending stress* ($\sigma_{\text{allowable}}$, see below), determine

$$S_{\text{req}} = \left(\frac{I}{c}\right)_{\text{req}} = \frac{M_{\text{max}}}{\sigma_{\text{allowable}}} \tag{8.5}$$

in which S is the *section modulus.*

5. In the handbook of steel members, "Manual of the American Institute of Steel Construction," called the *AISC Manual,* choose the lightest member having the required (or larger) S, the section modulus. In Table 4, p. 131, we have listed the properties of typical steel beams.

For shear design —

6. Knowing the *allowable shear stress* ($\tau_{\text{allowable}}$, see below), determine

$$\tau_{\text{actual}} = \frac{V}{td} \tag{8.6}$$

and check that this is equal to or less than the allowable shear stress value.

In connection with the above, we note the following points:

(a) In the *AISC Manual* there is a "specification" governing the design of steel members as approved by the AISC. This authoritative specification represents good design practice, and a design made in accordance with this specification will generally be considered satisfactory.

(b) In the section of the specifications titled "Unit Stresses," various allowable unit stresses are listed. In our present simplified treatment we are only interested in the allowable bending stress of 20,000 psi and the allowable shear stress of 13,000 psi. However, we emphasize that in an actual design *all* of the factors involved in a beam analysis should be considered.

(c) Although we consider, in our treatment, the shear stress as well as bending stress, it will be found (see later illustrative examples) that for the ordinary steel engineering beam — say one that is at least ten times as long as it is deep — the shear is not important, and only the bending must be considered. Shear is important in steel beam design only if the beam is an extremely short and deep beam.

If we are designing a timber beam, which is frequently rectangular in shape, we proceed as follows:

1. Determine the load on the beam. Include the weight of the beam itself, estimating this if necessary.
2. Draw the shear and moment curve for the beam and loading.
3. Determine the maximum shear and moment.

For bending moment design —

4. Knowing the *allowable bending stress* ($\sigma_{\text{allowable}}$, see below) determine

$$S_{\text{req}} = \left(\frac{I}{c}\right)_{\text{req}} = \frac{M_{\text{max}}}{\sigma_{\text{allowable}}} \tag{8.7}$$

5. Choose a cross-section which has the required (or larger) S. (See tabulation in the *AISC Manual* for example or Table 5, p. 139, in which we have listed the properties of typical wood beams.)

For shear design —

6. Knowing the *allowable shear stress* ($\tau_{\text{allowable}}$, see below) determine

$$A_{\text{req}} = \frac{3}{2}\frac{V}{\tau_{\text{allowable}}} \tag{8.8}$$

and check that this is equal to, or less than, the actual beam area.

In connection with the above we note the following points:

(a) Allowable stresses for timber are given in various handbooks and specifications. Typical allowable values of bending stress for timber are from 1000 psi to 2000 psi. Typical allowable values of shear stress for timber are from 50 psi to 150 psi.

(b) In general, design of timber beams is based entirely upon bending and shear. The additional factors which must be considered in steel beam design are ordinarily not important in timber beam design, although occasionally deflection of the beam governs the design.

(c) However, as opposed to the situation in steel beam design, the shear stress is extremely important in timber design and, in fact, frequently governs the design. We shall see this in the illustrative examples which follow.

8-3 Structural Analysis — Investigation and Design

There are, in general, two basically different problems involved in structural analysis:

1. The *investigation* of an existing structure to check its strength;
2. The *design* of a new structure.

Problems of group (1) are usually somewhat simpler than those in group (2). We shall solve both types in the illustrative examples which follow.

8-4 Steel Beams — Three Examples

The first two illustrative problems were chosen to illustrate the fact that the shear stress design problem is only important in very short, deep steel beams. Having made this point, we shall ignore shear stresses in all later problems involving rolled steel sections.

In connection with the investigation or design of steel beams, we point out that the tables of properties of the *AISC Manual* are essential. The properties of typical sections are included herein and are shown in Table 4.

In this table, I is the moment of inertia, S is the section modulus ($= I/(d/2)$), where d is the depth of the beam and r is the radius of gyration of the cross section for the moment of inertia shown.

TABLE 4

WF SHAPES

PROPERTIES FOR DESIGNING

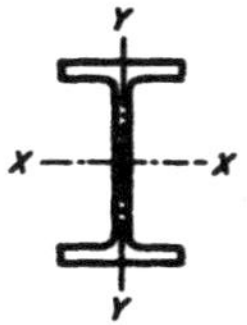

Nominal Size	Weight per Foot	Area	Depth	Flange Width	Flange Thickness	Web Thickness	Axis X-X I	Axis X-X S	Axis X-X r	Axis Y-Y I	Axis Y-Y S	Axis Y-Y r
In.	Lb	In.^{2}	In.	In.	In.	In.	In.^{4}	In.^{3}	In.	In.^{4}	In.^{3}	In.
36 x 12	194	57.11	36.48	12.117	1.260	.770	12103.4	663.6	14.56	355.4	58.7	2.49
	182	53.54	36.32	12.072	1.180	.725	11281.5	621.2	14.52	327.7	54.3	2.47
	170	49.98	36.16	12.027	1.100	.680	10470.0	579.1	14.47	300.6	50.0	2.45
	160	47.09	36.00	12.000	1.020	.653	9738.8	541.0	14.38	275.4	45.9	2.42
	150	44.16	35.84	11.972	.940	.625	9012.1	502.9	14.29	250.4	41.8	2.38
27 x 10	114	33.53	27.28	10.070	.932	.570	4080.5	299.2	11.03	149.6	29.7	2.11
	102	30.01	27.07	10.018	.827	.518	3604.1	266.3	10.96	129.5	25.9	2.08
	94	27.65	26.91	9.990	.747	.490	3266.7	242.8	10.87	115.1	23.0	2.04
18 x 7½	60	17.64	18.25	7.558	.695	.416	984.0	107.8	7.47	47.1	12.5	1.63
	55	16.19	18.12	7.532	.630	.390	889.9	98.2	7.41	42.0	11.1	1.61
	50	14.71	18.00	7.500	.570	.358	800.6	89.0	7.38	37.2	9.9	1.59
14 x 12	84	24.71	14.18	12.023	.778	.451	928.4	130.9	6.13	225.5	37.5	3.02
	78	22.94	14.06	12.000	.718	.428	851.2	121.1	6.09	206.9	34.5	3.00
12 x 8	50	14.71	12.19	8.077	.641	.371	394.5	64.7	5.18	56.4	14.0	1.96
	45	13.24	12.06	8.042	.576	.336	350.8	58.2	5.15	50.0	12.4	1.94
	40	11.77	11.94	8.000	.516	.294	310.1	51.9	5.13	44.1	11.0	1.94
10 x 8	45	13.24	10.12	8.022	.618	.350	248.6	49.1	4.33	53.2	13.3	2.00
	39	11.48	9.94	7.990	.528	.318	209.7	42.2	4.27	44.9	11.2	1.98
	33	9.71	9.75	7.964	.433	.292	170.9	35.0	4.20	36.5	9.2	1.94
8 x 6½	28	8.23	8.06	6.540	.463	.285	97.8	24.3	3.45	21.6	6.6	1.62
	24	7.06	7.93	6.500	.398	.245	82.5	20.8	3.42	18.2	5.6	1.61

TABLE 4 (Cont.)

AMERICAN STANDARD BEAMS
PROPERTIES FOR DESIGNING

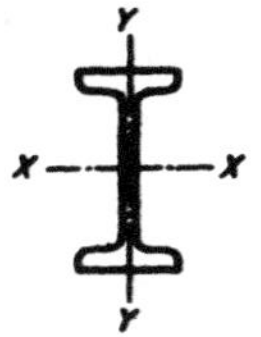

Nominal Size	Weight per Foot	Area	Depth	Flange		Web Thickness	AXIS X-X			AXIS Y-Y		
				Width	Thickness		I	S	r	I	S	r
In.	Lb	In.^{2}	In.	In.	In.	In.	In.^{4}	In.^{3}	In.	In.^{4}	In.^{3}	In.
20 x 6¼	75.0	21.90	20.00	6.391	.789	.641	1263.5	126.3	7.60	30.1	9.4	1.17
	65.4	19.08	20.00	6.250	.789	.500	1169.5	116.9	7.83	27.9	8.9	1.21
18 x 6	70.0	20.46	18.00	6.251	.691	.711	917.5	101.9	6.70	24.5	7.8	1.09
	54.7	15.94	18.00	6.000	.691	.460	795.5	88.4	7.07	21.2	7.1	1.15
15 x 5½	50.0	14.59	15.00	5.640	.622	.550	481.1	64.2	5.74	16.0	5.7	1.05
	42.9	12.49	15.00	5.500	.622	.410	441.8	58.9	5.95	14.6	5.3	1.08

AMERICAN STANDARD CHANNELS
PROPERTIES FOR DESIGNING

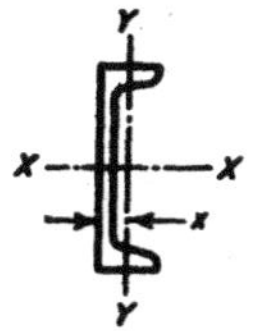

Nominal Size	Weight per Foot	Area	Depth	Flange		Web Thickness	AXIS X-X			AXIS Y-Y			
				Width	Average Thickness		I	S	r	I	S	r	x
In.	Lb	In.^{2}	In.	In.	In.	In.	In.^{4}	In.^{3}	In.	In.^{4}	In.^{3}	In.	In.
10 x 2⅝	30.0	8.80	10.00	3.033	.436	.673	103.0	20.6	3.42	4.0	1.7	.67	.65
	25.0	7.33	10.00	2.886	.436	.526	90.7	18.1	3.52	3.4	1.5	.68	.62
	20.0	5.86	10.00	2.739	.436	.379	78.5	15.7	3.66	2.8	1.3	.70	.61
	15.3	4.47	10.00	2.600	.436	.240	66.9	13.4	3.87	2.3	1.2	.72	.64

TABLE 4 (Cont.)

ANGLES — EQUAL LEGS

PROPERTIES FOR DESIGNING

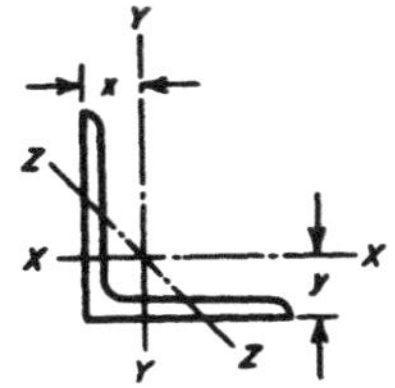

Size	Thickness	Weight per Foot	Area	AXIS X-X AND AXIS Y-Y				AXIS Z-Z
				I	S	r	x or y	r
In.	In.	Lb	In.^{2}	In.^{4}	In.^{3}	In.	In.	In.
3 x 3	1/2	9.4	2.75	2.2	1.1	.90	.93	.58
	7/16	8.3	2.43	2.0	.95	.91	.91	.58
	3/8	7.2	2.11	1.8	.83	.91	.89	.58
	5/16	6.1	1.78	1.5	.71	.92	.87	.59
	1/4	4.9	1.44	1.2	.58	.93	.84	.59
	3/16	3.71	1.09	.96	.44	.94	.82	.59

ANGLES — UNEQUAL LEGS

PROPERTIES FOR DESIGNING

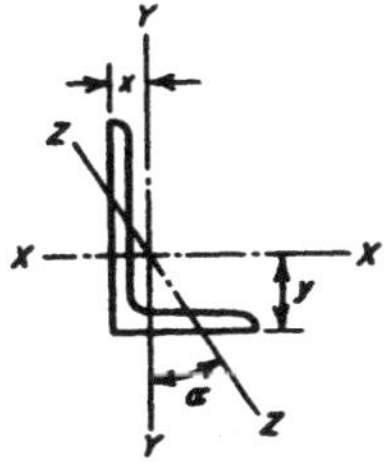

Size	Thick- ness	Weight per Foot	Area	AXIS X-X				AXIS Y-Y				AXIS Z-Z	
				I	S	r	y	I	S	r	x	r	Tan α
In.	In.	Lb	In.^{2}	In.^{4}	In.^{3}	In.	In.	In.^{4}	In.	In.	In.	In.	
6 x 3½	1/2	15.3	4.50	16.6	4.2	1.92	2.08	4.3	1.6	.97	.83	.76	.344
	3/8	11.7	3.42	12.9	3.2	1.94	2.04	3.3	1.2	.99	.79	.77	.350
	5/16	9.8	2.87	10.9	2.7	1.95	2.01	2.9	1.0	1.00	.76	.77	.352
	1/4	7.9	2.31	8.9	2.2	1.96	1.99	2.3	0.85	1.01	.74	.78	.355

Example 1 Given a 36 WF 194 (i.e., a 36-in.-deep, wide-flange beam weighing 194 lb/ft — this is the method of designation of wide flange beams), 10 ft long, under a total uniform load, including the weight of the beam, of 80 kip/ft. Determine the maximum shear stress and the maximum bending stress.

Solution The shear and moment curves are shown in Fig. 8.2.

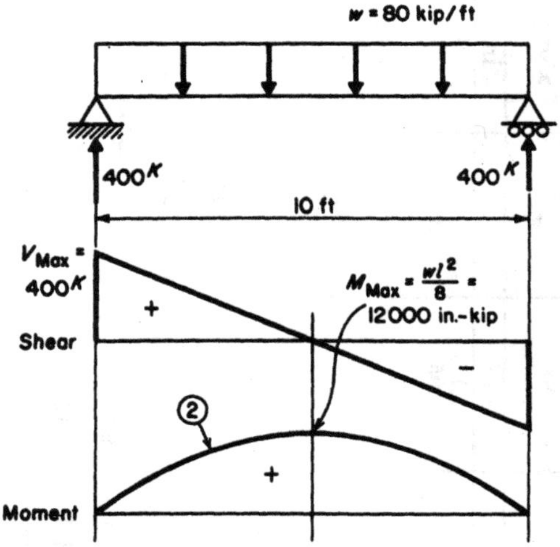

Fig. 8.2

Shear Stress Investigation

$$\tau_{max} = \frac{V_{max}}{td} = \frac{400{,}000}{(36.48)(.77)} \tag{8.9}$$

$$= 14{,}200 \text{ psi} \tag{8.10}$$

which is greater than the allowable shear stress of 13,000 psi.

Bending Stress Investigation — Bending About x-x Axis

$$\sigma_{max} = \frac{M_{max}}{S} = \frac{12{,}000{,}000}{663.6} \tag{8.11}$$

$$= 18{,}100 \text{ psi} \tag{8.12}$$

which is less than the allowable bending stress of 20,000 psi.

Thus, in the example given above, the shear stress is the governing stress in the design. But the student is asked to note particularly the dimensions of this beam. It is 10 ft long and 3 ft deep, so that $\frac{l}{d} = 3.33$. This is what we would call a "short, deep, steel beam" — one for which shear stresses would be expected to be important.

EXAMPLE 2 Consider the same 36 WF 194 as in the first problem. Now, however, the length is 30 ft and the total superposed load is 10 kip/ft. Determine the maximum shear stress and the maximum bending stress.

SOLUTION The shear and moment curves are shown in Fig. 8.3.

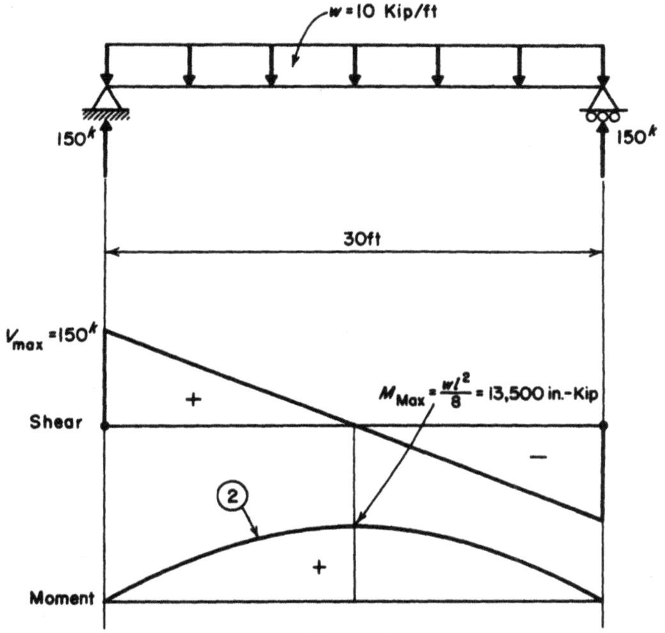

Fig. 8.3

Shear Stress Investigation

$$\tau_{max} = \frac{V_{max}}{td} = \frac{150{,}000}{(0.77)(36.48)} \tag{8.13}$$

$$= 5350 \text{ psi} \tag{8.14}$$

which is much less than the allowable shear stress of 13,000 psi.

Bending Stress Investigation

$$\sigma_{max} = \frac{M_{max}}{S} = \frac{13{,}500{,}000}{663.3} \tag{8.15}$$

$$= 20{,}200 \text{ psi} \tag{8.16}$$

which is just slightly over the allowable bending stress of 20,000 psi.

Thus, in the beam of this example, we see that the shear stress is much lower than the allowable when the bending stress is at the allowable value. This beam has an $l/d = 30/3 = 10$ and may be thought of as a typical engineering steel beam — one for which bending stress governs and the shear is not important.

Example 3 Let us now consider the *design* of a steel beam 20 ft long subjected to the loading shown in Fig. 8.4.

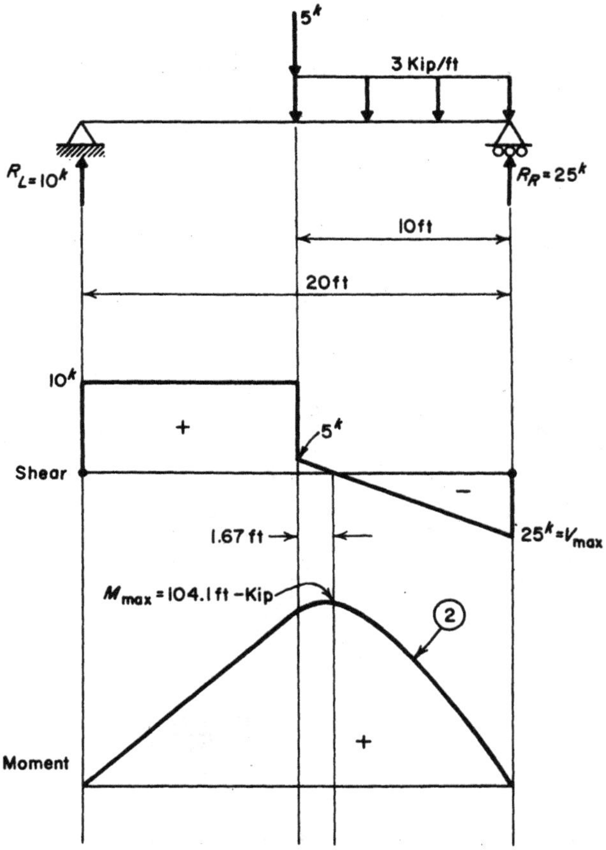

Fig. 8.4

Solution Following the procedure outlined in Art. 8.2, we shall number our steps in accordance with the numbering used therein.

1. The external load on the beam is given and shown on the figure. We will assume the weight of the beam is 50 lb/ft and will include this effect when we determine the maximum moment (see Step 3).

2. The shear and moment diagrams are shown on Fig. 8.4. To draw the shear

diagram we must first determine the reactions. Thus, taking moments about the right reaction, we have

$$R_L(20) = 5(10) + 3(10)(5) \tag{8.17}$$

or

$$R_L = 2.5 + 7.5 = 10^K \tag{8.18}$$

Then, taking $\Sigma F_V = 0$,

$$R_R = 5 + 30 - 10 \tag{8.19}$$

$$= 25^K \tag{8.20}$$

We may now draw the shear diagram, noting that the point of zero shear (and hence of maximum moment) is 11.67 ft from the left end. Having the shear curve, we now prepare the moment curve, using the methods previously described in Art. 7.5. The maximum moment value is obtained as follows:

$$M_{max} = 10(11.67) - 5(1.67) - \frac{3(1.67)(1.67)}{2} \tag{8.21}$$

$$= 116.7 - 8.35 - 4.20 \tag{8.22}$$

$$= 104.1 \text{ ft-kip} \tag{8.23}$$

3. Now, including the weight of beam effect, we have, for the moment due to the beam at the point of maximum determined in (2) above (and this may be assumed as the point of combined maximum with sufficient accuracy), see Fig. 8.5,

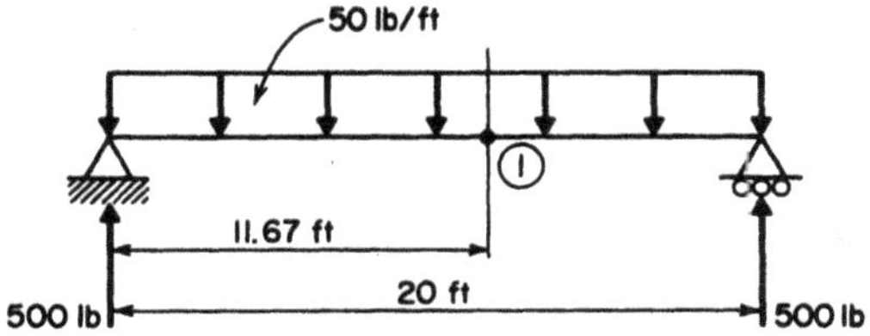

Fig. 8.5

$$M_1 = 500(11.67) - \frac{50(11.67)(11.67)}{2} \tag{8.24}$$

$$= 5835 - 3450 \tag{8.25}$$

$$= 2385 \text{ ft-lb} \tag{8.26}$$

or

$$= 2.4 \text{ ft-kip} \tag{8.27}$$

Hence, M_{max}, including the assumed weight of the beam, is given by

$$M_{max} = 104.1 + 2.4 \tag{8.28}$$

$$= 106.5 \text{ ft-kip} \tag{8.29}$$

$$= 106.5(12) \text{ in.-kip} \tag{8.30}$$

4. Using for the allowable bending stress the specification value of 20 ksi, we have, therefore, for the required S,

$$S_{req} = \frac{M_{max}}{\sigma_{allowable}} \tag{8.31}$$

$$= \frac{(106.5)(12)}{20} \tag{8.32}$$

$$= 64 \text{ in.}^3 \tag{8.33}$$

5. Referring to Table 4, we see that, of the sections listed the 12 WF 50, with an $S = 64.7$ in.^{3} will be acceptable. Also, the weight of this beam is just equal to the assumed weight, hence the beam is also acceptable in this respect.
   In any given case, if the assumed weight is much less than the required weight, and the difference in required S and actual S is not too great, then it may be necessary to redesign the beam using a larger assumed weight. But this is a matter of experience, and the designer very quickly learns how to handle this phase of the design procedure.

6. As pointed out above, shear is never critical in beams of relatively large length with respect to depth, and in this case shear need not be considered.

8-5 Timber Beams — Investigation and Design

For wood beams, just as for steel, there are, in general, the two separate phases of the analysis problem — investigation and design. Also, in connection with timber, as pointed out in Art. 8.2, we must consider shear as well as bending in our analysis.

We shall consider as illustrative examples one problem of each type. These may be considered as typical of all problems that may be encountered in ordinary practice.

As an aid to design and investigation, we include Table 5, which lists needed properties for typical standard timber beams. Note that actual sizes of dressed timber beams are less than nominal sizes, and the properties are based, of course, on actual sizes. The weights per foot listed are based on an assumed average weight of 40 lb/cu ft, a reasonable average for structural timber.

TABLE 5

TIMBER
AMERICAN STANDARD SIZES
PROPERTIES FOR DESIGNING
National Lumber Manufacturers Association

Nominal Size	American Standard Dressed Size	Area of Section	Weight per Foot	Moment of Inertia	Section Modulus
In.	In.	In.^{2}	Lb	In.^{4}	In.^{3}
2 x 4	1⅝ x 3⅝	5.89	1.64	6.45	3.56
6	5⅝	9.14	2.54	24.1	8.57
8	7½	12.2	2.39	57.1	15.3
10	9½	15.4	4.29	116	24.4
12	11½	18.7	5.19	206	35.8
14	13½	21.9	6.09	333	49.4
16	15½	25.2	6.99	504	65.1
18	17½	28.4	7.90	726	82.9
3 x 4	2⅝ x 3⅝	9.52	2.64	10.4	5.75
6	5⅝	14.8	4.10	38.9	13.8
8	7½	19.7	5.47	92.3	24.6
10	9½	24.9	6.93	188	39.5
12	11½	30.2	8.39	333	57.9
14	13½	35.4	9.84	538	79.7
16	15½	40.7	11.3	815	105
18	17½	45.9	12.8	1172	134

EXAMPLE 1 Determine the shear stress and the bending stress for the timber beam and loading shown. The uniform load includes the weight of the beam.

SOLUTION The shear and moment diagrams are drawn first. To prepare these we must determine the reactions. Taking moments about the right reaction, we have

$$R_L(16) - 50(16)(8) - 500(8) - 200(12) = 0 \tag{8.34}$$

or

$$R_L = 800 \text{ lb} \tag{8.35}$$

Then, taking

$$\Sigma F_V = 0 \tag{8.36}$$

$$R_L + R_R = 50(16) + 200 + 500 \tag{8.37}$$

$$R_R = 700 \text{ lb} \tag{8.38}$$

The reactions having been determined, the shear diagram is drawn as shown (Fig. 8.6).

The moment diagram is drawn next. Note that each segment of this curve is a second order parabola, with a discontinuity in slope at their junctions as shown. Also, the maximum moment, which occurs at the point of zero shear, is at the point of application of the 500-lb load. This moment is given by

$$M_{max} = 800(8) - 200(4) - 50(8)(4) \tag{8.39}$$

$$= 4000 \text{ ft lb} \tag{8.40}$$

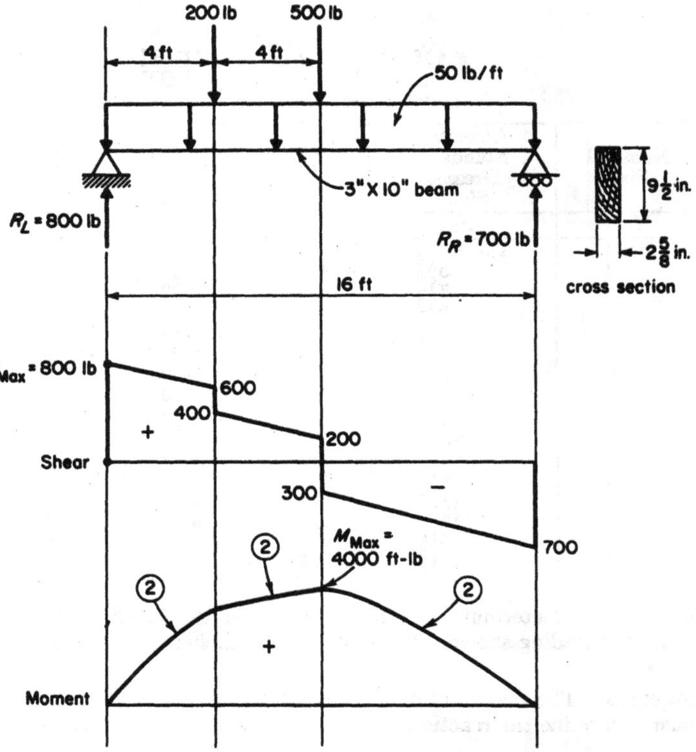

Fig. 8.6

Having the shear and moment, we can now determine the maximum shear stress and maximum bending stress in the following manner.

Shear Stress

$$\tau_{max} = \frac{3}{2}\frac{V_{max}}{A} \tag{8.41}$$

where

$$V_{max} = 800 \text{ lb} \tag{8.42}$$

(see the shear curve), and

(see Table 5)

$$A = 24.9 \text{ sq in.} \tag{8.43}$$

Hence

$$\tau_{max} = \frac{3}{2}\frac{(800)}{(24.9)} \tag{8.44}$$

$$= 48.3 \text{ psi} \tag{8.45}$$

Bending Stress

$$\sigma_{max} = \frac{M_{max}}{S} \tag{8.46}$$

where

$$M_{max} = 4000(12) \text{ in. lb} \tag{8.47}$$

(see moment curve)

$$S = 39.5 \text{ in.}^3 \tag{8.48}$$

(see Table 5) Hence

$$\sigma_{max} = \frac{4000(12)}{39.5} \tag{8.49}$$

$$= 1220 \text{ psi} \tag{8.50}$$

Both stresses — shear and bending — are very near the maximum allowables for timber beams. This emphasizes the point made before: for timber beams, shear stresses are frequently as important as bending stresses. Indeed, one of the more common failures of timber beams is a shear failure, which is indicated by a condition shown in Fig. 8.7. The split, which invariably occurs at the mid-height (point of maximum shear stress), generally starts at the support (point of maximum shear load) and extends some distance away from the support.

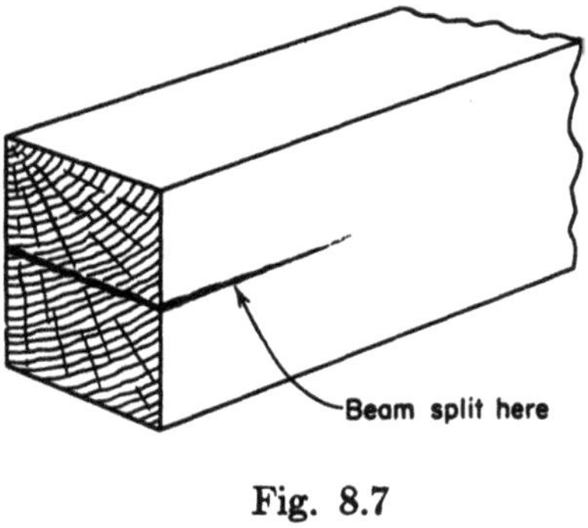

Fig. 8.7

EXAMPLE 2 Given the beam and loading shown in Fig. 8.8. If the allowable shear stress, τ, is equal to 100 psi and the allowable bending stress σ is 1500 psi, determine the size of beam required.

SOLUTION We draw first the shear and moment diagrams, neglecting the weight of the beam. To do this, we first obtain R_L and R_R. Taking moments about the right reaction,

$$R_L(20) = 20(20)(10) + 500(12) \tag{8.51}$$

$$R_L = 500 \text{ lb} \tag{8.52}$$

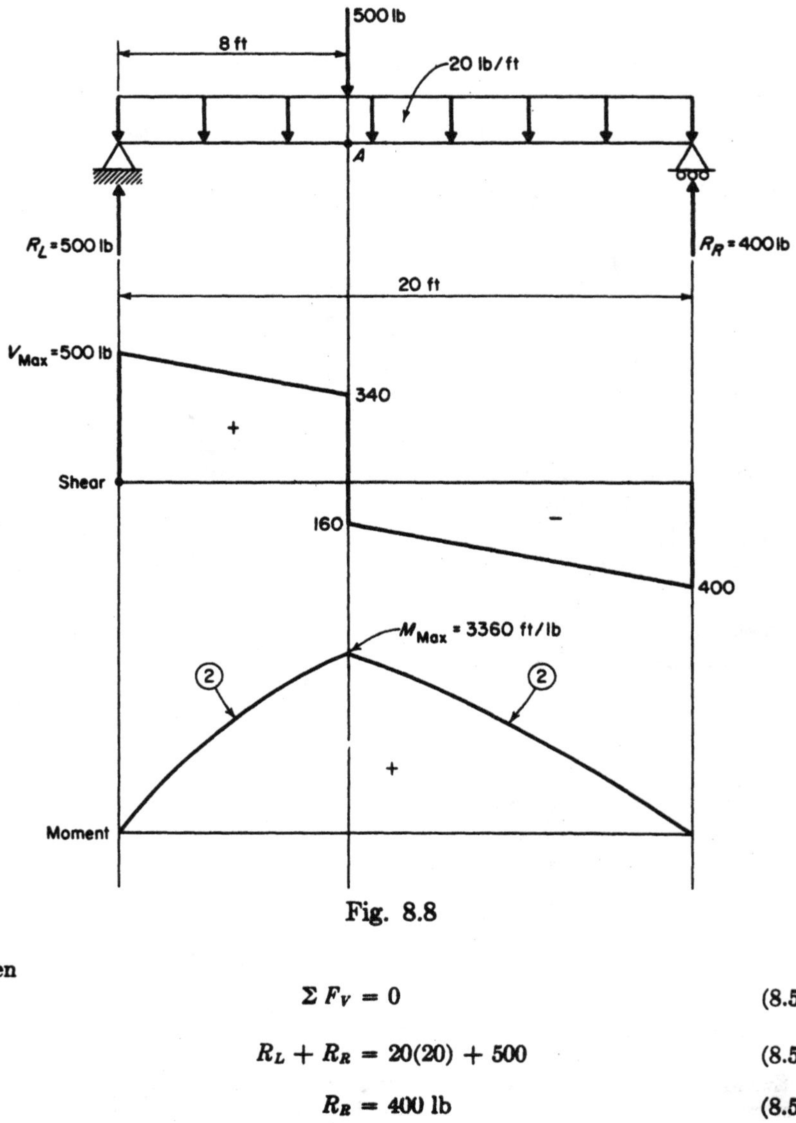

Fig. 8.8

Then

$$\Sigma F_V = 0 \tag{8.53}$$

$$R_L + R_R = 20(20) + 500 \tag{8.54}$$

or

$$R_R = 400 \text{ lb} \tag{8.55}$$

Having the reactions, the shear diagram and then the moment diagram are drawn, as shown in Fig. 8.8. Note that

$$M_{1_{max}} = 500(8) - 20(8)(4) \tag{8.56}$$

$$= 3360 \text{ ft lb} \tag{8.57}$$

If we assume the weight of the beam is 10 lb/ft, then there will be an additional moment at point A given by

$$100(8) - 10(8)(4) = 480 \text{ ft lb} \tag{8.58}$$

or

$$M_{max} = M_{1max} + 480 \tag{8.59}$$

$$= 3360 + 480 \tag{8.60}$$

$$= (3840)(12) \text{ in.-lb} \tag{8.61}$$

and

$$V_{max} = 500 \text{ lb} + 100 \text{ lb} = 600 \text{ lb} \tag{8.62}$$

Design for Shear

The area required for shear is determined from

$$\tau_{allow} = \frac{3}{2}\frac{V_{max}}{A_{req}} \tag{8.63}$$

or

$$A_{req} = \frac{3}{2}\left(\frac{600}{100}\right) = 9 \text{ sq in.} \tag{8.64}$$

Moment design

The required section modulus, S_{req}, is determined from

$$\sigma_{allow} = \frac{M_{max}}{S_{req}} \tag{8.65}$$

or

$$S_{req} = \frac{M_{max}}{\sigma_{allow}} = \frac{(3840)(12)}{1500} \tag{8.66}$$

$$= 30.7 \text{ in.}^3 \tag{8.67}$$

Hence we need a beam which has a minimum area of 9 sq in. and a minimum section modulus, $S = 30.7$ in.^{3}. Referring to Table 5, we see that the following two beams satisfy these requirements (also shown is the weight of each):

Beam	*Area*	*S*	*Weight*
2 × 12	18.7	35.8	5.19 lb/ft
3 × 10	24.9	39.5	6.93 lb/ft

Since the 2 × 12 beam is the lightest, it is also, in general, the least costly and hence would be the one chosen. Note also that the actual weight of 5.19 lb/ft is less than the assumed weight of 10 lb/ft; hence this choice is satisfactory.

8-6 Summary

The investigation and design procedures for steel and timber beams were explained in detail and illustrative problems of both kinds were solved for both materials.

By means of the illustrative examples it was verified that

(a) For steel beams, shear stresses are not important unless the beam is a short, deep one. In general bending stresses govern in the design of the engineering type of steel beam.

(b) For timber beams, shear and bending are of equal importance and either stress may govern the design of the beam.

It was also pointed out, however, that in steel beam design there are other factors besides simple shear load and bending load which must be considered. These are described in the appropriate specification — as, for example, the specifications of the American Institute of Steel Construction, which are given in the *AISC Manual*

Problems

In the problems which follow use the following values as allowable stresses for steel and timber:

	Steel	*Timber*
Bending	20,000 psi	1500 psi
Shear	13,000 psi	150 psi

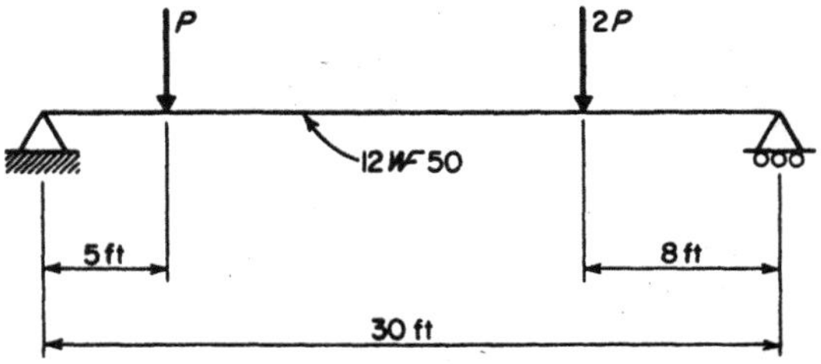

1. Determine the allowable load, P, for the beam and loading shown.

2. Determine the allowable load, w, per unit length for the beam and loading shown.

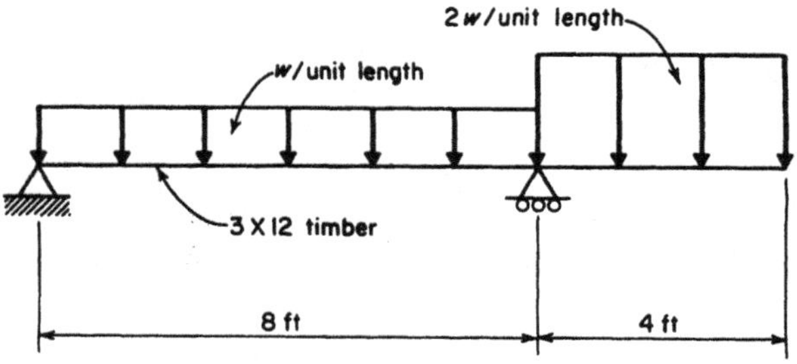

3. If cover plates are attached to the top and bottom flanges of the beam as shown, so that the composite structure acts as a unit, determine the load, P, which the beam can withstand.

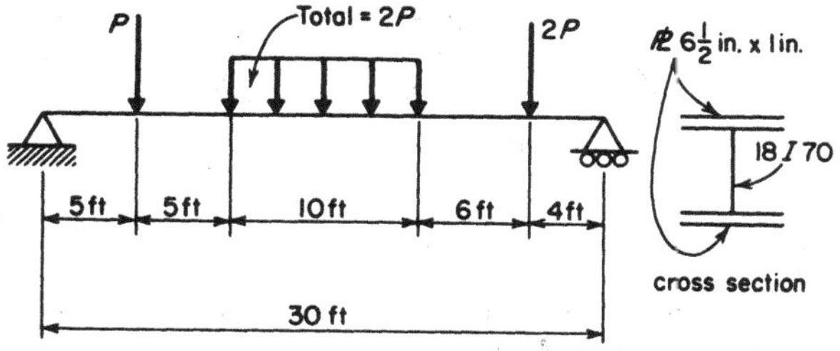

4. A 10 [25 is loaded as shown. Determine the maximum allowable value of P.

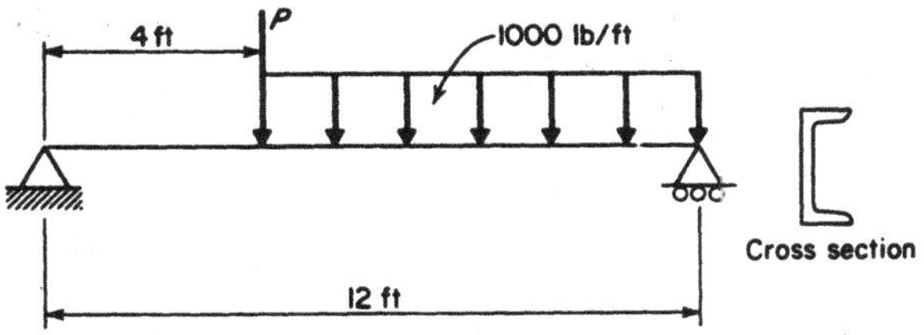

5. Given the loading as shown, determine the required size of steel beam.

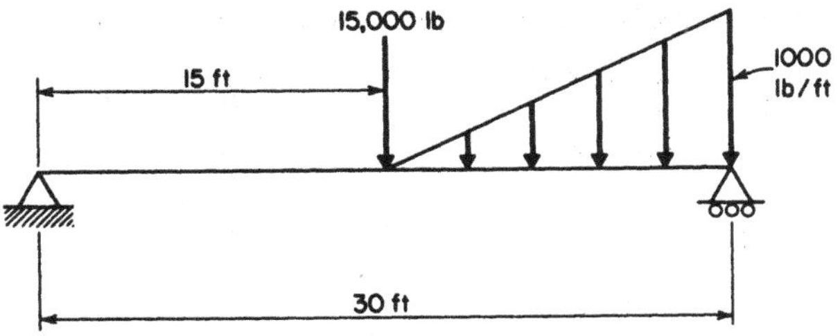

6. Determine a suitable timber beam for the load and span shown.

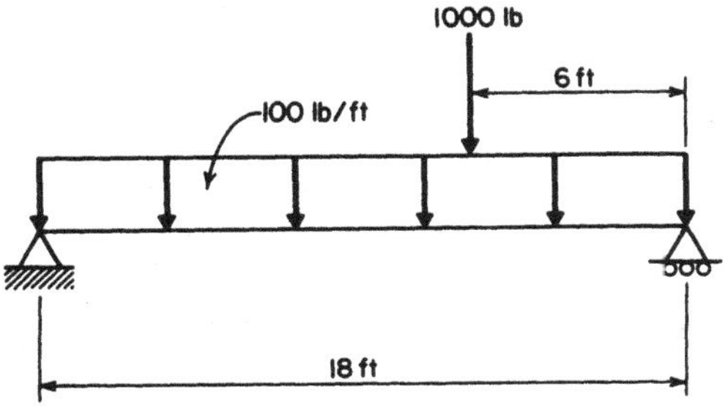

7. A steel beam is subjected to moment loadings as indicated. Determine the required sise of beam.

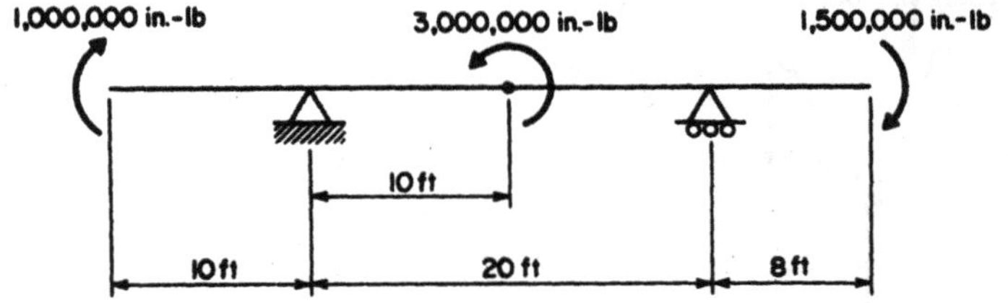

8. Determine the maximum length of span for a 3 × 12 timber beam if the only load on the beam is its own weight. It is supported at its ends.

9. Determine the maximum length of span for a 20 I 75 if the only load on the beam is its own weight. It is supported at its ends.

10. Determine the maximum span, *l*, for a 2 × 14 timber beam supported as shown if its only load is its own weight.

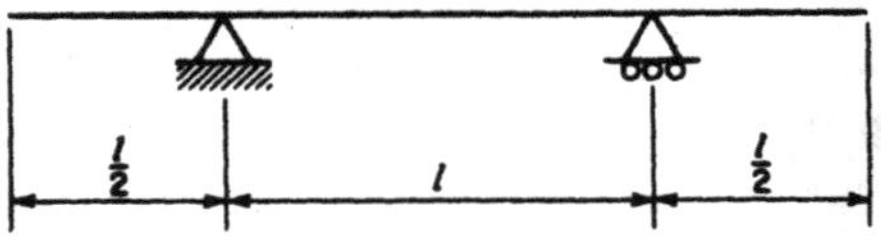

11. If the beam shown in Prob. 10 is a 36 WF 170, what is its maximum span, *l*?

12. The figure shows a typical steel framing plan for a building. Members marked *C* are columns, members marked *B* are beams, and members marked *G* are girders. Loads in the different areas, including weights of floor, are

I — 150 lb/sq ft	IV — 150 lb/sq ft
II — 50 lb/sq ft	V — 50 lb/sq ft
III — 70 lb/sq ft	VI — 200 lb/sq ft

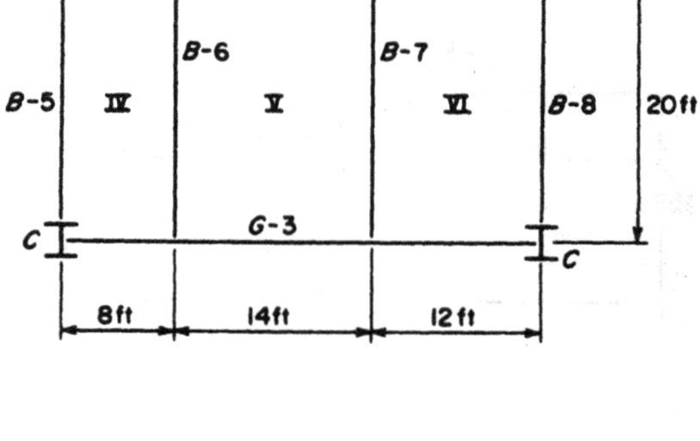

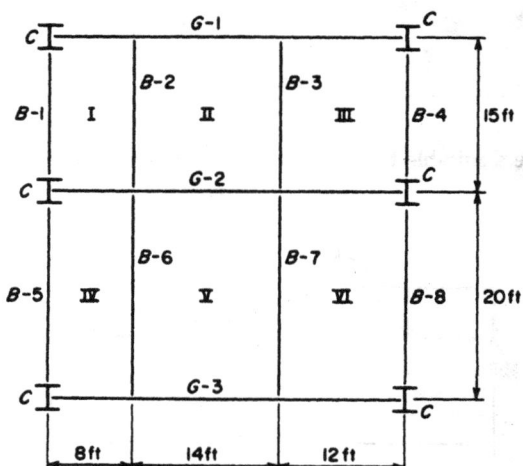

Assume the floor load divides among the beams in accordance with tributary areas. Assume beam loads are applied to girders as concentrated loads. Do not neglect the weights of the beams. Design the beams and girders.

13. The figure shows a typical timber framing plan. Members marked *C* are columns, members marked *J* are joists, and members marked *G* are girders. The superposed live load and floor dead load may be taken as 100 lb/sq ft.
(a) Design the members *J* and *G*.

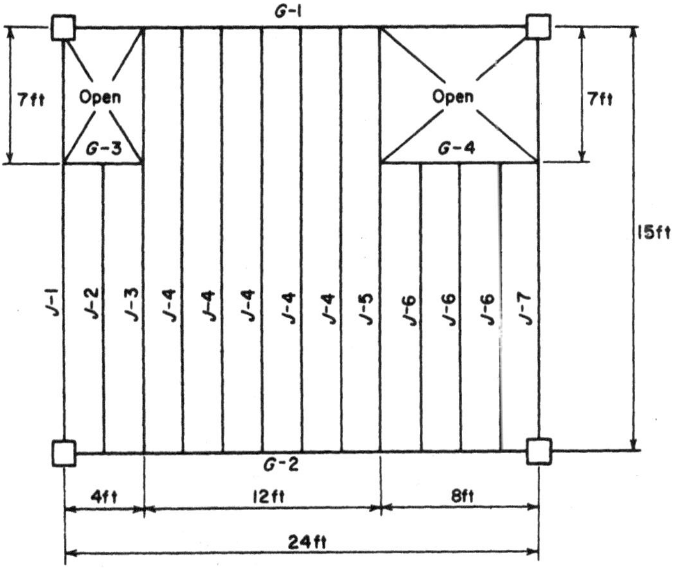

Chapter 9

THE DEFLECTION OF BEAMS

9-1 Introduction

In the last several chapters we discussed various topics which are important in the design of the engineering beam for shear and moment loadings. In this chapter we shall discuss another topic which is extremely important in the overall problem of the analysis of beams: *the deflection of beams.*

We are interested in beam deflections for two reasons:

1. For the deflection itself. In many cases it is important to limit the maximum deflection of a beam. Indeed, in some cases, the amount of permissible deflection governs the design of the beam. This is especially true when we have plaster or other brittle material coverings on the beam. Experience indicates, in these cases, that unsightly cracking can be eliminated if the maximum deflection of the beam is limited to a certain percentage of its length, generally in the order of 0.3%.

2. The analysis of deflections is the key step in the solution of the statically indeterminate beam (which we will discuss in the next chapter). A statically indeterminate beam is one which can not be completely solved for shear and moment using the equations of statics alone (Eq. 4.89–4.92). For these beams, in addition to the equations of statics, we must use the compatibility of strain relations in their engineering form. This requires a knowledge of the methods for calculating beam deflections.

In this chapter, therefore, we shall discuss several different methods for determining the deflections of beams. We shall describe

(a) the double integration method,
(b) the moment-area method,
(c) the conjugate beam method,
(d) the finite difference method.

Methods (a), (b) and (c) are typical of "classical methods" for determining beam deflections.

Method (d) is typical of advanced modern techniques which utilize electronic computers.†

9-2 The Double Integration Method

It was pointed out in Eq. 6.46 that if the deflection of the beam is small,

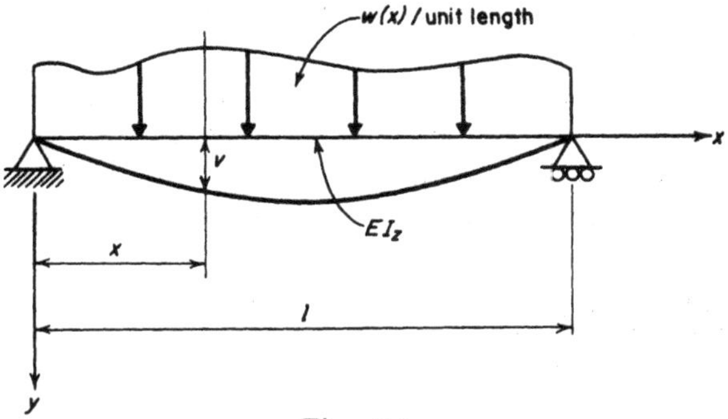

Fig. 9.1

the following Bernoulli-Euler relation holds for the engineering beam (see Fig. 9.1‡)

$$\frac{d^2v}{dx^2} = -\frac{M(x)}{EI_z} \tag{9.1}$$

in which

v is the deflection in the y direction (positive downward) of the neutral plane of the beam — hence, in the engineering approximation, of the beam itself.

$M(x)$ is the bending moment at any point x on the beam.

EI_z is the "beam stiffness," since as we shall see, the smaller EI_z is, the larger is the deflection, and vice versa.

†Another general technique for determining deflections of beams and beam-type structures is the "energy method". Many engineers prefer one or another of the various energy methods and, in fact, many problems not otherwise solvable can be approximately solved using an energy method. Among the different (although very similar) energy solutions are the method of least work, Castigliano's Theorems, conservation of energy, minimum potential energy, and others. Examples of different applications of these will be found in Refs. 9, 10, 20, 25.

‡In this figure, and in all later figures in which deflected structures are shown, we shall for clarity indicate all deflections in an exaggerated manner. The student must bear in mind, however, that except for the elastica solution of Chap. 11, *all* deflections are assumed to be very small and the theories which are given hold only for small deflections.

This equation can be integrated twice to give an expression for v, the deflection. Thus, a first integration gives

$$\frac{dv}{dx} = -\int \frac{M(x)\,dx}{EI_z} + C_1 \tag{9.2}$$

and integrating once more,

$$v = -\int \left[\int \frac{M(x)\,dx}{EI_z} \right] dx + C_1 x + C_2 \tag{9.3}$$

in which C_1 and C_2 are constants of integration determined from the boundary conditions. In any given case, $M(x)$ is usually a simple polynomial in x so that the integrations can be performed without difficulty.

Two simple examples will suffice to illustrate the application of this method.

Example 1 Let us determine the equation of the deflection curve for the beam and loading of Fig. 9.2.

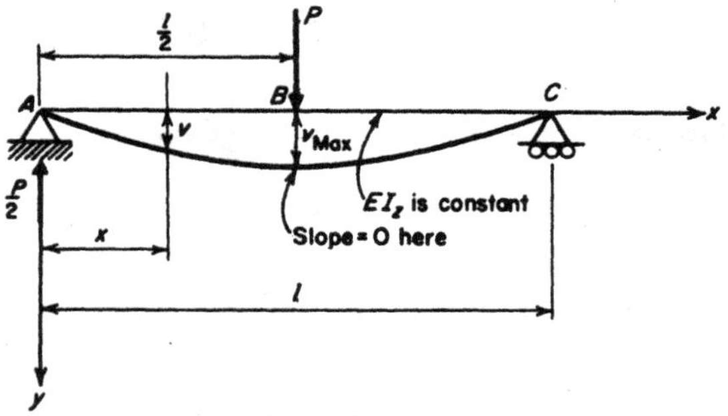

Fig. 9.2

Solution In this problem we can simplify the solution by utilising the properties of symmetry. For example, at the point B on the deflected beam, the slope must be zero. Also, the portion AB is obviously the same as the portion CB, insofar as deformations are concerned. Hence, if we solve only half the beam — say the portion AB, — then we have, in effect, a solution for the entire beam. We shall do this.

$$\frac{d^2v}{dx^2} = -\frac{M(x)}{EI_z} \tag{9.4}$$

and

$$M(x) = \frac{Px}{2}, \quad 0 \le x \le \frac{l}{2} \tag{9.5}$$

or

$$\frac{d^2v}{dx^2} = \frac{-Px}{2EI_z} \tag{9.6}$$

Integrating once, we get†

$$\frac{dv}{dx} = \frac{-Px^2}{4EI_z} + C_1 \tag{9.7}$$

and, integrating once more,

$$v = \frac{-Px^3}{12EI_z} + C_1x + C_2 \tag{9.8}$$

We determine C_1 and C_2 by utilizing the boundary conditions, which are

$$\left.\begin{aligned} v &= 0 \text{ at } x = 0 \\ \frac{dv}{dx} &= 0 \text{ at } x = \frac{l}{2} \end{aligned}\right\} \tag{9.9}$$

Using the second of these in Eq. 9.7, we have

$$0 = \frac{-P\left(\frac{l^2}{4}\right)}{4EI_z} + C_1 \tag{9.10}$$

or

$$C_1 = \frac{Pl^2}{16EI_z} \tag{9.11}$$

Therefore, from Eq. 9.8,

$$v = \frac{-Px^3}{12EI_z} + \frac{Pl^2x}{16EI_z} + C_2 \tag{9.12}$$

and since, from the first boundary condition, $v = 0$ at $x = 0$, it follows that

$$C_2 = 0 \tag{9.13}$$

Thus, the equation of the deflection curve is

$$v = -\frac{Px^3}{12EI_z} + \frac{Pl^2x}{16EI_z}, \quad 0 \leq x \leq \frac{l}{2} \tag{9.14}$$

To obtain v_{max}, we substitute $x = l/2$ in the above. This gives

$$v_{max} = -\frac{Pl^3}{96EI_z} + \frac{Pl^3}{32EI_z} \tag{9.15}$$

$$= +\frac{Pl^3}{48EI_z} \tag{9.16}$$

The positive sign indicates the deflection is in the direction of the positive y-axis.

†Note that if I_z is a variable, expressible as a function of x, it may be handled without difficulty. If I_z is a variable not exactly expressible as a simple function of x, we may still very often approximate it as a simple function.

Similarly, the slope at the left end, the maximum slope is given by (see Eq. 9.7)

$$\left(\frac{dv}{dx}\right)_{x=0} = -\frac{Px^2}{4EI_z} + \frac{Pl^2}{16EI_z}, \quad x = 0 \tag{9.17}$$

or

$$\left(\frac{dv}{dx}\right)_{x=0} = +\frac{Pl^2}{16EI_z} \tag{9.18}$$

and the positive sign indicates a positive (i.e., clockwise, in this case) slope.

Example 2 As a second illustrative problem, let us solve for the deflection curve for the beam and loading shown in Fig. 9.3.

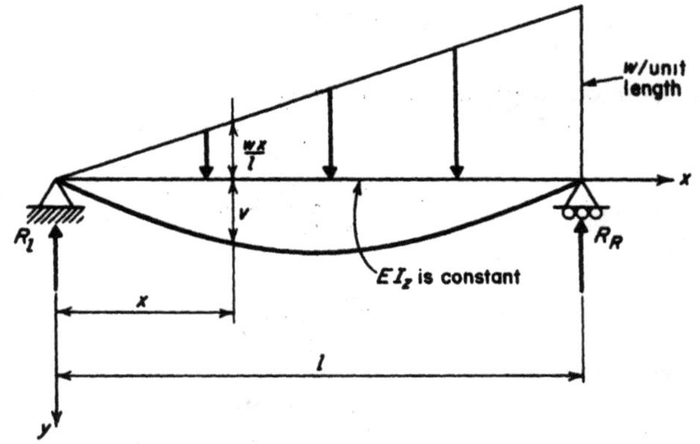

Fig. 9.3

Solution We must find R_L in order to determine M_z. Taking moments about R_R, we have

$$R_L l - w\left(\frac{l}{2}\right)\left(\frac{l}{3}\right) = 0 \tag{9.19}$$

or

$$R_L = \frac{wl}{6} \tag{9.20}$$

Therefore (see Fig. 9.3),

$$M_z = R_L x - \frac{wx}{l}\cdot\frac{x}{2}\cdot\frac{x}{3} \tag{9.21}$$

$$= \frac{wlx}{6} - \frac{wx^3}{6l} \tag{9.22}$$

Now, applying the differential relation,

$$\frac{d^2v}{dx^2} = -\frac{M(x)}{EI_z} \tag{9.23}$$

we get

$$EI_z\frac{d^2v}{dx^2} = -\frac{wlx}{6} + \frac{wx^3}{6l} \tag{9.24}$$

Integrating once, we have

$$EI_z \frac{dv}{dx} = -\frac{wlx^2}{12} + \frac{wx^4}{24l} + C_1 \tag{9.25}$$

and integrating again we obtain

$$EI_z v = -\frac{wlx^3}{36} + \frac{wx^5}{120l} + C_1 x + C_2 \tag{9.26}$$

The boundary conditions which will enable us to determine C_1 and C_2 are

$$\left.\begin{aligned} \text{at } x &= 0, v = 0 \\ \text{at } x &= l, \ v = 0 \end{aligned}\right\} \tag{9.27}$$

Substituting these in the equation we get, from the first,

$$C_2 = 0 \tag{9.28}$$

and from the second,

$$0 = -\frac{wl^4}{36} + \frac{wl^4}{120} + C_1 l \tag{9.29}$$

or

$$C_1 = +\frac{7}{360} wl^3 \tag{9.30}$$

and therefore the equation of the deflection curve is given by

$$v = \frac{1}{EI_z}\left(-\frac{wlx^3}{36} + \frac{wx^5}{120l} + \frac{7wl^3x}{360}\right) \tag{9.31}$$

In anticipation of work to be done later in this chapter, we determine the deflections at the points $l/4$, $l/2$ and $3l/4$ as given by this equation. These are

$$v_{l/4} = 0.00442 \frac{wl^4}{EI_z} = \frac{27.15}{6144} \frac{wl^4}{EI_z} \tag{9.32}$$

$$v_{l/2} = 0.0065 \frac{wl^4}{EI_z} = \frac{39.92}{6144} \frac{wl^4}{EI_z} \tag{9.33}$$

$$v_{3l/4} = 0.00485 \frac{wl^4}{EI_z} = \frac{29.60}{6144} \frac{wl^4}{EI_z} \tag{9.34}$$

9-3 The Moment-Area Method†

The moment-area method is essentially a procedure for determining the constants of integration, of the deflection equation, in a graphical manner. We first state the two moment-area relations; then we derive them. Finally we shall solve the same problems considered in the last section, using the moment-area method.

†The moment-area method (and the essentially equivalent conjugate beam method described in Art. 9-4) appear to have been developed in different countries at about the same time. In the United States, one of the early users (about 1873) was Professor Charles E. Greene of the University of Michigan. Otto Mohr, the German engineer, outlined the conjugate beam method in 1868.

Given a beam AB under any transverse loading, as shown in Fig. 9.4. Also shown are the deflection curve, and the $M(x)/(EI_z)$ curve; the latter is a curve whose ordinates are given by the moment at each point divided

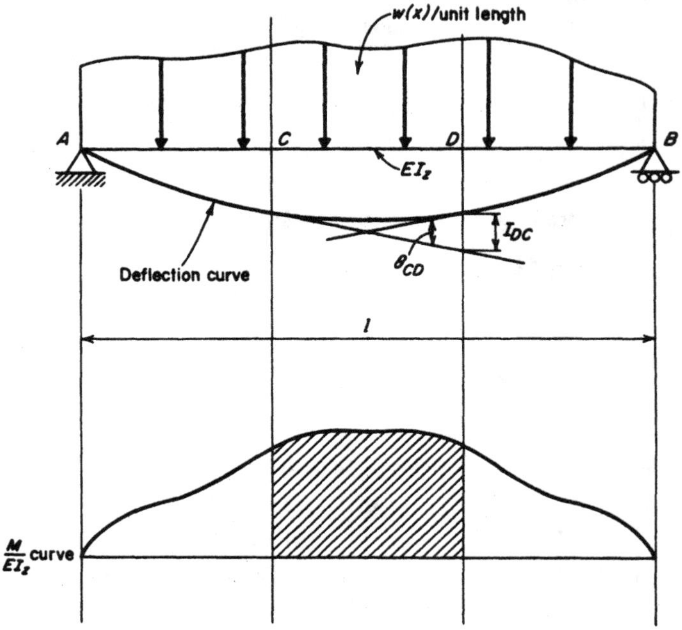

Fig. 9.4

by EI_z at the corresponding points. Consider any two points, such as C and D. Draw tangents to the deflected beam at these two points. Then, the two moment-area relations are as follows:

I The angle between the tangents drawn at C and at D is equal to the area under the $M(x)/(EI_z)$ curve between C and D.

II The intercept at D, between the tangents drawn at C and D, in a direction normal to the undeflected position of the beam, called I_{DC}, is given by the moment, about D, of the area under the $M(x)/(EI_z)$ curve between C and D.

In connection with two relations, we point out the following:

(a) Both expressions give relations between tangents, to the deflected beam, drawn at two points. Thus, in applying the moment-area method the first step, *always*, is to draw two tangents at the proper points.

(b) In using this method it is necessary to have an $M(x)/(EI_z)$ curve. As noted, this is simply the M curve divided by the EI_z at each point. If EI_z is constant, then the *shape* of the $M(x)/(EI_z)$ curve is the same as the shape of the M curve. If EI_z is not constant then the M curve is altered accordingly.

(c) Note the order of subscripts in the expression I_{DC}. The first subscript represents the point at which the intercept is taken; the second subscript represents the other point of tangency.

(d) In the second moment-area relation, the moment of the area under the $M(x)/(EI_z)$ curve is *always* taken about the point where the intercept is taken. Also note that, in general,

$$I_{DC} \neq I_{CD} \tag{9.35}$$

and so the point about which the moment of the $M(x)/(EI_z)$ area is taken determines whether we are getting I_{DC} or I_{CD}.

We now prove the two moment-area relations. Refer to Fig. 9.5, showing a typical beam and loading. Also shown is the $M(x)/(EI_z)$ curve.

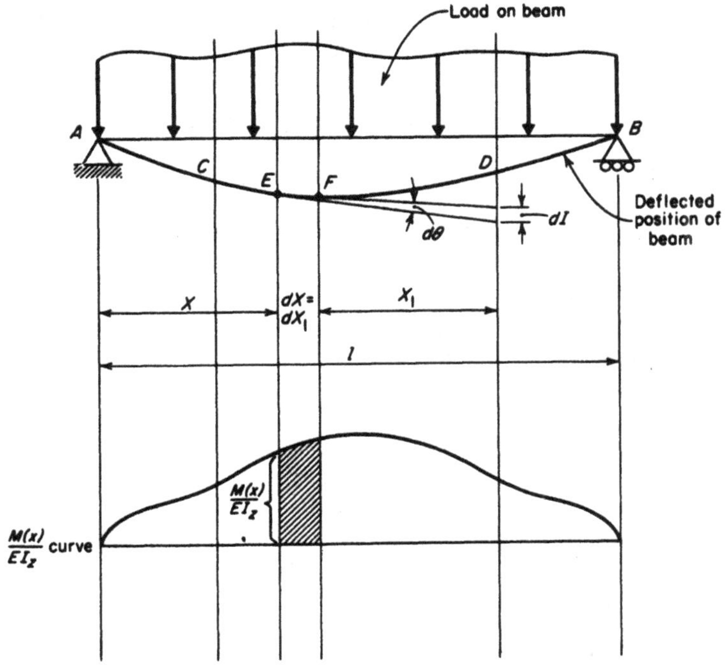

Fig. 9.5

We start with the fundamental Bernoulli-Euler equation:

$$\frac{d^2v}{dx^2} = -\frac{M(x)}{EI_z} \tag{9.36}$$

or

$$d\left(\frac{dv}{dx}\right) = -\frac{M(x)}{EI_z}\,dx \tag{9.37}$$

and recalling (see Eq. 6.45) that, for small deflections,

$$\frac{dv}{dx} = \tan\theta \cong \theta \tag{9.38}$$

the above becomes

$$d\theta = -\frac{M(x)}{EI_z}\,dx \tag{9.39}$$

That is, the differential angle between tangents drawn at E and F is given by the shaded area of the $M(x)/(EI_z)$ curve shown. Hence, if we wish to sum up the effect of all of these differential angles between C and D — or, otherwise stated, if we want the angle between tangents drawn at C and at D — we integrate this expression, obtaining

$$\int_C^D d\theta = -\int \frac{M(x)}{EI_z}\,dx \tag{9.40}$$

which is given by

$$\theta_{CD} = \text{area under the } \frac{M(x)}{EI_z} \text{ curve between } C \text{ and } D \tag{9.41}$$

This proves Relation I.

Now return to Fig. 9.5. Note that $dx = dx_1$. Also, since we are dealing with small deflections, it follows that

$$dI = x_1 d\theta \tag{9.42}$$

and, since

$$d\theta = \frac{-M(x)\,dx}{EI_z} \tag{9.43}$$

it follows that

$$dI = \frac{-M(x)x_1\,dx_1}{EI_z} \tag{9.44}$$

which says, in words, that the differential vertical intercept, at D, between tangents drawn at E and F is equal to the moment, about D, of the differential area of the $M(x)/(EI_z)$ curve at E and F.

Hence, if we want the intercept at D between tangents drawn at C and D, we would integrate this expression between these two points, and this would then give

$$I_{DC} = \text{the moment, about } D\text{, of the area under the } \frac{M(x)}{EI_z} \text{ curve between } C \text{ and } D \tag{9.45}$$

This proves Relation II.

In order to indicate how these relations are utilized in solving deflection problems, we now solve the same illustrative examples considered earlier in this chapter, using the moment-area method.

EXAMPLE 1 Let us determine the equation of the deflection curve for the beam and loading of Fig. 9.6.

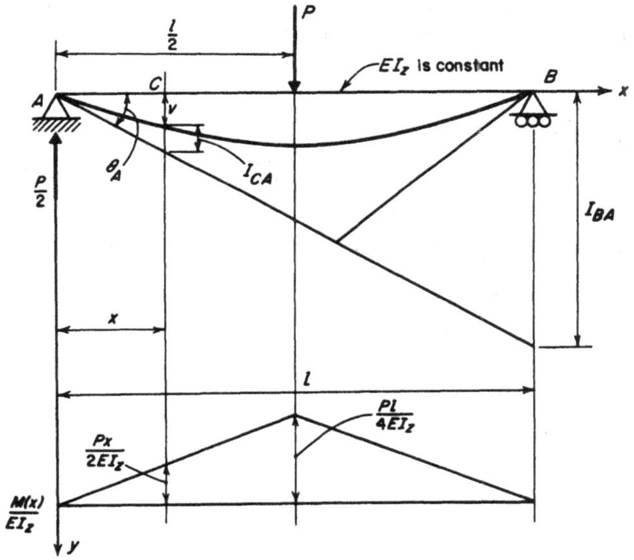

Fig. 9.6

SOLUTION The $M(x)/(EI_z)$ curve, for the beam with constant EI_z is the same shape as the M curve. The deflection at any point, x, is given by (see Fig. 9.6)

$$v = \frac{x}{l} I_{BA} - I_{CA}, \quad 0 \leq x \leq \frac{l}{2} \tag{9.46}$$

In this expression, I_{BA} is the moment of the complete $M(x)/(EI_z)$ diagram about B, and I_{CA} is the moment about C, of the $M(x)/(EI_z)$ diagram between A and C.

Therefore,

$$I_{BA} = \frac{Pl}{4EI_z}\left(\frac{l}{4}\right)\left[\frac{2}{3}\frac{l}{2} + \left(\frac{l}{2} + \frac{1}{3}\frac{l}{2}\right)\right] \tag{9.47}$$

$$= \frac{Pl^3}{16EI_z} \tag{9.48}$$

and

$$I_{CA} = \frac{Px}{2EI_z}\cdot\frac{x}{2}\cdot\frac{x}{3} \tag{9.49}$$

$$= \frac{Px^3}{12EI_z} \tag{9.50}$$

so that

$$v = \frac{Pl^2x}{16EI_z} - \frac{Px^3}{12EI_z}, \quad 0 \le x \le \frac{l}{2} \tag{9.51}$$

Compare this with Eq. 9.14.

If we wanted the slope of the beam at the left end, we would obtain this by noting

$$\theta_A = \frac{I_{BA}}{l} \tag{9.52}$$

or (using Eq. 9.48)

$$\theta_A = \frac{Pl^2}{16EI_z} \tag{9.53}$$

Other slopes or deflections would be obtained in a similar manner.

Example 2 Let us obtain the deflection curve for the beam and loading shown in Fig. 9.7.

Solution The $M(x)/(EI_z)$ diagram is drawn by parts, starting from the left end.

The complete curve consists of the triangle, positive, due to R_L and the third order parabola, negative, due to the load.

Then

$$v = \frac{x}{l} I_{BA} - I_{CA} \tag{9.54}$$

In the above,

$$I_{BA} = \frac{wl^2}{6EI_z}\cdot\frac{l}{2}\cdot\frac{l}{3} - \frac{wl^2}{6EI_z}\cdot\frac{l}{4}\cdot\frac{l}{5} \tag{9.55}$$

$$= \frac{7}{360}\frac{wl^4}{EI_z} \tag{9.56}$$

and (see Fig. 9.7)

$$I_{CA} = \frac{wlx}{6EI_z}\cdot\frac{x}{2}\cdot\frac{x}{3} - \frac{wx^3}{6lEI_z}\cdot\frac{x}{4}\cdot\frac{x}{5} \tag{9.57}$$

$$= \frac{wlx^3}{36EI_z} - \frac{wx^5}{120lEI_z} \tag{9.58}$$

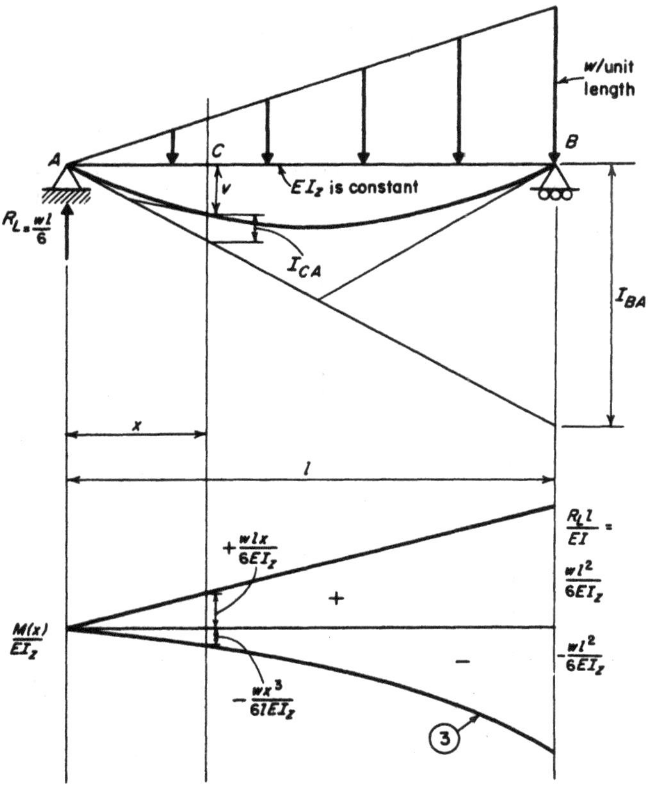

Fig. 9.7

Therefore, from Eq. 9.54,

$$v = \frac{7}{360}\frac{wl^3x}{EI_z} - \frac{wlx^3}{36EI_z} + \frac{wx^5}{120lEI_z} \tag{9.59}$$

Compare this result with Eq. 9.31.

The student should note particularly, in the illustrative examples, how similar the term-by-term expressions are to those obtained previously using the double integration method. That this is so is not surprising since, as pointed out before, the moment-area method is, fundamentally, merely a graphical interpretation of the formal integration process, in which the constants of integration are also obtained graphically.

In the next section we discuss still another "interpretive method" — the *conjugate beam method*. This method also will give — term for term — identical results to those obtained by the moment-area method. This is so, because the conjugate beam method and moment-area method are, fundamentally, identical methods. They are merely different representations of the integration process for the Bernoulli-Euler relation.

9-4 The Conjugate Beam Method

While the conjugate beam method is, as noted in the previous paragraph, essentially the same as the moment-area method,† it does have certain advantages for our particular applications. These advantages stem from the fact that we can utilize a very simple sign convention for the conjugate beam method. This convention, which will determine the sign and direction of the slopes and deflections, will utilize the convention previously described and used for shear and moments. Hence it fits in very nicely with the previous work done in beam analysis.

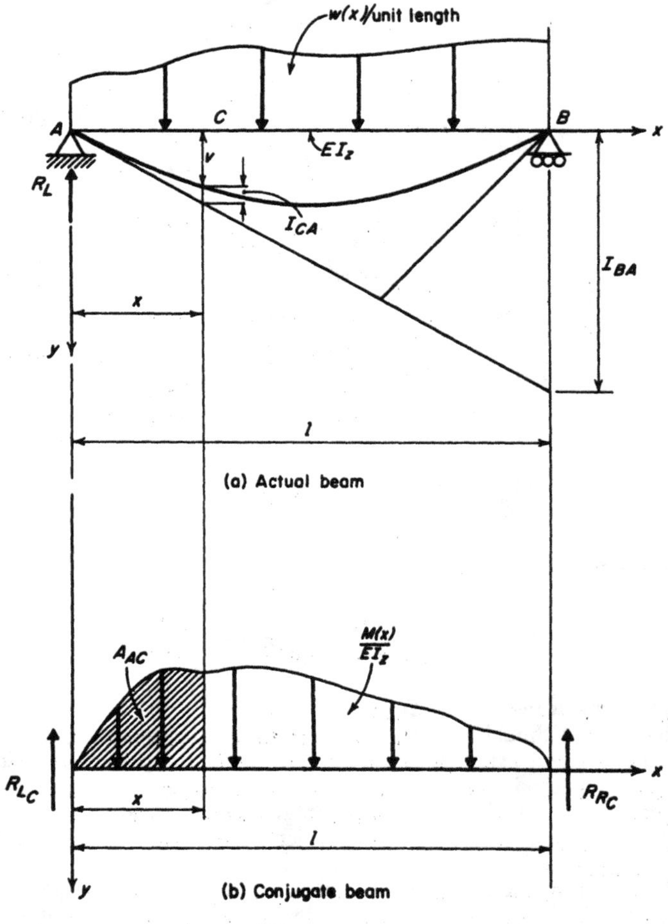

Fig. 9.8

†The student should compare the solution given herein for the same problems using the moment-area method and the conjugate beam method, and should note that these solutions are identical, term for term.

Consider the beam and loading shown in Fig. 9.8a. In applications of the conjugate beam method, we shall call this beam the *actual beam*.

Draw the $M(x)/(EI_z)$ curve for this actual beam, and consider another beam, of the same length as the actual beam, but loaded with the $M(x)/(EI_z)$ diagram of the actual beam *as a load*.

This load is in the direction of positive y, i.e., down if $M(x)$ is positive and up if $M(x)$ is negative. The beam loaded with the $M(x)/(EI_z)$ load is called the *conjugate beam* (see Fig. 9.8b).

Then, we shall prove the following two relations:

I The *shear* at any point, x, on the conjugate beam is equal — in sign and value — to the *slope* at that point on the actual beam.

II The *moment* at any point, x, on the conjugate beam is equal — in sign and value — to the *deflection* at that point on the actual beam.

The proof of I follows.

See Fig. 9.9, which shows that

$$\theta_C = \theta_A - \theta_{CA} \tag{9.60}$$

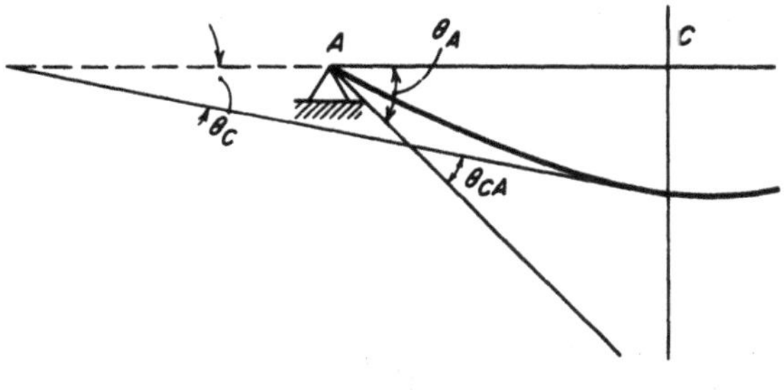

Fig. 9.9

We shall show, using the moment-area method, that θ_C is just equal to, in sign and value, the shear at C on the conjugate beam.

We note first that

$$\theta_A = \frac{I_{BA}}{l} \tag{9.61}$$

$$= \frac{\text{moment of area of entire } M(x)/(EI_z) \text{ curve about point } B}{l} \tag{9.62}$$

We observe in passing that this is just equal to R_{LC}, the left reaction on the conjugate beam, since the right-hand side of the above is just the expression which would be obtained when R_{LC} is solved for in the usual manner.

Also

$$\theta_{CA} = \text{area under } M(x)/(EI_s) \text{ curve between } A \text{ and } C \tag{9.63}$$

$$= A_{AC} \tag{9.64}$$

and therefore,

$$\theta_C = R_{LC} - A_{AC} \tag{9.65}$$

But this is just equal to the shear at C on the conjugate beam. Also, we see by inspection that the conjugate beam shear is positive when the slope is positive.

We observe, too, that $R_{LC} + R_{RC}$ is just equal to the total load of the $M(x)/(EI_s)$ diagram, so that the conjugate beam is in equilibrium under its loading and supports (which are R_{LC} and R_{RC}).

All of the above proves Relation I.

The proof of Relation II follows.

The deflection, v, at any point x (see Fig. 9.8a) is given by

$$v = \frac{x}{l} I_{BA} - I_{CA} \tag{9.66}$$

In Eq. 9.61, we showed that

$$R_{LC} = \frac{I_{BA}}{l} \tag{9.67}$$

hence

$$R_{LC}x = \frac{x}{l} I_{BA} \tag{9.68}$$

or, the first term on the right-hand side of Eq. 9.66 is just the moment at C, on the conjugate beam due to R_{LC}.

Similarly, I_{CA} is equal to the moment, about C, of the load on the conjugate beam between A and C.

Hence,

$$v = \frac{x}{l} I_{BA} - I_{CA} \tag{9.69}$$

is just equal to the moment at C, on the conjugate beam. Furthermore, we see that this deflection is positive (i.e., down) when the moment on the conjugate beam is positive.

This proves Relation II.

Before going on to illustrative examples, we discuss the special support conditions of the conjugate beam method.

9-5 Special Support Conditions of the Conjugate Beam Method

The conjugate beam support conditions shown in Fig. 9.8b indicate reactions R_{LC} and R_{RC} at the ends of the conjugate beam. These reactions are to be expected, since the actual beam is hinged at the ends A and B, and therefore we have *slopes* at A and B on the actual beam — the amount and direction of these slopes being just equal to the shears (i.e., R_{LC} and R_{RC}) on the conjugate beam.

However, if the actual beam has a free end (Fig. 9.10a) then it follows

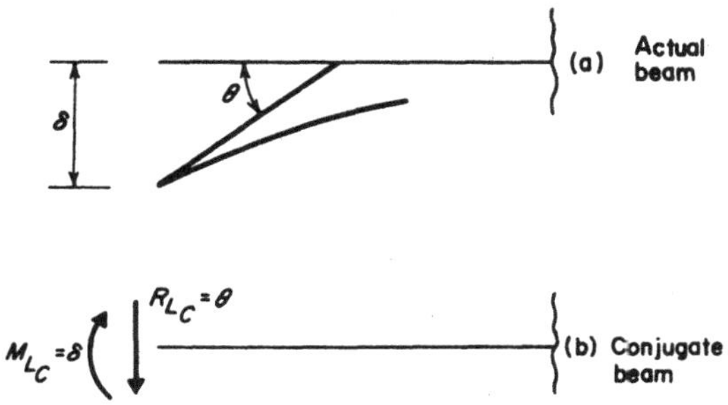

Fig. 9.10

that the conjugate beam, at this end, must have a reaction (shear) and moment (Fig. 9.10b). This must be so, since the actual beam has a slope and deflection at the free end, and this requires a shear and moment at the end of the conjugate beam.

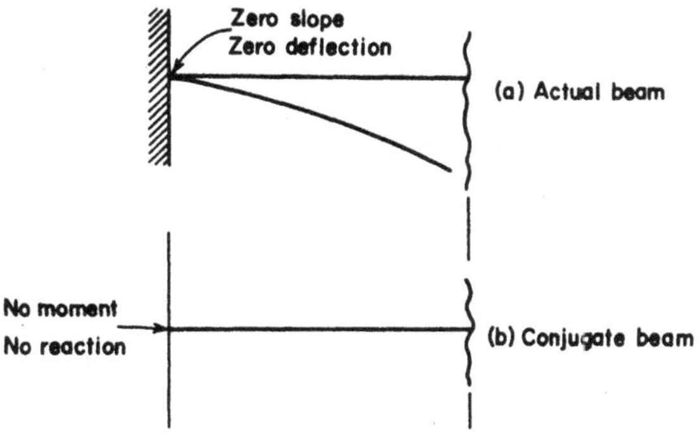

Fig. 9.11

If the actual beam has a built-in end (Fig. 9.11a), then, because it has zero slope and zero deflection at this point, it follows that the conjugate beam must have zero shear and zero moment at this end. This means that there is no concentrated reaction or concentrated moment at this end on the conjugate beam (Fig. 9.11b).

Finally, if we have a continuous beam, i.e., a beam with interior supports, such as at A in Fig. 9.12a, then, because the beam is continuous, there is a

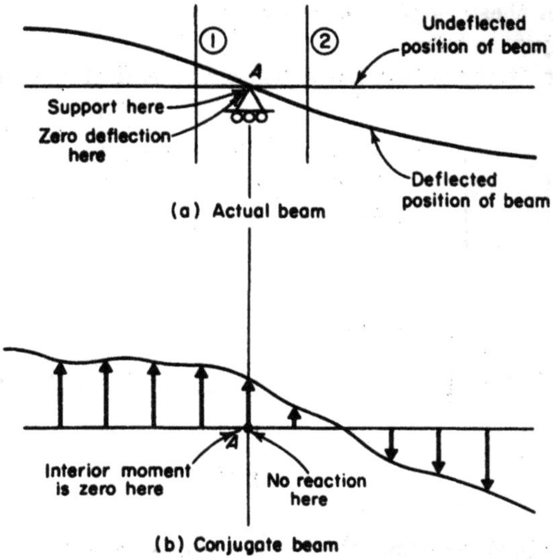

Fig. 9.12

continuous slope of the beam at A. That is, the slope at 1, a differential distance to the left of A, is the same as the slope at 2, a differential distance to the right of A. This means that there can be no discontinuity in shear at A on the conjugate beam — and this means that there can be no reaction at A on the conjugate beam (Fig. 9.12b).

Also, the moment at A on the conjugate beam is zero because the deflection at A on the actual beam is zero.

We now solve the two examples using the conjugate beam method.

EXAMPLE 1 Given the beam and loading shown in Fig. 9.13, determine the equation of the deflection curve.

SOLUTION The $M(x)/(EI_z)$ diagram is drawn, and the supports, R_{LC} and R_{RC} are indicated for the conjugate beam. The loading is a down load since the moment is a positive moment. R_{LC} and R_{RC} were assumed upward. However they could have been assumed acting downward and the solution would have indicated the proper direction.

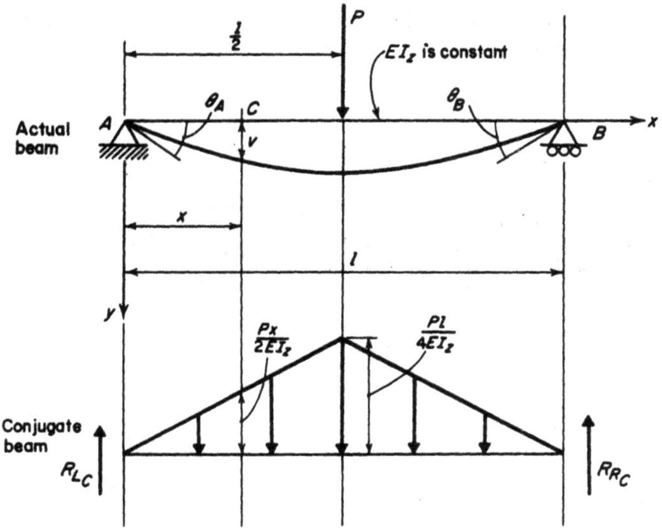

Fig. 9.13

Referring to the conjugate beam, we see that, because of symmetry,

$$R_{LC} = R_{RC} \tag{9.70}$$

and taking

$$\Sigma F_{VC} = 0$$

we have

$$R_{LC} + R_{RC} - \frac{Pl}{4EI_z} \cdot \frac{l}{2} = 0 \tag{9.71}$$

so that

$$R_{LC} = R_{RC} = \frac{Pl^2}{16EI_z} \tag{9.72}$$

This indicates that R_{LC} and R_{RC} act in the directions shown on Fig. 9.13. This means that the shear at A on the conjugate beam is positive; hence, θ_A is positive. Also, this shows that the shear at B on the conjugate beam is negative; hence θ_B is negative. Both are shown in Fig. 9.13.

The moment at C on the conjugate beam is given by

$$M_{CC} = R_{LC}x - \frac{Px}{2EI_z} \cdot \frac{x}{2} \cdot \frac{x}{3} \tag{9.73}$$

$$= \frac{Pl^2x}{16EI_z} - \frac{Px^3}{12EI_z}, \quad 0 \leq x \leq \frac{l}{2} \tag{9.74}$$

and this *positive* quantity indicates a positive deflection, i.e., a down deflection.

Example 2 Determine the equation of the deflection curve for the beam and loading shown in Fig. 9.14.

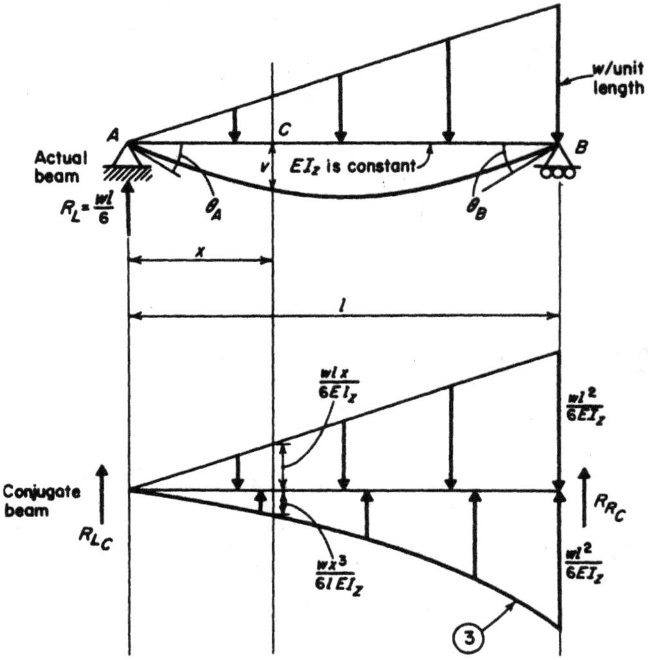

Fig. 9.14

Solution The $M(x)/(EI_z)$ diagram is drawn by parts. The part due to R_L (the triangle) is a down load, corresponding the positive moment; the part due to the triangular load on the beam, a third order parabola, acts upward, since this is a negative moment. We also show R_{LC} and R_{RC}, the reactions at the ends of the conjugate beam.

We determine R_{LC} first by taking moments about the right reaction. This gives

$$R_{LC}(l) + \frac{wl^2}{6EI_z}\cdot\frac{l}{4}\cdot\frac{l}{5} - \frac{wl^2}{6EI_z}\cdot\frac{l}{2}\cdot\frac{l}{3} = 0 \tag{9.75}$$

or

$$R_{LC} = \frac{wl^3}{36EI_z} - \frac{wl^3}{120EI_z} \tag{9.76}$$

$$= \frac{7}{360}\frac{wl^3}{EI_z} \tag{9.77}$$

Hence R_{LC} acts as shown on Fig. 9.14; this being a positive shear means that θ_A is positive and equal to $\frac{7}{360}\frac{wl^3}{EI_z}$.

The deflection at C is given by

$$v = M_{CC} = R_{LC}x + \frac{wx^3}{6lEI_z}\cdot\frac{x}{4}\cdot\frac{x}{5} - \frac{wlx}{6EI_z}\cdot\frac{x}{2}\cdot\frac{x}{3} \tag{9.78}$$

$$= \frac{7}{360}\frac{wl^3x}{EI_z} + \frac{wx^5}{120lEI_z} - \frac{wlx^3}{36EI_z} \tag{9.79}$$

Because this is a positive quantity, it means that v is positive, or downward.

9-6 Deflections by the Finite Difference Method

In Chapter 2 we presented the fundamentals of the finite difference method. It was pointed out there that the differential equation

$$\left(\frac{d^2v}{dx^2}\right)_n = -\left(\frac{M(x)}{EI_z}\right)_n \tag{9.80}$$

can be given, in finite difference notation as (see Eq. 2.105)

$$v_{n+1} - 2v_n + v_{n-1} = -\left[\frac{M(x)}{EI_z}\right]_n [\Delta x]^2 \tag{9.81}$$

in which the v's are the deflections at points $n + 1$, n and $n - 1$.

We shall use this form of the finite difference relations to solve the two problems being considered in this chapter.

EXAMPLE 1 Determine the deflections at the quarter points and also the center of the beam for the loading shown in Fig. 9.15.

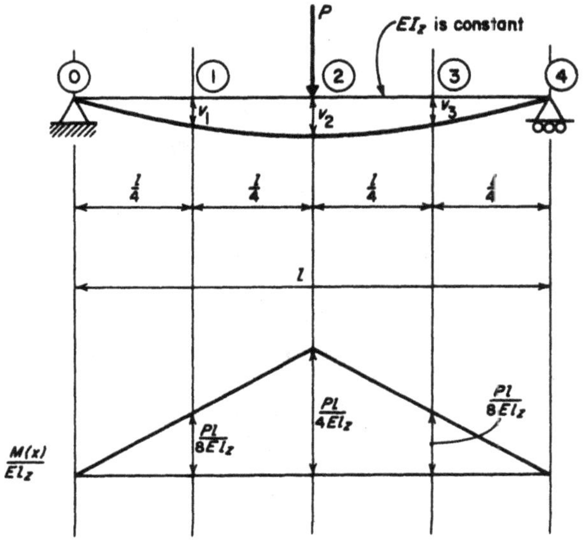

Fig. 9.15

SOLUTION Assume the beam is divided into four equal sections as shown. Hence $\Delta x = l/4$. We shall assume further that at each point, 1, 2, and 3, the right-hand side term is just equal to the $M(x)/(EI_z)$ at that point. Actually, we should use a *weighted* $M(x)/(EI_z)$ at each point, since this quantity does vary from point to point on the beam. There are various ways to weight the right-hand side function, some of which are shown in Ref. 7. Since we are primarily interested in illustrating a typical application of the method, however, we shall use the simpler procedure described above. What we are doing, essentially, is shown in Fig. 9.16.

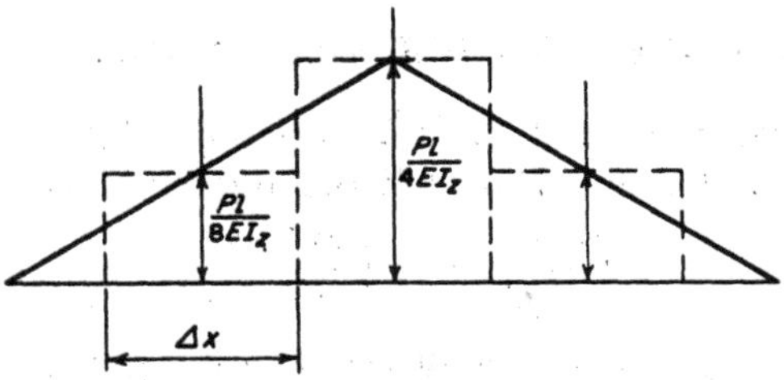

Fig. 9.16

We are replacing the actual $M(x)/(EI_z)$ curve (full lines) by an approximate curve as shown by the dotted lines. It is obvious that the smaller we make Δx, the more closely the approximation will approach the true curve. But, of course, the smaller we make Δx, the larger the computational task. Hence, a compromise must be made between desired accuracy and the amount of permissible labor.

In our case, therefore, we have at point 1, considering points 0, 1 and 2, using Eq. 9.81, and noting that $\Delta x = l/4$, the following:

$$v_2 - 2v_1 + v_0 = -\frac{l^2}{16}\left(\frac{Pl}{8EI_z}\right) \tag{9.82}$$

at point 2, considering points 1, 2, and 3:

$$v_3 - 2v_2 + v_1 = -\frac{l^2}{16}\left(\frac{Pl}{4EI_z}\right) \tag{9.83}$$

and finally, at point 3, considering points 2, 3 and 4:

$$v_4 - 2v_3 + v_2 = -\frac{l^2}{16}\left(\frac{Pl}{8EI_z}\right) \tag{9.84}$$

Now, noting that

$$v_0 = v_4 = 0 \tag{9.85}$$

we obtain from the above the following three equations in terms of the three unknown deflections, v_1, v_2 and v_3

$$v_2 - 2v_1 = -\frac{Pl^3}{EI_z}\left(\frac{1}{128}\right) \tag{9.86}$$

$$v_3 - 2v_2 + v_1 = -\frac{Pl^3}{EI_z}\left(\frac{1}{64}\right) \tag{9.87}$$

$$-2v_3 + v_2 = -\frac{Pl^3}{EI_z}\left(\frac{1}{128}\right) \tag{9.88}$$

These may be solved for v_1, v_2 and v_3 without difficulty, giving

$$v_1 = v_3 = \frac{Pl^3}{64EI_z}\left(\text{vs. } \frac{\frac{11}{12}Pl^3}{64EI_z}, \text{ see Eq. 9.14}\right) \tag{9.89}$$

and

$$v_2 = \frac{\frac{9}{8}Pl^3}{48EI_z}\left(\text{vs. } \frac{Pl^3}{48EI_z}, \text{ see Eq. 9.16}\right) \tag{9.90}$$

We see, therefore, that the approximate results, even for the relatively coarse grid chosen, give values which are not too different from the more accurate engineering beam results obtained by other means.

EXAMPLE 2 Determine the deflections at the quarter points and center of the beam for the beam and loading shown in Fig. 9.17, using finite difference methods.

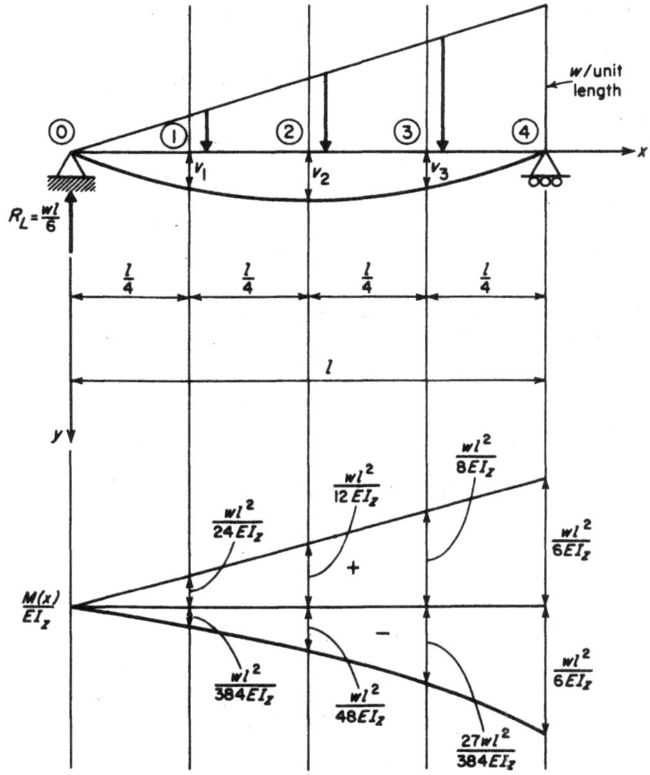

Fig. 9.17

SOLUTION Again we assume the beam divided into four equal spaces, so that $\Delta x = l/4$.

The $M(x)/(EI_s)$ diagram is drawn by parts, as shown. Also shown are the quarter-point ordinates of the two curves, recalling that the equation of the third degree parabola is $\frac{wx^3}{6lEI_s}$.

Therefore we have

$$\left[\frac{M(x)}{EI_s}\right]_{x=l/4} = \frac{wl^2}{24EI_s} - \frac{wl^2}{384EI_s} = \frac{15}{384}\frac{wl^2}{EI_s} \tag{9.91}$$

$$\left[\frac{M(x)}{EI_s}\right]_{x=l/2} = \frac{wl^2}{12EI_s} - \frac{wl^2}{48EI_s} = \frac{24}{384}\frac{wl^2}{EI_s} \tag{9.92}$$

$$\left[\frac{M(x)}{EI_s}\right]_{x=3/4l} = \frac{wl^2}{8EI_s} - \frac{27}{384}\frac{wl^2}{EI_s} = \frac{21}{384}\frac{wl^2}{EI_s} \tag{9.93}$$

and proceeding as before, we obtain at point 1, using Eq. 9.81:

$$v_2 - 2v_1 + v_0 = -\frac{l^2}{16}\left(\frac{15}{384}\frac{wl^2}{EI_s}\right) \tag{9.94}$$

at point 2 :

$$v_3 - 2v_2 + v_1 = -\frac{l^2}{16}\left(\frac{24}{384}\frac{wl^2}{EI_s}\right) \tag{9.95}$$

and finally, at point 3 :

$$v_4 - 2v_3 + v_2 = -\frac{l^2}{16}\left(\frac{21}{384}\frac{wl^2}{EI_s}\right) \tag{9.96}$$

Now, noting that $v_0 = v_4 = 0$, we obtain from the above the following three equations in terms of the three unknown deflections v_1, v_2 and v_3:

$$v_2 - 2v_1 = -\frac{15}{6144}\frac{wl^4}{EI_s} \tag{9.97}$$

$$v_3 - 2v_2 + v_1 = -\frac{24}{6144}\frac{wl^4}{EI_s} \tag{9.98}$$

$$-2v_3 + v_2 = -\frac{21}{6144}\frac{wl^4}{EI_s} \tag{9.99}$$

These may be solved for v_1, v_2 and v_3 without difficulty, giving

$$v_1 = \frac{28.5}{6144}\frac{wl^4}{EI_s}\left(\text{vs. } \frac{27.15}{6144}\frac{wl^4}{EI_s}, \text{ see Eq. 9.32}\right) \tag{9.100}$$

$$v_2 = \frac{42.0}{6144}\frac{wl^4}{EI_s}\left(\text{vs. } \frac{39.92}{6144}\frac{wl^4}{EI_s}, \text{ see Eq. 9.33}\right) \tag{9.101}$$

$$v_3 = \frac{31.5}{6144}\frac{wl^4}{EI_s}\left(\text{vs. } \frac{29.60}{6144}\frac{wl^4}{EI_s}, \text{ see Eq. 9.34}\right) \tag{9.102}$$

The results given above show once more how relatively accurate the finite difference approximation is, in the form used above.

9-7 A Finite-Difference Solution Using $EI_z \dfrac{d^4v}{dx^4} = w(x)$

We shall consider one more finite difference solution in order to bring out an additional factor which occasionally arises in these solutions. The entire discussion can be presented best by considering an illustrative example.

EXAMPLE Determine the deflections at the quarter point and center of the beam for the given loading, using finite difference methods and the relation, see Eq. 6.53

$$EI_z \frac{d^4v}{dx^4} = w(x) \tag{9.103}$$

The beam and loading are shown in Fig. 9.18.

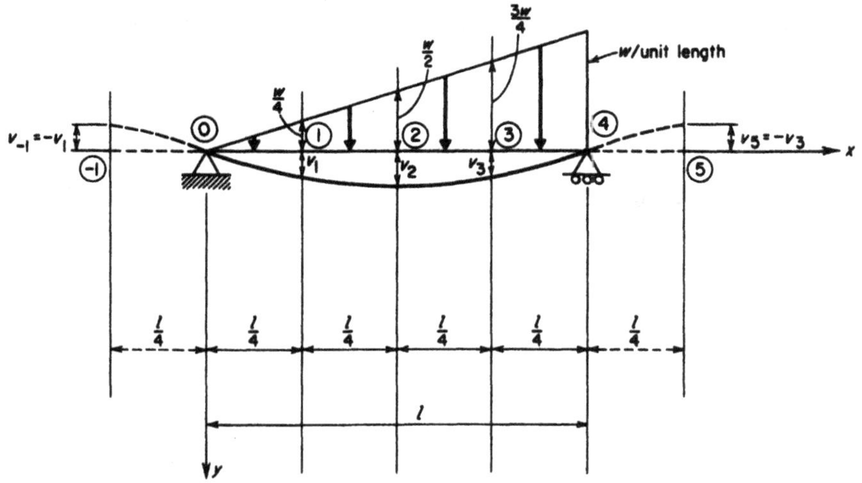

Fig. 9.18

SOLUTION Assume the beam divided into four equal sections, as shown in Fig. 9.18. Hence $\Delta x = l/4$.

In Eq. 2.103, we showed that

$$\left(\frac{d^4v}{dx^4}\right)_n = \frac{v_{n+2} - 4v_{n+1} + 6v_n - 4v_{n-1} + v_{n-2}}{(\Delta x)^4} \tag{9.104}$$

Since point 1 is the first point which can be considered in this relation, and because (as we see) two points on either side of 1 must appear in the equation, it follows that we will need an auxiliary point -1 for the beam, as shown in Fig. 9.18. The same argument applied to the other end of the beam indicates that an auxiliary point 5 will be needed.

Now, in the finite difference approximation, for a beam on a hinged support, the introduction of an auxiliary point such as -1 implies simply an extension of the beam past point 0 with the same slope as it has between points 0 and 1; see Fig. 9.19. This means simply that at point -1 we have a deflection, v_{-1}, given by

$$v_{-1} = -v_1 \tag{9.105}$$

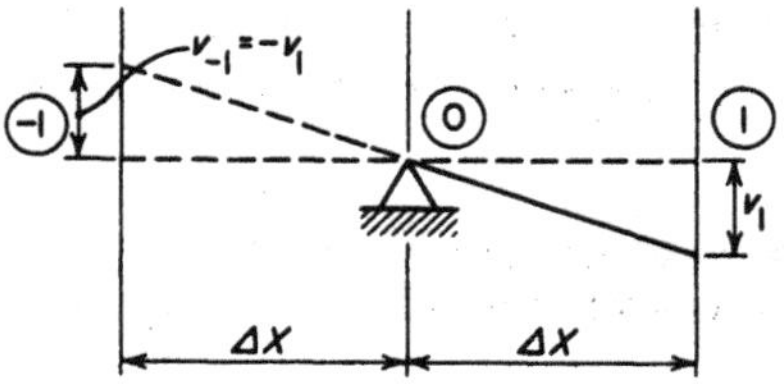

Fig. 9.19

and this fact enables us to proceed with the solution in the usual way.

Thus, in the present case, using for any w_n the value of the ordinate at the point n (although in more accurate solutions a weighted value of w can be used), we obtain for point 1, using -1, 0, 1, 2, 3, and noting that $\Delta x = l/4$:

$$v_3 - 4v_2 + 6v_1 - 4v_0 + v_{-1} = +\frac{l^4}{256}\cdot\frac{w}{4EI_z} \tag{9.106}$$

for point 2, using 0, 1, 2, 3, 4:

$$v_4 - 4v_3 + 6v_2 - 4v_1 + v_0 = +\frac{l^4}{256}\cdot\frac{w}{2EI_z} \tag{9.107}$$

and finally, for point 3, using 1, 2, 3, 4 and 5:

$$v_5 - 4v_4 + 6v_3 - 4v_2 + v_1 = +\frac{l^4}{256}\cdot\frac{3w}{4EI_z} \tag{9.108}$$

In the above equations we have

$$v_{-1} = -v_1 \tag{9.109}$$

$$v_5 = -v_3 \tag{9.110}$$

and

$$v_0 = v_4 = 0 \tag{9.111}$$

Substituting the values from Eqs. 9.109 through 9.111 in Eqs. 9.106 through 9.108, we get

$$v_3 - 4v_2 + 5v_1 = \frac{1}{1024}\frac{wl^4}{EI_z} \tag{9.112}$$

$$-4v_3 + 6v_2 - 4v_1 = \frac{2}{1024}\frac{wl^4}{EI_z} \tag{9.113}$$

$$5v_3 - 4v_2 + v_1 = \frac{3}{1024}\frac{wl^4}{EI_z} \tag{9.114}$$

Solving the above, we get for v_1, v_2 and v_3 the following (also shown are the more exact values obtained using the other methods):

$$v_1 = \frac{28.5}{6144}\frac{wl^4}{EI_z}\left(\text{vs. } \frac{27.15}{6144}\frac{wl^4}{EI_z}, \text{ see Eq. 9.32}\right) \tag{9.115}$$

$$v_2 = \frac{42}{6144}\frac{wl^4}{EI_z}\left(\text{vs. } \frac{39.92}{6144}\frac{wl^4}{EI_z}, \text{ see Eq. 9.33}\right) \tag{9.116}$$

$$v_3 = \frac{31.5}{6144}\frac{wl^4}{EI_z}\left(\text{vs. } \frac{29.60}{6144}\frac{wl^4}{EI_z}, \text{ see Eq. 9.34}\right) \tag{9.117}$$

The above results indicate once more how closely this particular finite difference formulation approximates the more accurate direct integration solution of the differential equation. It would appear that the finite difference solution will give results which are sufficiently accurate for ordinary engineering purposes.

9-8 Summary

We discussed the significance of beams deflections in engineering elasticity theory and then showed how beam deflections can be obtained using the following methods:

(a) double integration,
(b) moment area,
(c) conjugate beam,
(d) finite difference.

In order to bring out most clearly and forcefully the similarities and differences of the various methods, the same two basic problems were solved using all the methods.

Problems

In the problems which follow use these values: for steel, $E = 30{,}000{,}000$ psi; for wood, $E = 1{,}500{,}000$ psi.

1. Determine the deflection and slope at points A and B for the beams and loadings shown. Use (a) the moment-area method, (b) the conjugate beam method, and (c) the finite difference method (see Fig. 10.17 for the built-in end condition).

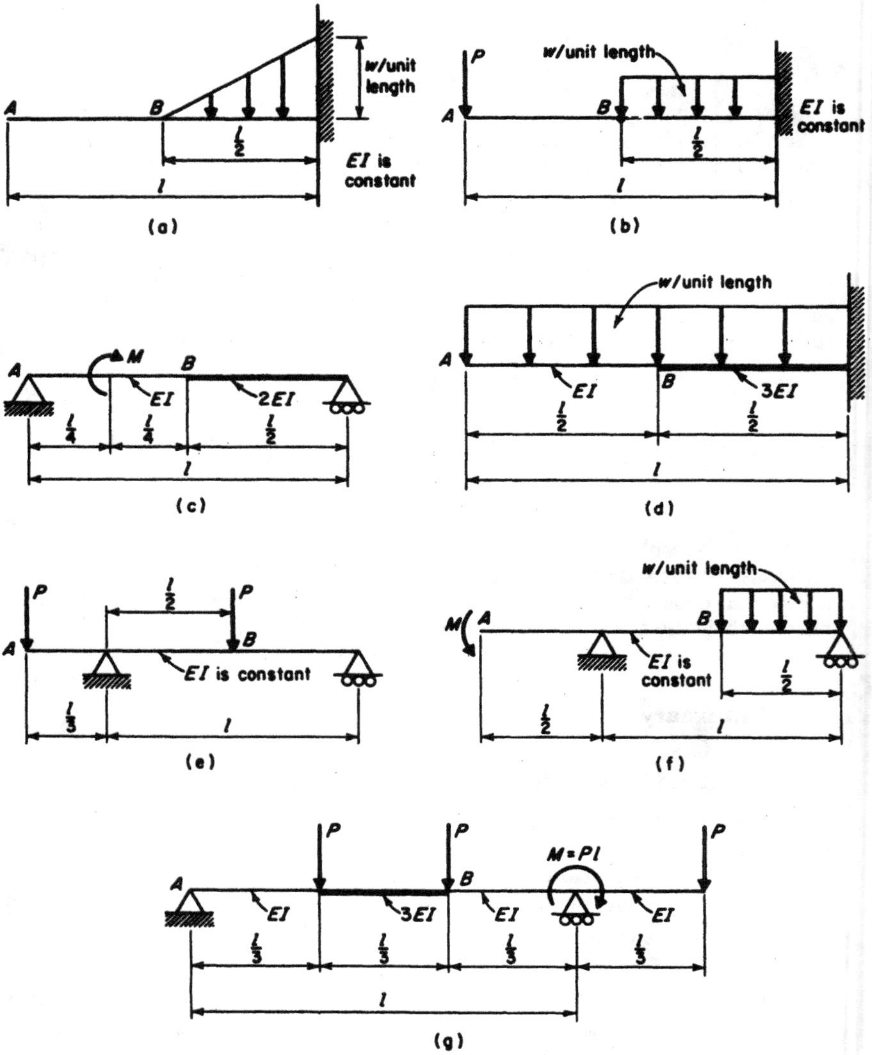

2. Determine, using the finite difference method, the deflection and slope at A for the beam and loading shown.

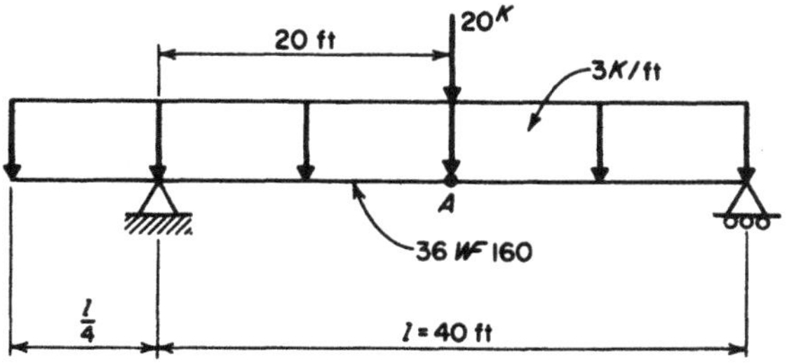

3. If the deflection at A is limited to $l/360$, what is the maximum allowable value of P? Does deflection or stress govern the design of this beam if the allowable bending stress is 20,000 psi?

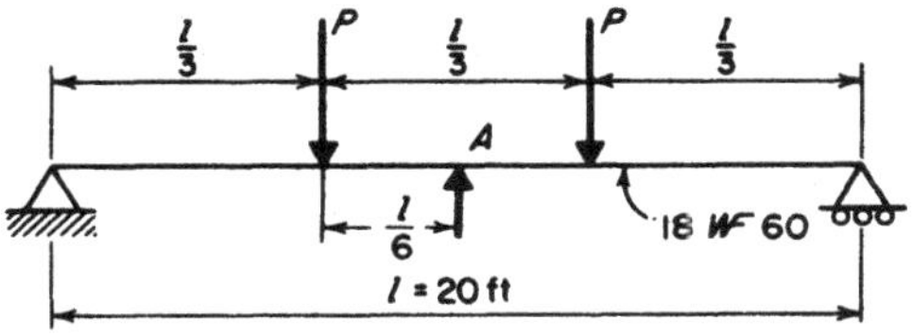

4. A timber floor beam is loaded as shown. If the maximum deflection is $l/360$, what is the allowable value of W? If the allowable bending stress is 1200 psi and allowable shear stress is 75 psi, does deflection or stress govern the design of this beam?

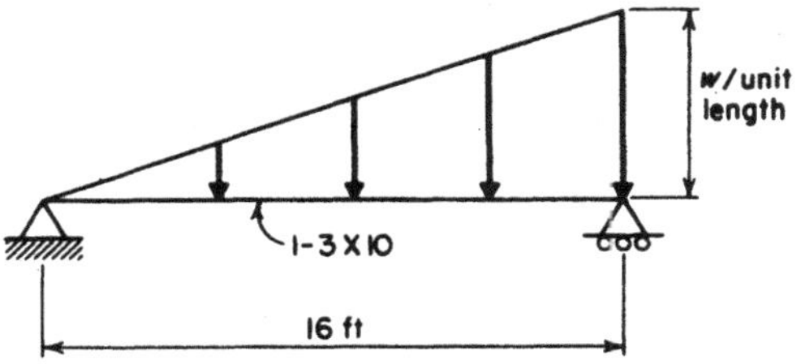

5. Two 3 × 18 wood beams are placed side by side to support the load shown. If the allowable bending stress is 1800 psi and the allowable shear stress is 150 psi, what will be the maximum value of P in order that the deflection may not exceed 1 in.?

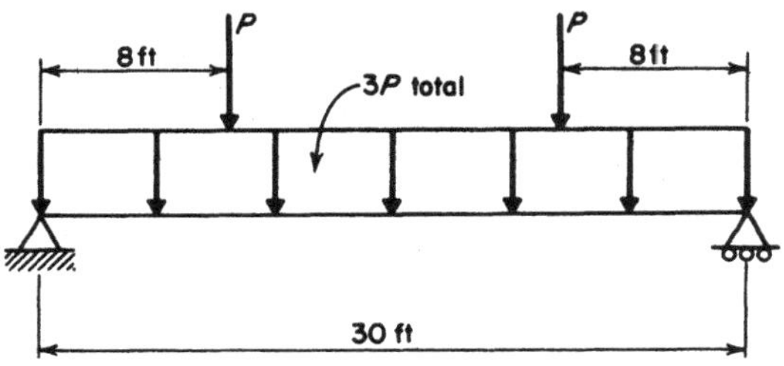

Chapter 10

STATICALLY INDETERMINATE BEAMS

10-1 Introduction

In this chapter we shall discuss the statically indeterminate beam and two of the methods for solving this structure.

We begin with a discussion of the difference between the statically determinate and statically indeterminate beam. We describe in some detail a simple method for determining how manifoldly indeterminate a beam is.

Following this, we show how statically indeterminate beams are solved†, using

(a) the conjugate beam method,

(b) the finite difference method.

Method (a) above may be taken as typical of the "classical" methods which have been developed for solving this problem. We emphasize it especially in this book because, as pointed out before, it is a technique with a very simple sign convention — and the matter of signs is frequently a stumbling block to the beginning student. In addition, once the two basic conjugate beam relations are mastered and understood, the method is a simple, straightforward one.

Method (b) is typical of a modern application for the solution of the beam problem which utilizes electronic computers.‡

10-2 Statically Determinate Beams

It was pointed out in Art. 4.7, that the equations of static equilibrium for the engineering beam are

$$\Sigma F_H = 0 \tag{10.1}$$

$$\Sigma F_V = 0 \tag{10.2}$$

$$\Sigma M_{\text{about any point}} = 0 \tag{10.3}$$

†Just as beam deflections may be determined by means of the various energy methods, so also may statically indeterminate beams be solved in this manner. All of the methods mentioned in the footnote on p. 149 may be used for this purpose. There are, in addition, many others available. See the references cited for some of these.

‡It may be pointed out here that the finite difference formulation of problems is the basis of the so-called "relaxation method" which is extremely useful throughout mathematical physics and engineering. See Ref. 6 for some of the different applications of the relaxation method and Ref. 25 for the application of this method to the beam problem.

For the usual beam these represent just three independent equations, and they will permit us to solve for three independent unknowns. The unknowns in beam analysis are the reactions; it follows, therefore, that for the ordinary beam there are three reactions which can be obtained using the equations of static equilibrium.†

For example, consider the beam of Fig. 10.1.

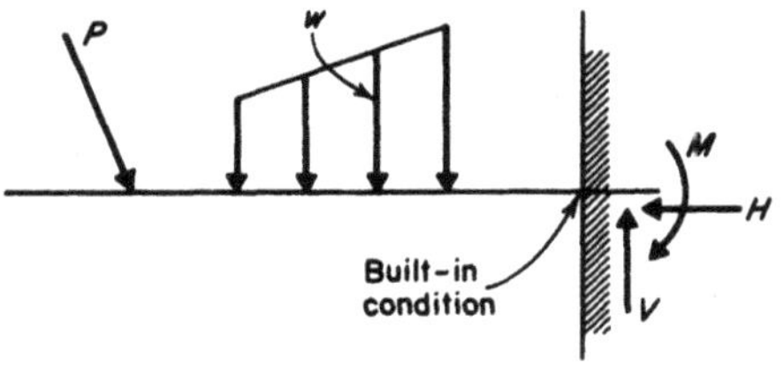

Fig. 10.1

In this beam, P and w are given (hence known), and there are three unknown reactions, V, H and M. These can be determined quite easily by utilizing the three equations given above. The beam of Fig. 10.1 is therefore said to be "statically determinate."

Now consider the beam of Fig. 10.2. In this figure, P, w_1 and w_2 are

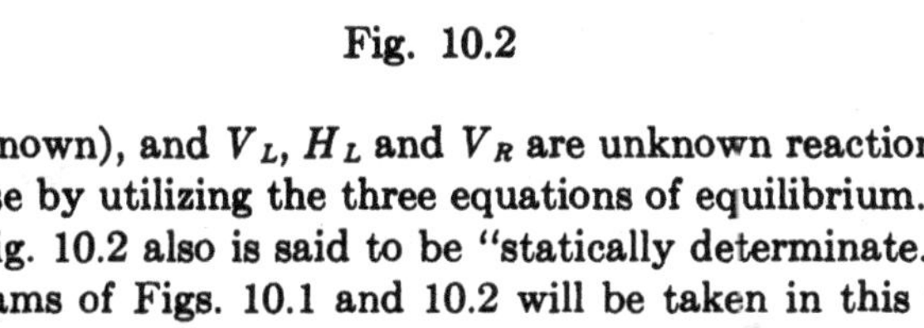

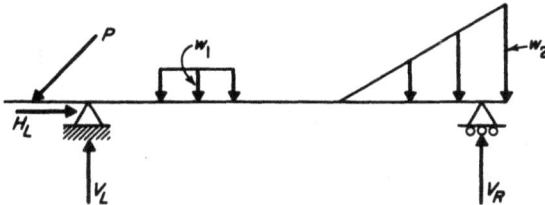

Fig. 10.2

given (hence known), and V_L, H_L and V_R are unknown reactions. We can determine these by utilizing the three equations of equilibrium. Therefore the beam of Fig. 10.2 also is said to be "statically determinate."

The two beams of Figs. 10.1 and 10.2 will be taken in this text as the basic statically determinate beams. In the next section we shall determine the degree of indeterminateness of beams by using these two as guides.

†It is possible to introduce artificial restraints into beams (and other structures) which essentially introduce additional independent equations and hence permit one to statically solve beams which have more than three reactions. We are excluding these cases in this text.

10-3 Statically Indeterminate Beams

A statically indeterminate beam is one whose unknown reactions can not be determined using the three equations of statics. For example, consider

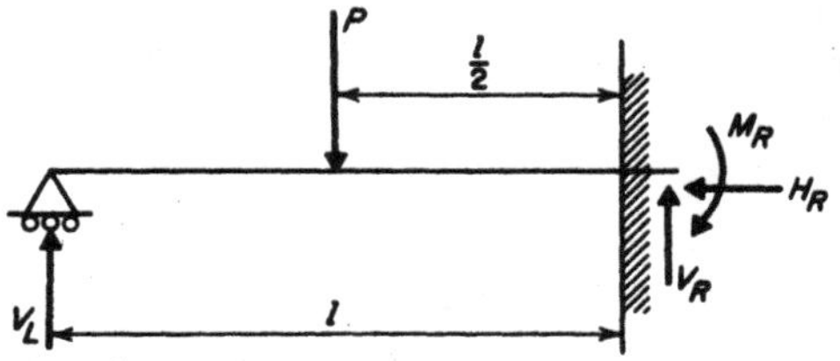

Fig. 10.3

the beam of Fig. 10.3. The student should convince himself that it is impossible to obtain V_L, V_R, M_R and H_R using

$$\Sigma F_V = 0 \tag{10.4}$$

$$\Sigma F_H = 0 \tag{10.5}$$

$$\Sigma M_{\text{about any point}} = 0 \tag{10.6}$$

This beam is said to be "statically indeterminate;" it can only be solved by making use of a strain compatibility condition in addition to the three equations of equilibrium. This point will be discussed further presently. For the moment, it may be noted that the beam shown has one extra reaction, over and above the number which is required for statical determinateness. Therefore we speak of this beam as being

1. statically indeterminate in the first degree, or as having
2. a redundant reaction, or as being
3. redundant in the first degree.

All of these descriptions apply and are equivalent.

The redundancy description is the one which enables us to determine very simply the degree of indeterminateness of more complicated beams, as follows:

We simply determine how many reactions must be removed in order that we have remaining either of the basic beams shown in Fig. 10.1 and 10.2. This number of reactions which must be removed is just equal to the degree of redundancy.

As an example, consider the beam of Fig. 10.3. We see that if V_L is removed, we have Fig. 10.4, which is the basic beam of Fig. 10.1. Hence,

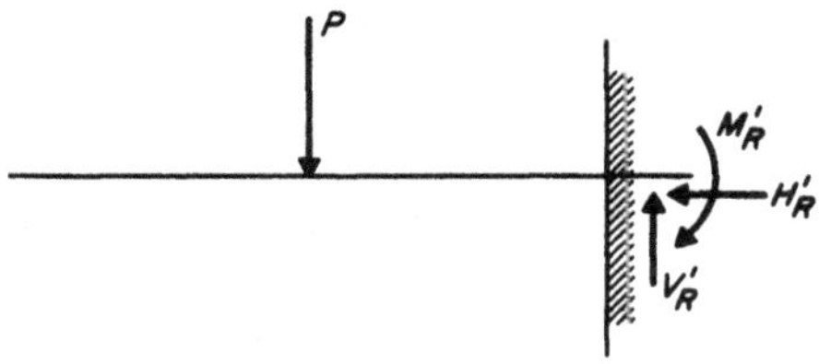

Fig. 10.4

the beam of Fig. 10.3 is indeterminate in the first degree. Note particularly that even though H'_R is equal to zero, it *is* a beam reaction, and one of the equations of statics must be utilized to establish its zero value.

Consider the beam of Fig. 10.5. How many degrees indeterminate is the

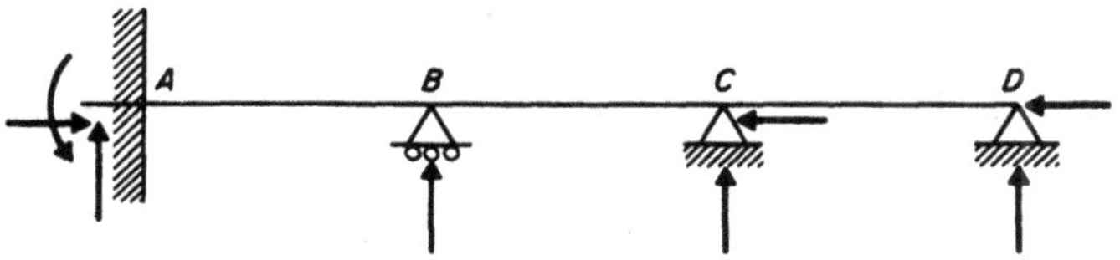

Fig. 10.5

beam? The answer is that this beam has five redundant reactions. We prove this by noting that if reactions B, C and D are removed we have Fig. 10.6, which is the basic determinate structure. Also we removed five reactions — one at B, two at C and two at D.

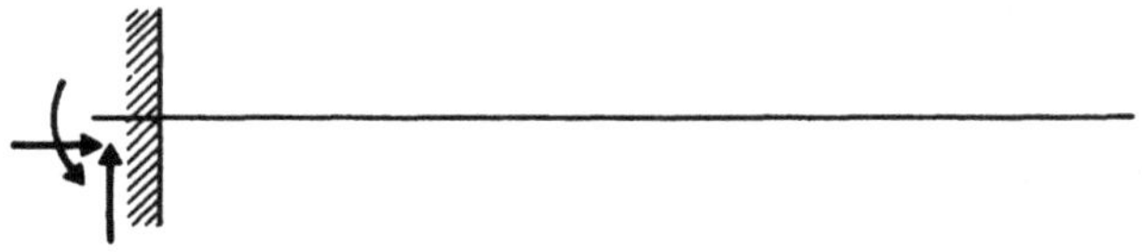

Fig. 10.6

Or we could have analyzed this beam by removing A and C, leaving B and D, which corresponds again to the basic beam. Also, A has three reactions and C has two so that once more we have removed five reactions and the beam is indeterminate in the fifth degree.

It was pointed out earlier in this section that the key to solving the indeterminate beam is the requirement of strain compatibility. We are now in a position to discuss this point further.

10-4 Strain or Deformation Compatibility for Indeterminate Beams

Let us consider the beam of Fig. 10.7. This beam, we know, has one redundant.

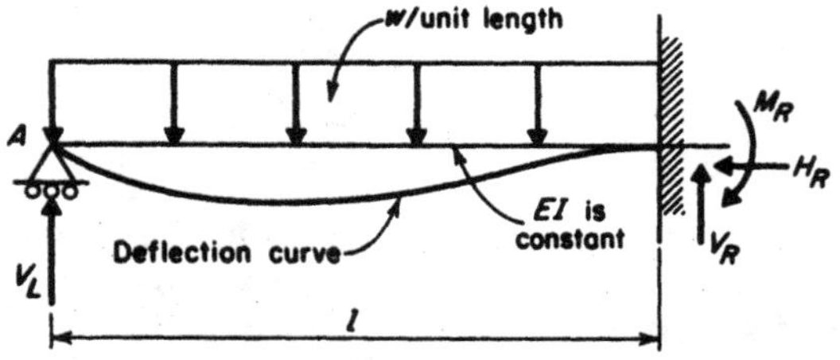

Fig. 10.7

Let us assume V_L is this redundant, and let us further assume it is removed. Then we have the statically determinate beam shown in Fig. 10.8; furthermore, we see that point A of this beam deflects an amount δ_A.

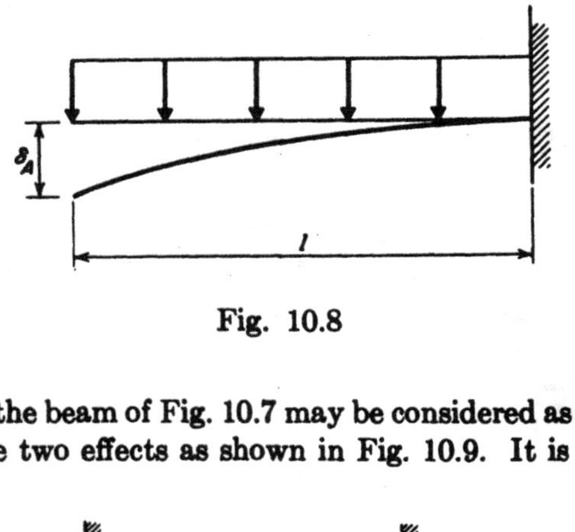

Fig. 10.8

It is clear that the beam of Fig. 10.7 may be considered as being subjected to the sum of the two effects as shown in Fig. 10.9. It is also clear that,

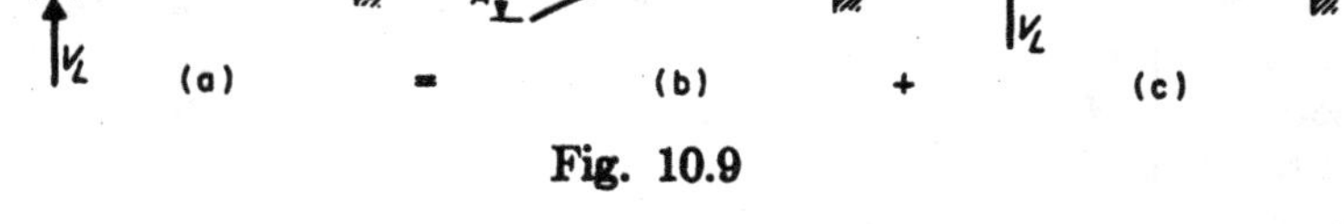

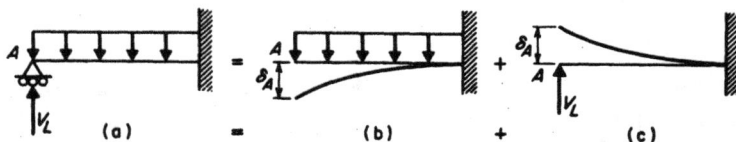

Fig. 10.9

since the support of A in the actual beam of Fig. 10.7 cannot move vertically, the reaction V_L, as shown in Fig. 10.9c, must have such a value that it causes a deflection just equal to δ_A of Fig. 10.9b. This is, in fact, a strain compatibility condition for the beam in question, that is, the vertical deflection at A must equal zero.

We could have analyzed this problem in another way. The beam of Fig. 10.7 can also be thought of as being equivalent to the sum of the two beams shown in Figs. 10.10b and 10.10c. In this case, we are essentially

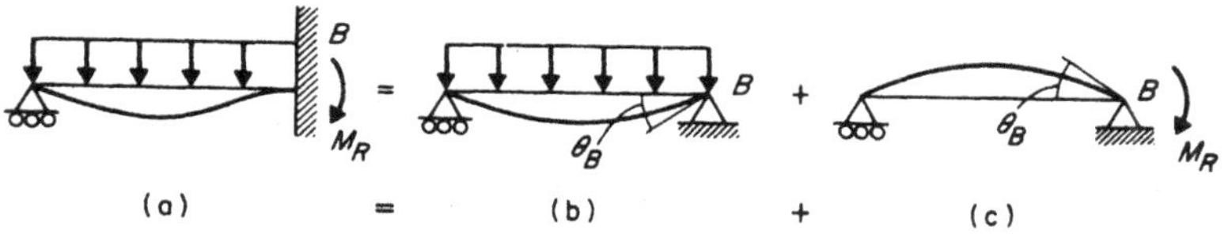

Fig. 10.10

considering M_R as the redundant, and we are, in effect saying that, since the beam is built-in at B, it can not rotate at this point. Hence M_R must have such a value that it causes a rotation at B just equal and opposite the rotation caused by the loading of Fig. 10.10b. It is this fact which enables us to determine M_R.

In both cases considered, having obtained V_L or M_R, we can then by statics determine the other reactions of the beam. Hence we can draw the shear and moment diagrams. In other words, we have solved the beam.

We emphasize once more that in all indeterminate structures we must utilize the requirements of strain compatibility in order to solve for the redundant reactions.

Refer back to Figs. 10.5 and 10.6. In this structure the strain compatibility requirements are

$$\delta_{V_B} = 0 \tag{10.7}$$

$$\delta_{V_C} = 0 \tag{10.8}$$

$$\delta_{H_C} = 0 \tag{10.9}$$

$$\delta_{V_D} = 0 \tag{10.10}$$

$$\delta_{H_D} = 0 \tag{10.11}$$

These five requirements will enable us to solve for the five redundants.

Or, if we consider A and C removed, as suggested, we have as the five strain compatibility requirements:

$$\delta_{H_A} = 0 \tag{10.12}$$

$$\delta_{V_A} = 0 \tag{10.13}$$

$$\delta_{\theta_A} = 0 \tag{10.14}$$

$$\delta_{V_C} = 0 \tag{10.15}$$

$$\delta_{H_C} = 0 \tag{10.16}$$

and so on for any other combination of redundants.

We are now ready to solve for the redundants of indeterminate beams.

10-5 Indeterminate Beam Solution by the Conjugate Beam Method

We may best describe the procedure for solving indeterminate beams using the conjugate beam method by solving several simple problems and explaining in detail all steps involved in the solution.

Example 1 Determine the reactions for the beam and loading shown in Fig. 10.11.

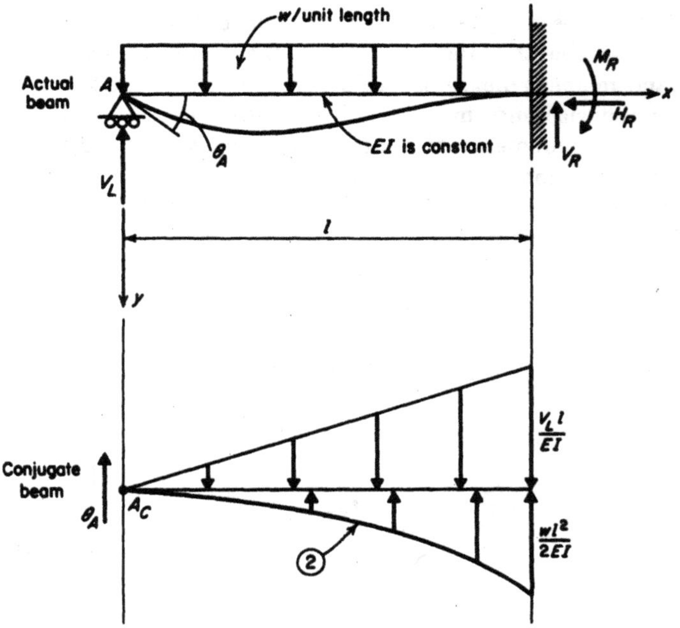

Fig. 10.11

SOLUTION I The first step, in all cases, is to determine the degree of redundancy and to decide on which reactions are to be considered as redundants. In this solution (I) let us consider V_L as the one redundant.

Then draw an M/EI diagram, *in terms of the given load* (*w in this case*) *and the redundants.* This curve can generally be drawn best by parts.

Now set up the conjugate beam. Recall that positive moments are down loads, negative moments are up loads.

Also — *and this is where the strain compatibility requirement is used* — show the supports on the conjugate beam. In the present case, this consists only of θ_A, representing the slope at the left end of the beam. There are no supports at the right end of the conjugate beam, since this is a built-in end (see Art. 9.5). Also, there is no moment at the left end of the conjugate beam, since point A on the actual beam does not deflect vertically.

Because the conjugate beam is in equilibrium, we can take moments about point A_C and obtain at once

$$\frac{V_L l}{EI}\cdot\frac{l}{2}\cdot\frac{2l}{3} - \frac{wl^2}{2EI}\cdot\frac{l}{3}\cdot\frac{3l}{4} = 0 \tag{10.17}$$

or

$$V_L = \tfrac{3}{8}wl \tag{10.18}$$

Now, using the equations of statics, for the actual beam

$$\Sigma F_V = 0 \tag{10.19}$$

or

$$V_L + V_R - wl = 0 \tag{10.20}$$

$$\therefore \quad V_R = \tfrac{5}{8}wl \tag{10.21}$$

Also,

$$\Sigma F_H = 0 \tag{10.22}$$

$$\therefore \quad H_R = 0 \tag{10.23}$$

and

$$\Sigma M_A = 0 \tag{10.24}$$

or

$$M_R - V_R l + \frac{wl^2}{2} = 0 \tag{10.25}$$

$$\therefore \quad M_R = \frac{wl^2}{8} \tag{10.26}$$

Having determined the reactions, we can now draw the shear and moment diagrams. These are shown in Fig. 10.12.

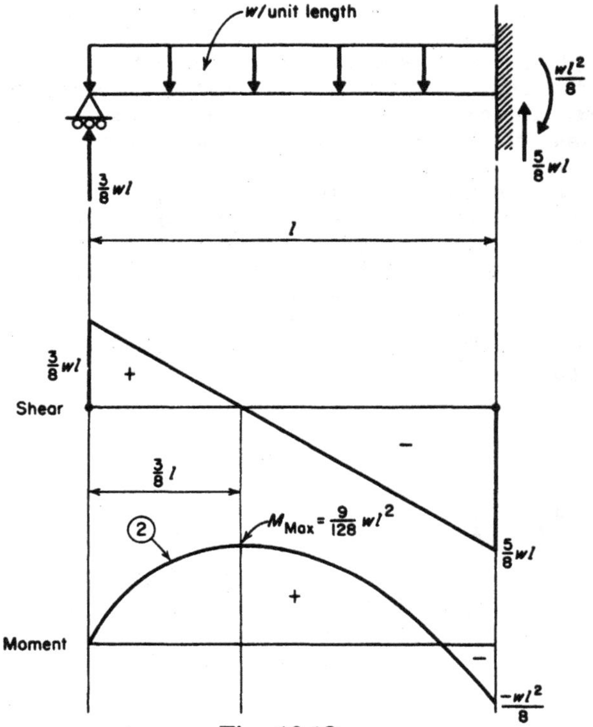

Fig. 10.12

SOLUTION II If we assume M_R as the redundant, we may proceed as follows; see Fig. 10.13. Note that the M/EI diagram consists of the two separate curves

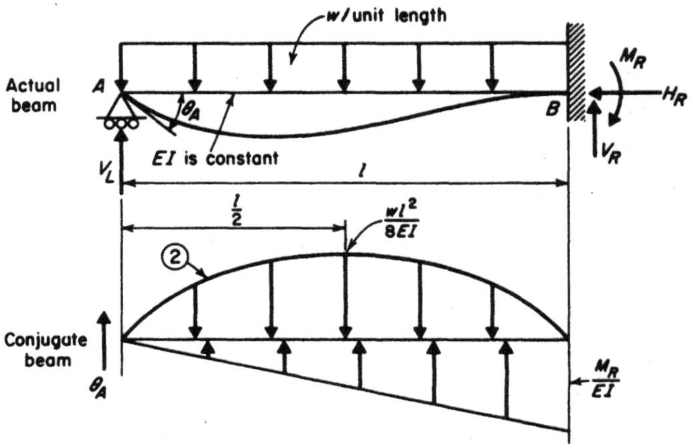

Fig. 10.13

corresponding to the two separate loadings of Fig. 10.10b and 10.10c. One of these curves is in terms of the known given loading and the other is in terms of the assumed redundant, M_R. Once again, on the conjugate beam we have the single reaction θ_A.

Now, taking moments about the left support on the conjugate beam, we have

$$\frac{2}{3}\left(\frac{wl^2}{8EI}\right)(l)\left(\frac{l}{2}\right) - \frac{M_R}{EI}\left(\frac{l}{2}\right)\left(\frac{2l}{3}\right) = 0 \tag{10.27}$$

Solving this for M_R, we obtain

$$M_R = \frac{wl^2}{8} \tag{10.28}$$

Following this, statics will give for the actual beam

$$V_L = \tfrac{3}{8}wl \tag{10.29}$$

$$V_R = \tfrac{5}{8}wl \tag{10.30}$$

$$H_R = 0 \tag{10.31}$$

and the problem, again, is completely solved.

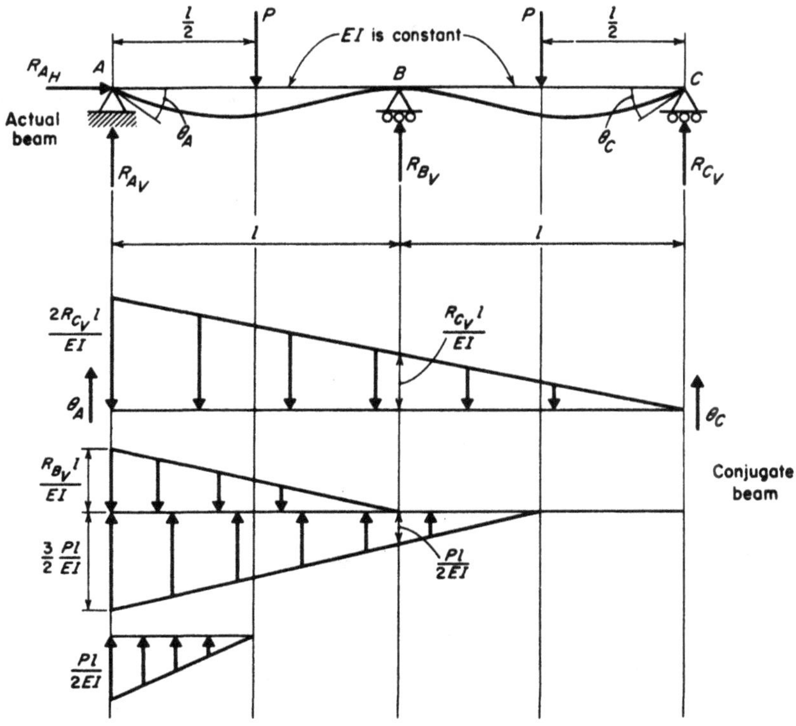

Fig. 10.14

If desired, the moment diagram could have been drawn from the right end to the left end. In any case, the use of the conjugate beam with the proper support conditions determines the redundant. Having this, we use the equations of static equilibrium to find the other reactions.

EXAMPLE 2 Determine the reactions for the beam and loading shown in Fig. 10.14.

SOLUTION This beam has one redundant. Let us assume R_{CV} to be the redundant. Then the M/EI diagram, drawn by parts from the right to the left, is as shown in Fig. 10.14.

Note that since there is a slope at end A on the actual beam, there is a shear (i.e., reaction) at this point on the conjugate beam. This is shown as θ_A. Similarly at the right end of the conjugate beam we have a reaction (i.e., shear) of value θ_C.

Also, by virtue of the special support conditions as described in Art. 9.5, there is no reaction at point B on the conjugate beam.

We can utilize symmetry to solve this problem, since at B, the center support, the beam must be horizontal. That is,

$$\theta_B = 0 \tag{10.32}$$

— the shear at B on the conjugate beam must be zero. This means that

$$-\frac{R_{CV}l}{EI}\cdot\frac{l}{2}+\frac{Pl}{2EI}\cdot\frac{l}{4}+\theta_C = 0 \tag{10.33}$$

Also, since the deflection at B on the actual beam is zero, we have $M_{B\,\text{conj}} = 0$, or

$$\theta_C l+\frac{Pl}{2EI}\cdot\frac{l}{4}\cdot\frac{l}{6}-\frac{R_{CV}l}{EI}\cdot\frac{l}{2}\cdot\frac{l}{3} = 0 \tag{10.34}$$

Substituting the value of θ_C from Eq. 10.33 in the above and solving for R_{CV}, we find that

$$\frac{R_{CV}l^3}{2EI}-\frac{Pl^3}{8EI}+\frac{Pl^3}{48EI}-\frac{R_{CV}l^3}{6EI} = 0 \tag{10.35}$$

or

$$R_{CV} = \tfrac{5}{16}P \tag{10.36}$$

Then, using statics, we find that

$$R_{AV} = \tfrac{5}{16}P \tag{10.37}$$

$$R_{BV} = \tfrac{11}{8}P \tag{10.38}$$

and

$$R_{AH} = 0 \tag{10.39}$$

Having these, we can draw the shear and moment diagrams, as shown in Fig. 10.15.

The two solutions given above are typical of the method by which redundant structures are solved using the conjugate beam method.

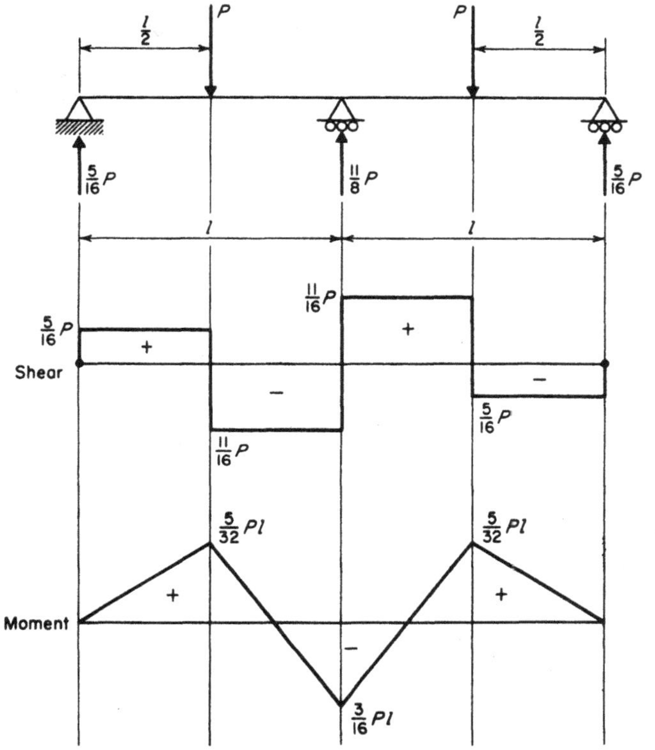

Fig. 10.15

10-6 The Indeterminate Beam Solved by the Finite Difference Method

We may best describe the application of the finite difference method in the solution of statically indeterminate beams by solving in detail the same problems that were solved in the last section by the conjugate beam method.

Example 1 Determine the reactions for the beam and loading of Fig. 10.16.

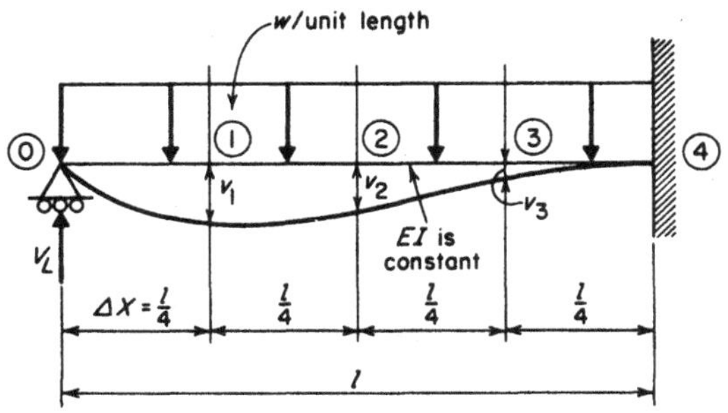

Fig. 10.16

SOLUTION We divide the beam length into four equal parts so that $\Delta x = l/4$.

Before proceeding, it is necessary that we discuss the technique for handling a built-in end in the finite difference method. A built-in end, physically, is merely a means for preventing rotation of the end of the beam. In other words, at point 4 we have a zero slope. In finite difference terminology (see Fig. 10.17), this is entirely

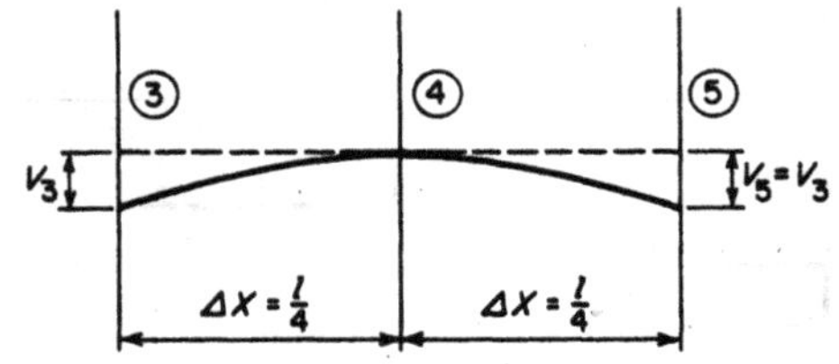

Fig. 10.17

equivalent to the effect which would be introduced if the beam were extended a distance past 4 (to point 5) and if at point 5 the deflection was equal to the deflection at point 3.

This is how we handle the built-in end. We introduce a fictitious point 5, $l/4$ distance past point 4, having a deflection equal to $v_5 = v_3$.

With this done, our beam and moment diagram, drawn by parts from the left end, is as shown in Fig. 10.18.

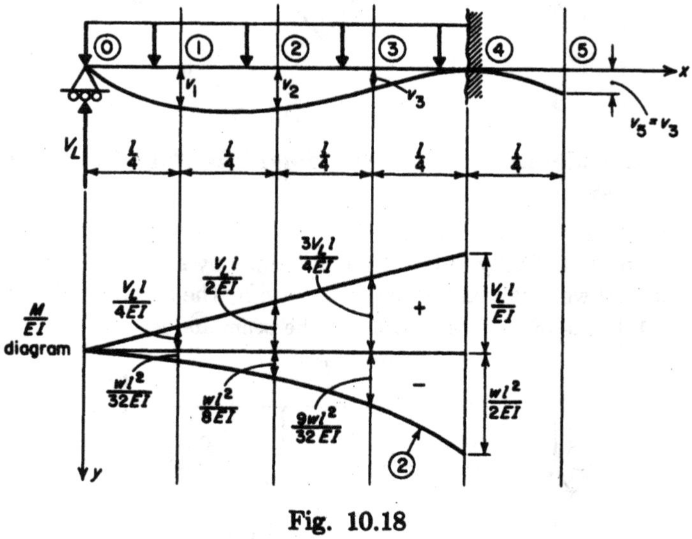

Fig. 10.18

The Bernoulli-Euler relation is given by

$$\frac{d^2v}{dx^2} = -\frac{M(x)}{EI} \tag{10.40}$$

or, in finite difference notation, at any point, n (see Eq. 2.105):

$$v_{n+1} - 2v_n + v_{n-1} = -(\Delta x)^2 \frac{M(x)}{EI} \tag{10.41}$$

Thus, at points 1 , 2 , 3 , and 4 , we have

$$v_0 - 2v_1 + v_2 = -\frac{l^2}{16}\left(\frac{V_L l}{4EI} - \frac{wl^2}{32EI}\right) \tag{10.42}$$

$$v_1 - 2v_2 + v_3 = -\frac{l^2}{16}\left(\frac{V_L l}{2EI} - \frac{wl^2}{8EI}\right) \tag{10.43}$$

$$v_2 - 2v_3 + v_4 = -\frac{l^2}{16}\left(\frac{3}{4}\frac{V_L l}{EI} - \frac{9wl^2}{32EI}\right) \tag{10.44}$$

$$v_3 - 2v_4 + v_5 = -\frac{l^2}{16}\left(\frac{V_L l}{EI} - \frac{wl^2}{2EI}\right) \tag{10.45}$$

Now, recalling that $v_4 = v_0 = 0$, and that $v_5 = v_3$ we see that these four equations have four unknowns: v_1, v_2, v_3 and V_L.

We solve these for V_L and obtain

$$V_L = \frac{69}{22}\frac{wl}{8}\left(\text{vs. } \frac{66}{22}\frac{wl}{8}, \text{ see Eq. 10.29}\right) \tag{10.46}$$

We see, therefore, that the finite difference method gives a result which is within 2% of the value obtained by more accurate solutions of the differential equation. This is probably sufficiently accurate for most engineering purposes. But, more important perhaps, is the fact that we have in this method a technique which can be applied directly to the most complicated beams and loadings, including those problems which can not be readily solved by the usual means. The finite difference formulation, furthermore, will permit us to set up the problem for simple electronic computer solution.

The student should also note that, in the finite difference formulation for the redundant beam, the *redundants* as well as the deflections are the unknowns. There will always be as many finite difference equations as there are unknown deflections and redundants.

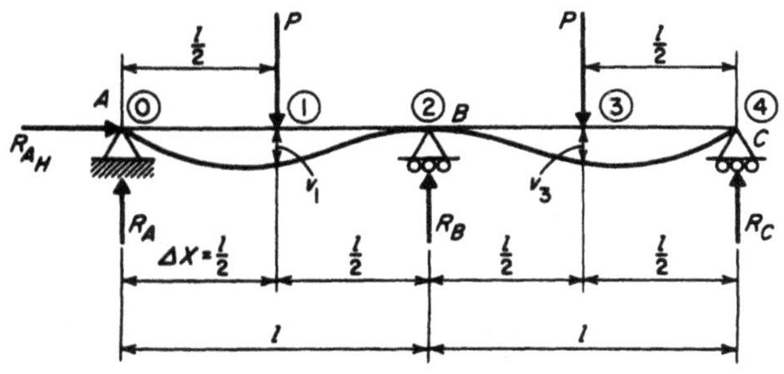

Fig. 10.19

One more example will be solved in order to further familiarize the reader with the procedure. We shall solve the second illustrative problem considered earlier in this chapter.

EXAMPLE 2 Determine the reactions for the beam and loading of Fig. 10.19.

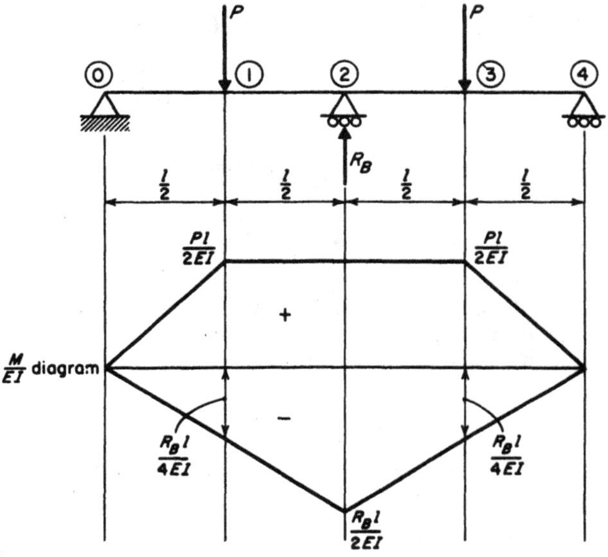

Fig. 10.20

SOLUTION We assume R_B to be the redundant. In terms of this redundant and the applied loads, the M/EI diagram can be given as the sum of the two curves shown in Fig. 10.20.

The Bernoulli-Euler equation, in finite difference form,

$$v_{n+1} - 2v_n + v_{n-1} = -(\Delta x)^2 \frac{M(x)}{EI} \tag{10.47}$$

is applied to the points 1, 2 and 3 giving

$$v_0 - 2v_1 + v_2 = -\frac{l^2}{4EI}\left(\frac{Pl}{2} - \frac{R_B l}{4}\right) \tag{10.48}$$

$$v_1 - 2v_2 + v_3 = -\frac{l^2}{4EI}\left(\frac{Pl}{2} - \frac{R_B l}{2}\right) \tag{10.49}$$

$$v_2 - 2v_3 + v_4 = -\frac{l^2}{4EI}\left(\frac{Pl}{2} - \frac{R_B l}{4}\right) \tag{10.50}$$

Noting that $v_0 = v_2 = v_4 = 0$ and that, because of symmetry, $v_1 = v_3$, we obtain from the above

$$R_B = \tfrac{4}{3}P (\text{vs. } \tfrac{11}{8}P, \text{ see Eq. 10.38}) \tag{10.51}$$

Thus, even for the relatively coarse net chosen to solve this problem, we obtain for R_B a value which is only 3% in error, compared to the more accurate solution.

10-7 Summary

In this chapter we discussed the statically indeterminate beam. We began with a descriptive discussion of the differences between the determinate and indeterminate beams, and presented a simple method for determining the degree of indeterminateness for ordinary beams.

Following this it was pointed out that the key to the solution of the indeterminate beam is the requirement that the engineering deformation (or deflections) be compatible ones. This was emphasized in the illustrative examples, which were used to show how the conjugate beam method and the finite difference method can be applied to the problem of solving the indeterminate beam.

Although there are other methods for solving indeterminate beams available (see Refs. 25, 28) of the two methods chosen in this text, one (the conjugate beam method) is fairly typical of other methods, and the second (the finite difference method) is an example of the use — in our present field, the study of the beam — of a technique which has wide application and is extremely useful in many other fields.

Problems

In the problems which follow, the solution may be either by the conjugate beam method, by the finite difference method, or by both methods.

1. Draw the shear and moment diagrams for the beam and loading shown.

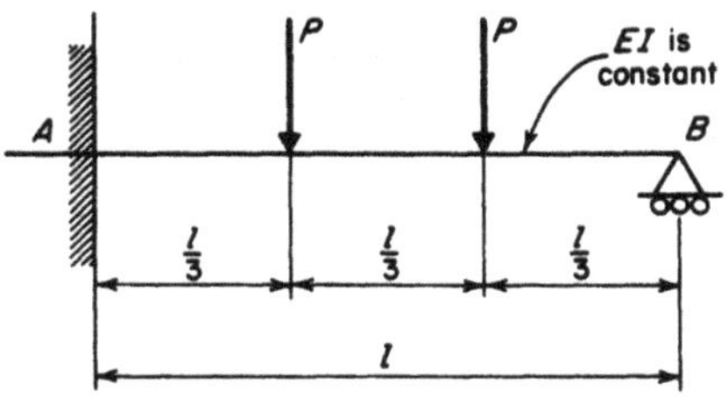

2. Draw the shear and moment diagrams for the beam of Prob. 1 if $l = 25$ ft, $P = 5000$ lb, the beam is a 27 WF 94, and support B moves down ½ in. Compare with the solution given in Prob. 1.

3. Draw the shear and moment diagrams for the beam and loading shown.

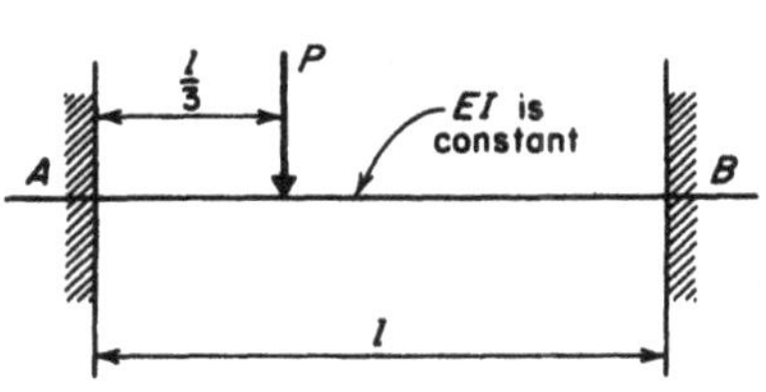

4. Draw the shear and moment diagrams for the beam of Prob. 3 if $l = 10$ ft, $P = 1000$ lb, the beam is a 3×16 timber, $E = 1{,}200{,}000$ psi, and support B rotates through 0.001 radian. Compare with the solution given in Prob. 3.

5. Draw the shear and moment diagrams for the beams and loadings shown.

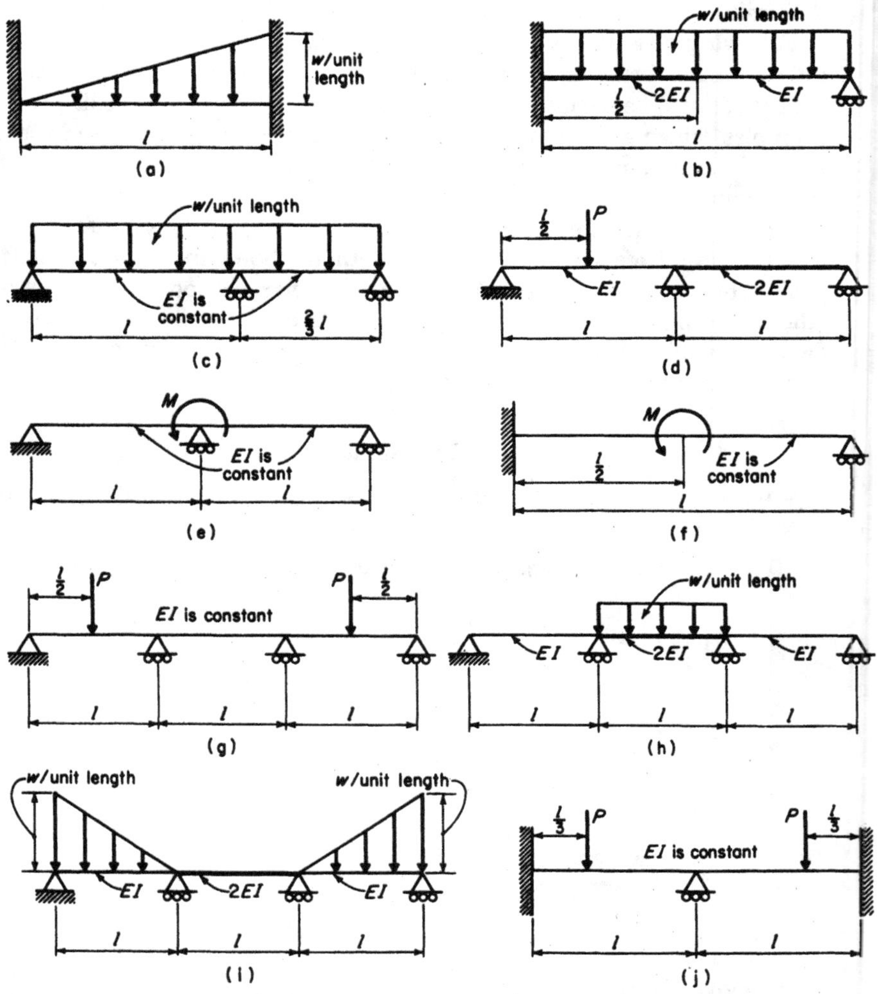

Chapter 11

BENDING INSTABILITY — THE DESIGN OF COLUMNS

11-1 Introduction

In the last several chapters we discussed in some detail various portions of the over-all beam bending problem. The ultimate aim of all the previous discussion was to proportion a member so that its maximum bending stress (or associated shear stress) under the given loading was less than some (in general) constant predetermined value. The analyses presented in these chapters represent typical bending-shear stress analyses for beams — or, otherwise stated, for members subjected to so-called *beam action*.

In this chapter we describe *bending instability*, a fundamentally different type of action. It is still connected with bending stress effects, but the structural behaviour which we consider now is basically different from beam behaviour in the type of failure associated with the structure.

Consider Fig. 11.1a, showing a typical beam and loading, P. If the

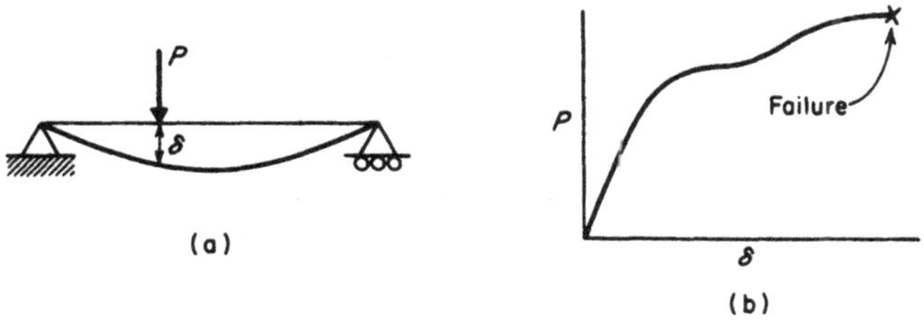

Fig. 11.1

load P is gradually increased, and if we measure, at intervals, the deflections under the varying load, a plot of the load deflection behaviour will be as shown in Fig. 11.1b. This is characteristic of *beam* action.

Now consider the beam loaded with the centrally applied loads shown in Fig. 11.2a. Once more assume that the load P is gradually increased. If the beam was initially quite straight, it will be found that at a certain *critical load*, P_{CR}, the beam buckles suddenly; for all practical purposes the load P_{CR} is the maximum that the beam can support without failing due to

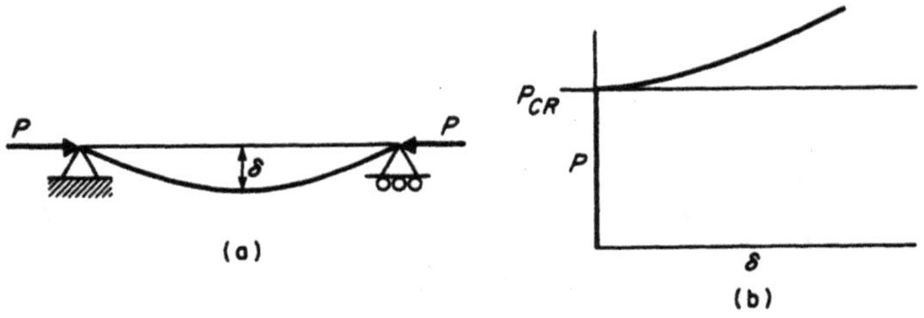

Fig. 11.2

excessive stress or excessive deflection. The situation shown in Fig. 11.2b is representative of this behaviour.

The phenomenon described above and illustrated in Fig. 11.2 is a classical instability failure in the following sense:

For P below P_{CR}, the structure (which is called a *column*) does not deflect laterally, but merely compresses. At $P = P_{CR}$ we obtain a sudden catastrophic deflection corresponding, for all practical purposes, to failure. This sudden, catastrophic behaviour is characteristic of many different types of engineering and physical instabilities.†

In this chapter we shall discuss first the classical solution of columns as given by Euler. We shall see that the solution is based upon the engineering form of the beam differential equation. Hence it represents a small-deflection approximation. We shall see that an inconsistency is introduced in this form of the solution.

Various column support conditions are also considered. It will be shown that allowable column loads for all boundary conditions of engineering interest can be obtained by considering the one basic solution obtained above.

Following the Euler solution, we shall obtain a more exact solution, based upon the nonlinear form of the Bernoulli-Euler relation. We shall see that the Euler solution appears as a special (or limit) case of this more exact solution, and our knowledge of various types of column behaviour will be extended by virtue of this more exact solution.

We shall then discuss the forms which the Euler solution takes when we are in a stress range past the linear part of the stress-strain curve.

After this we will discuss the rather arbitrary modifications made to the Euler formula by various specifications which enable engineers to design steel columns and also columns of other materials.

†There are other instabilities associated with beam action than the one considered in this chapter. The present discussion is confined to *compression, or column, instability*. An equally important instability may occur in members subjected to torsional loads which is characterized by a sudden twisting type of failure. We shall not discuss this instability here. The interested student will find a rather complete treatment of this topic in Ref. 20.

Finally we shall indicate how critical buckling loads may be obtained by the use of finite differences.

11-2 The Euler Solution for a Hinged-Hinged Column

Consider a column hinged at both ends and loaded as shown in Fig. 11.3a. This is a *hinged-hinged column.*

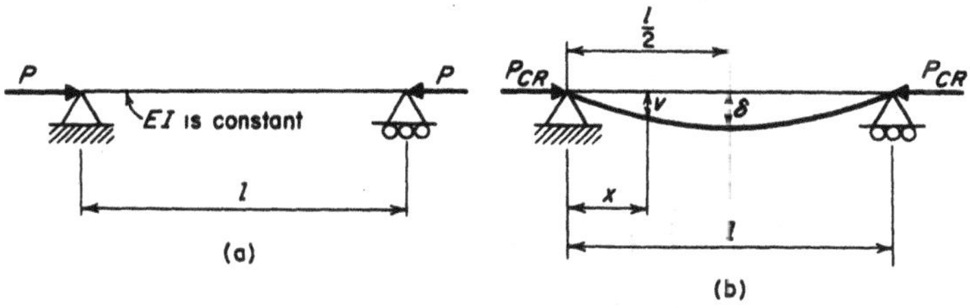

Fig. 11.3

In the Euler analysis, the critical buckling load, P_{CR}, (or P_{Euler}, as we shall occasionally call it) is taken as the load which will just cause the bar to bend. Stated in another manner, the load P_{CR} will keep the bar in a slightly deflected position, as shown by Fig. 11.3b. In this figure, if the load is reduced below P_{CR}, the bar will straighten; if the load is increased above P_{CR}, the bar will fail. Hence P_{CR} is the *critical* value of the load. At the center of the beam, δ is a small, arbitrary deflection.

The Bernoulli-Euler bending relation, which holds for small deflections and for bars in which stress is proportional to strain, is (see Eq. 6.46)

$$\frac{d^2v}{dx^2} = -\frac{M(x)}{EI_z} \tag{11.1}$$

In this case,

$$M(x) = P_{CR}v \tag{11.2}$$

or

$$\frac{d^2v}{dx^2} = -\frac{P_{CR}v}{EI_z} \tag{11.3}$$

Let us call

$$\frac{P_{CR}}{EI_z} = K^2 \tag{11.4}$$

Then the equation becomes

$$\frac{d^2v}{dx^2} + K^2v = 0 \tag{11.5}$$

Now we may verify by substitution in Eq. 11.5 that

$$v = A \sin Kx \tag{11.6}$$

satisfies the equation. A is a constant, not equal to zero.

Also, the boundary conditions are

$$v = 0 \quad \text{at} \quad x = 0 \tag{11.7}$$

and

$$v = 0 \quad \text{at} \quad x = l \tag{11.8}$$

The first of these conditions is automatically satisfied by the solution of Eq. 11.6. The second boundary condition requires that

$$0 = A \sin Kl \tag{11.9}$$

and this condition will be satisfied only if

$$Kl = n\pi, \quad n = 1, 2, 3, \ldots . \tag{11.10}$$

or if

$$\left(\sqrt{\frac{P_{CR}}{EI_z}}\right) l = n\pi, \quad n = 1, 2, 3, \ldots . \tag{11.11}$$

This means that for the column of Fig. 11.3

$$P_{CR} = n^2 \frac{\pi^2 EI_z}{l^2} \tag{11.12}$$

Note also that A, the constant of Eq. 11.6, is just equal to δ.

It is clear that the smallest value of P_{CR} will occur for $n = 1$. However, the other values of n also represent solutions to the column problem; physically they correspond to the solutions for the different buckling patterns shown in Fig. 11.4. It is clear also that unless the column is

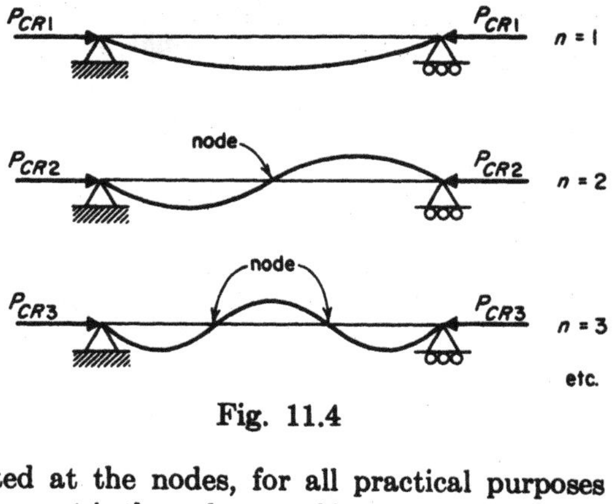

Fig. 11.4

specially supported at the nodes, for all practical purposes the solution corresponding to $n = 1$ is the only one of interest, since the column will fail at this load and the higher loads will not be reached.

So, for a hinged-hinged column, as shown in Fig. 11.3, the Euler buckling load is given by

$$P_{CR} = P_{\text{Euler}} = \frac{\pi^2 EI_z}{l^2} \tag{11.13}$$

Also it is obvious that I_z must be the *minimum* moment of inertia for the column cross section. That is, the critical axis for an unsupported column is the axis about which the moment of inertia is a minimum. In all column formulas used in this chapter, I_z will be this minimum moment of inertia.†

11-3 The Euler Solution for Other End Conditions

We can utilize the solution obtained in the last section, the hinged-hinged solution, to obtain solutions for other support conditions, in the following manner.

Consider the *fixed-free ended column* shown in Fig. 11.5a. It has a

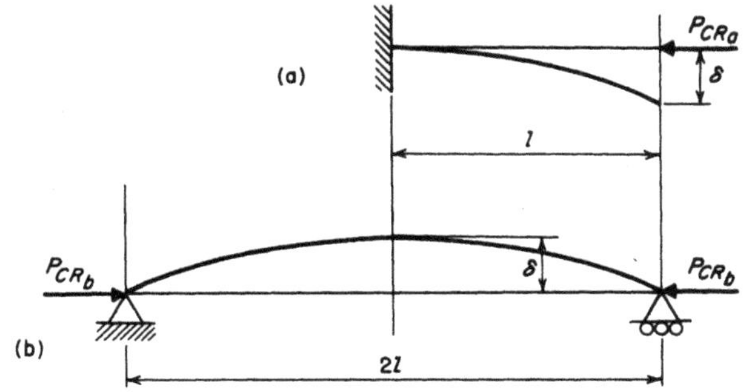

Fig. 11.5

length l and a constant EI_z. In Fig. 11.5b, we show a hinged-hinged column of length $2l$ and with the same stiffness constant, EI_z, as the fixed-free ended column. It can be seen that one-half of the hinged-hinged column is identical to the fixed-free column. Hence

$$P_{CR_b} = P_{CR_a} \tag{11.14}$$

or, since

$$P_{CR_b} = \frac{\pi^2 EI_z}{(2l)^2} \tag{11.15}$$

we have

$$P_{CR_a} = \frac{\pi^2 EI_z}{4l^2} \tag{11.16}$$

which is one-quarter the critical buckling load for a hinged-hinged column of the same length and stiffness.

†It is possible for a column to have intermediate supports which effectively prevent buckling about a particular axis. Thus, a longer length with its corresponding larger moment of inertia may be more critical than the shorter length and its smaller moment of inertia. See Prob. 3 at the end of this chapter.

Now consider a column built-in at both ends, the so-called *fixed-fixed column*, Fig. 11.6a. When this column buckles, it is found that the points of

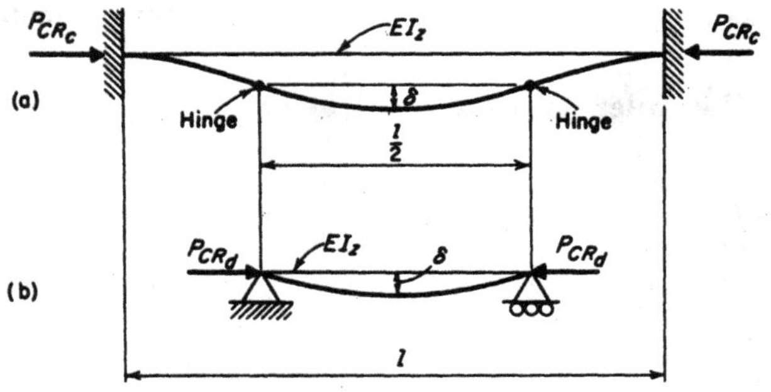

Fig. 11.6

zero moment (i.e., the hinges) occur at approximately the quarter points. Hence the center portion of the column is, in effect, a hinged-hinged column, as shown by Fig. 11.6b. This being so, it follows that

$$P_{CR_d} = P_{CR_c} = \frac{\pi^2 EI_z}{(l/2)^2} \tag{11.17}$$

or

$$P_{CR_c} = \frac{4\pi^2 EI_z}{l^2} \tag{11.18}$$

that is, the critical buckling load for the fixed-fixed column is four times the buckling load for a hinged-hinged column of the same length and stiffness.

Finally, consider the *fixed-hinged column* as shown in Fig. 11.7a. In this case, it is found that a hinge occurs at approximately $0.7l$, as shown in

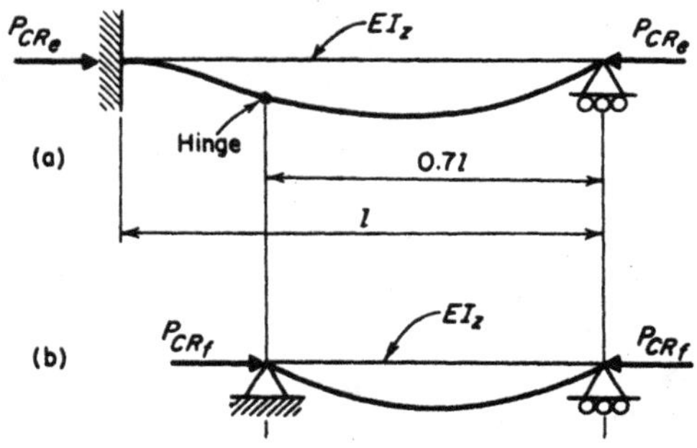

Fig. 11.7

Fig. 11.7a. Hence the portion of length $0.7l$ is, in effect, a hinged-hinged column, and we have for this case

$$P_{CR_f} = P_{CR_e} = \frac{\pi^2 EI_z}{(0.7l)^2} \tag{11.19}$$

or

$$P_{CR_e} = \frac{2\pi^2 EI_z}{l^2} \tag{11.20}$$

Thus, for a fixed-hinged column, the critical buckling load is twice the critical load for a hinged-hinged column.

All of the above is summarized in Table 6.

TABLE 6

Type of Column	Sketch	$P_{CR} = P_{Euler}$
Fixed-free	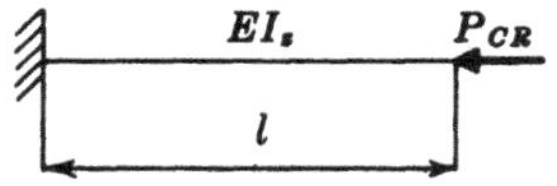	$P_{CR} = \frac{\pi^2 EI_z}{4l^2}$
Hinged-hinged	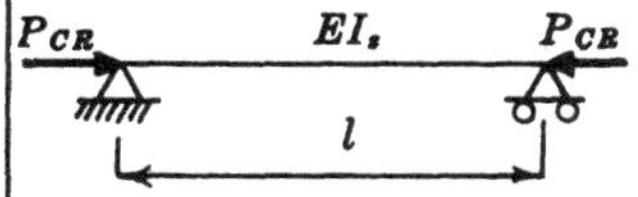	$P_{CR} = \frac{\pi^2 EI_z}{l^2}$
Fixed-hinged	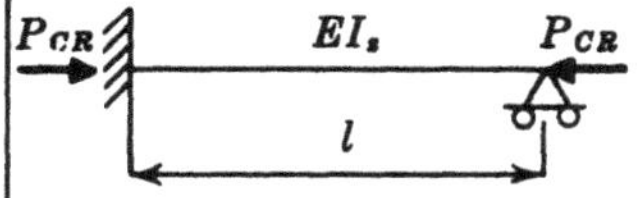	$P_{CR} = \frac{2\pi^2 EI_z}{l^2}$
Fixed-fixed	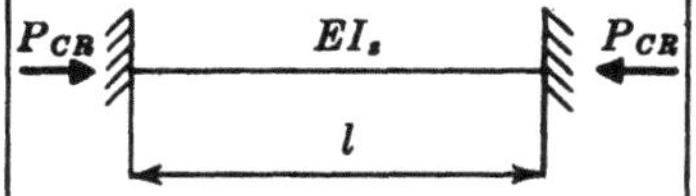	$P_{CR} = \frac{4\pi^2 EI_z}{l^2}$

11-4 The Inconsistency in the Euler Solution

If we examine the basic derivation of the Euler critical buckling load as given in Fig. 11.3, and the equations which refer to this figure, we note an inconsistency in that the deflection δ was not specified. That is, the solution apparently holds for *any* value of δ, just so long as it is "small." This facet of the Euler solution has been the subject of considerable discussion and has cast doubt, in some quarters, on the validity of the derivation, although the Euler result can be verified experimentally for the engineering-type column.

In the next section we examine a more precise formulation of the buckling problem and one, which we shall see, removes the inconsistency in the undefined deflection of the buckled column. In fact, this more exact solution will even permit us to solve those column buckling cases which occur with large deflection. But even more important, perhaps, we shall see that the solution of the next section effectively verifies the approximate result given by the Euler solution.

11-5 The Elastica — a Column With Large Deflections

In this article we shall speak of engineering-type columns and of elastica-type columns.

The engineering column is one whose length-least dimension ratio, for example, is of the order of 20 to 50. The elastica column, on the other hand, is one whose length-least dimension ratio is of the order of 200 to 500.

The two columns are fundamentally different in their response to central loads. Thus, the engineering column, Fig. 11.8a, effectively fails when the

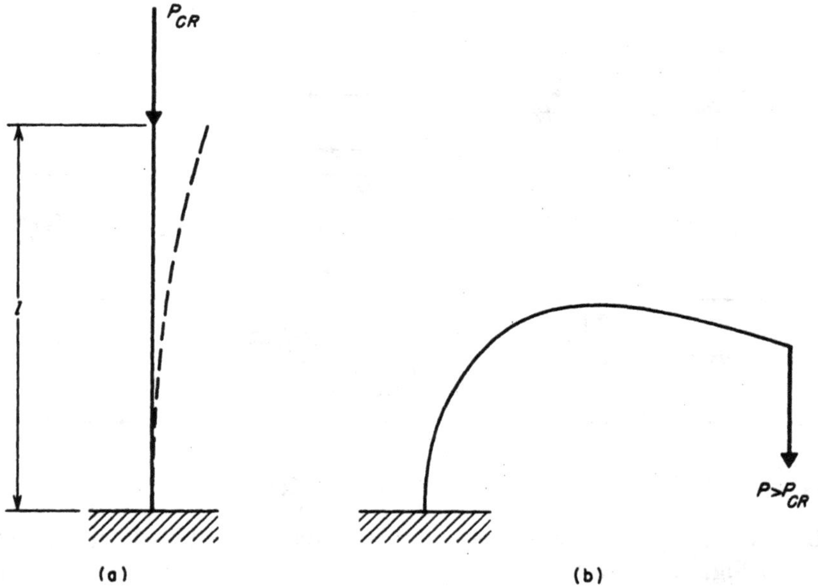

Fig. 11.8

Euler critical load, P_{CR}, is reached. On the other hand, the elastica column, Fig. 11.8b, as we shall see, can sustain loads past the critical load and will deform in a manner shown in the figure. It is clear, therefore, that if we wish to analyze the elastica column, we must examine the hypotheses of the Bernoulli-Euler relation and determine whether or not they apply to the given case.

It was pointed out in Art. 6.3 that, for the engineering beam, the approximate Bernoulli-Euler relation and the requirement that the beam bending stresses be within the proportional limit (i.e., stresses proportional to strains) lead to the following relation:

$$\frac{1}{\rho} = -\frac{M(x)}{EI_z} \tag{11.21}$$

In Eq. 11.21, ρ is the radius of curvature of the neutral plane at the point where the moment is $M(x)$ and the stiffness is EI_z.

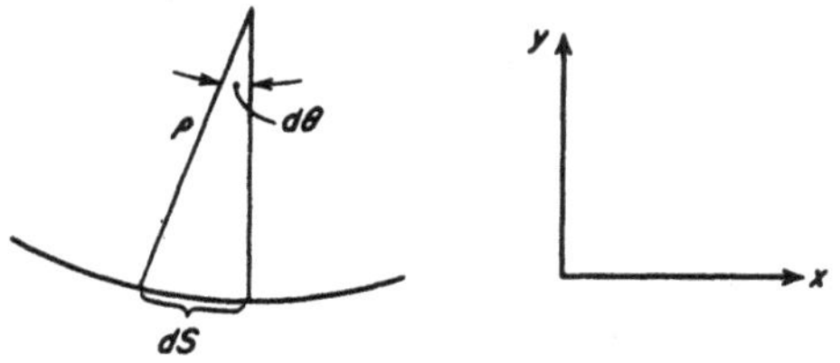

Fig. 11.9

The exact relation for ρ is given by (see Fig. 11.9)

$$\rho d\theta = ds \tag{11.22}$$

or

$$\frac{1}{\rho} = \frac{d\theta}{ds} \tag{11.23}$$

and, in Cartesian coordinates,

$$\frac{1}{\rho} = \frac{\dfrac{d^2y}{dx^2}}{\left(1 + \left(\dfrac{dy}{dx}\right)^2\right)^{3/2}} \tag{11.24}$$

As noted previously, if the slope dy/dx is much less than unity, then

$$\left(\frac{dy}{dx}\right)^2 \ll 1 \tag{11.25}$$

and approximately

$$\frac{1}{\rho} = \frac{d^2y}{dx^2} \tag{11.26}$$

which is the form of the expression used in all of the Bernoulli-Euler developments in the last several chapters.

However, for the column of Fig. 11.8b, we *cannot* assume that

$$\frac{dy}{dx} < 1 \tag{11.27}$$

It follows that we can not use, for $1/\rho$, the expression given in Eq. 11.26 above. If we wish to use the curvature form of the Bernoulli-Euler equation to solve the column of Fig. 11.8b, then we must use the equation†

$$\frac{1}{\rho} = \frac{d\theta}{ds} = -\frac{M(x)}{EI_z} \tag{11.28}$$

We shall now determine the solution for the buckling behaviour of the elastica column of Fig. 11.10. We shall list each step in the derivation. Although quite lengthy, it is not difficult, and it is suggested that the reader follow through the derivation step by step.

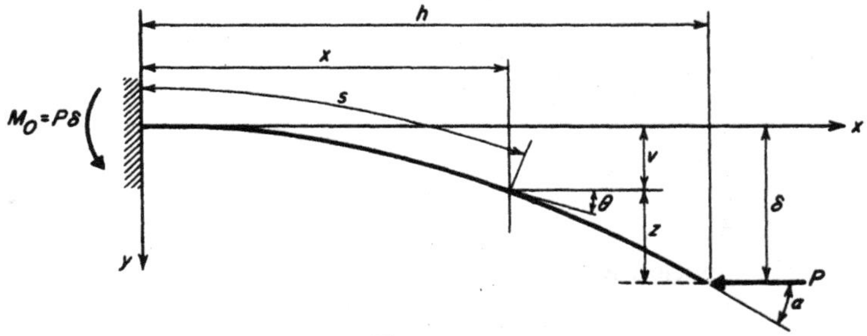

Fig. 11.10

The structure and loading are shown in Fig. 11.10.

$$EI_z \frac{d\theta}{ds} = P(\delta - v) \tag{11.29}$$

Calling

$$z = \delta - v \tag{11.30}$$

and

$$K^2 = \frac{P}{EI_z} \tag{11.31}$$

this becomes

$$\frac{d\theta}{ds} = K^2 z \tag{11.32}$$

†The Bernoulli-Euler relation is an approximate one even if the exact value of ρ is used as in Eq. 11.28. However, as pointed out in Art. 6.3, if we make the assumption (approximation) that, in general for beams, plane sections before bending are plane after bending, then Eq. 11.21 holds. A *more accurate form of this expression* is one which assumes the exact value for ρ, rather than the small deflection approximation for this term. This is what the elastica solution does, and in this respect it may be looked upon as a solution one step more exact than the Euler column solution.

Now we introduce some trigonometric identities and relations which hold for the structure.

$$\frac{dv}{ds} = \frac{dv}{d\theta}\frac{d\theta}{ds} = \sin\theta \tag{11.33}$$

or, using Eq. 11.32 above,

$$\sin\theta = \frac{dv}{d\theta} K^2 z \tag{11.34}$$

and since, from Eq. 11.30,

$$dz = -dv \tag{11.35}$$

Eq. 11.34 becomes

$$\sin\theta\, d\theta = -K^2 z\, dz \tag{11.36}$$

Let us integrate this, to obtain

$$\cos\theta = \frac{K^2 z^2}{2} + C \tag{11.37}$$

Noting the boundary conditions

$$\left.\begin{aligned} v &= \delta \\ z &= 0 \end{aligned}\right\} \text{ when } \theta = \alpha \tag{11.38}$$

this gives

$$\cos\alpha = C \tag{11.39}$$

or

$$2(\cos\theta - \cos\alpha) = K^2 z^2 \tag{11.40}$$

Since

$$\left.\begin{aligned} v &= 0 \\ z &= \delta \end{aligned}\right\} \text{ when } \theta = 0 \tag{11.41}$$

this gives

$$1 - \cos\alpha = \frac{K^2\delta^2}{2} \tag{11.42}$$

$$= 2\sin^2\frac{\alpha}{2} \tag{11.43}$$

or

$$\delta = \frac{2}{K}\sin\frac{\alpha}{2} \tag{11.44}$$

and this gives a relation between δ and α.

We also have the following well-known trigonometric identities:

$$\cos\theta = 1 - 2\sin^2\frac{\theta}{2} \tag{11.45}$$

$$\cos\alpha = 1 - 2\sin^2\frac{\alpha}{2} \tag{11.46}$$

Subtract the second from the first and get

$$\cos\theta - \cos\alpha = 2\left(\sin^2\frac{\alpha}{2} - \sin^2\frac{\theta}{2}\right) \tag{11.47}$$

which is also

$$= 2\sin^2\frac{\alpha}{2}\left(1 - \frac{\sin^2(\theta/2)}{\sin^2(\alpha/2)}\right) \tag{11.48}$$

Let us define a new angle, ϕ, by the relation

$$\sin\phi = \frac{\sin(\theta/2)}{\sin(\alpha/2)} \tag{11.49}$$

Then Eq. 11.48 becomes

$$\cos\theta - \cos\alpha = 2\sin^2\frac{\alpha}{2}\cos^2\phi \tag{11.50}$$

or

$$\sqrt{2(\cos\theta - \cos\alpha)} = 2\sin\frac{\alpha}{2}\cos\phi \tag{11.51}$$

If we differentiate Eq. 11.49, we obtain, noting that ϕ and θ are the variables,

$$\cos\phi\, d\phi = \frac{1}{2}\frac{\cos(\theta/2)\, d\theta}{\sin(\alpha/2)} \tag{11.52}$$

or

$$\frac{d\phi}{\cos(\theta/2)} = \frac{d\theta}{2\sin(\alpha/2)\cos\phi} \tag{11.53}$$

Now, returning to Eq. 11.32, we have

$$\frac{d\theta}{K^2z} = ds \tag{11.54}$$

and, from Eq. 11.40,

$$\frac{1}{K^2z} = \frac{1}{K\sqrt{2(\cos\theta - \cos\alpha)}} \tag{11.55}$$

or

$$ds = \frac{d\theta}{K\sqrt{2(\cos\theta - \cos\alpha)}} \tag{11.56}$$

and, from Eq. 11.51, this is also equal to

$$ds = \frac{d\theta}{2K\sin(\alpha/2)\cos\phi} \tag{11.57}$$

But, from Eq. 11.53,

$$\frac{d\phi}{\sqrt{1 - \sin^2 (\theta/2)}} = \frac{d\theta}{2 \sin (\alpha/2) \cos \phi} \tag{11.58}$$

which, since (Eq. 11.49),

$$\sin \frac{\theta}{2} = \sin \phi \sin \frac{\alpha}{2} \tag{11.59}$$

is also given by

$$\frac{d\phi}{\sqrt{1 - \sin^2 (\alpha/2) \sin^2 \phi}} = \frac{d\theta}{2 \sin (\alpha/2) \cos \phi} \tag{11.60}$$

so that

$$ds = \frac{1}{K} \frac{d\phi}{\sqrt{1 - \sin^2 (\alpha/2) \sin^2 \phi}} \tag{11.61}$$

We integrate this equation and get

$$Ks = \int_0^{\phi_1} \frac{d\phi}{\sqrt{1 - \sin^2 (\alpha/2) \sin^2 \phi}} \tag{11.62}$$

This expression is an *elliptic integral*, and it has been evaluated for various values of the parameters α and ϕ_1. These evaluations are shown in tabular form in, for example, "A Short Table of Integrals," by B. O. Peirce, published by Ginn and Co., p. 122. In Peirce's notation, then

$$Ks = F(k, \phi) \tag{11.63}$$

If we take s equal to l, the straight length of the bar, then $\theta = \alpha$, and

$$\sin \phi_1 = \frac{\sin (\theta/2)}{\sin (\alpha/2)} = 1 \tag{11.64}$$

or

$$\phi_1 = \frac{\pi}{2} \tag{11.65}$$

Hence, the expression becomes

$$Kl = \int_0^{\pi/2} \frac{d\phi}{\sqrt{1 - \sin^2 (\alpha/2) \sin^2 \phi}} \tag{11.66}$$

and, again in Peirce's notation (p. 121),

$$Kl = \mathcal{K} \tag{11.67}$$

This means that

$$K^2l^2 = \mathcal{K}^2 \tag{11.68}$$

But

$$K^2l^2 = \frac{Pl^2}{EI_z} \tag{11.69}$$

or, noting that, for this column (see Eq. 11.16)

$$P_{\text{Euler}} = \frac{\pi^2 EI_z}{4l^2} \tag{11.70}$$

the above relation gives

$$\frac{P}{P_{\text{Euler}}} = \left(\frac{2}{\pi}\mathcal{K}\right)^2 \tag{11.71}$$

In Eq. 11.44, we had

$$\delta = \frac{2}{K}\sin\frac{\alpha}{2} \tag{11.72}$$

Therefore,

$$\frac{\delta}{l} = \frac{2}{Kl}\sin\frac{\alpha}{2} \tag{11.73}$$

and, from Eq. 11.67,

$$\frac{\delta}{l} = \frac{2}{\mathcal{K}}\sin\frac{\alpha}{2} \tag{11.74}$$

Similarly, from Eq. 11.40,

$$K^2z^2 = 2(\cos\theta - \cos\alpha) \tag{11.75}$$

or

$$\frac{z}{l} = \frac{\sqrt{2(\cos\theta - \cos\alpha)}}{Kl} \tag{11.76}$$

and using Eq. 11.51 and Eq. 11.67,

$$\frac{z}{l} = \frac{2}{\mathcal{K}}\sin\frac{\alpha}{2}\cos\phi \tag{11.77}$$

We have also,

$$\frac{dx}{ds} = \cos\theta \tag{11.78}$$

But, from Eq. 11.61, and since

$$\cos\theta = 1 - 2\sin^2\frac{\theta}{2} \tag{11.79}$$

Eq. 11.78 becomes

$$dx = \frac{1}{K}\frac{(1 - 2\sin^2(\theta/2)\,d\phi}{\sqrt{1 - \sin^2(\alpha/2)\sin^2\phi}} \tag{11.80}$$

Now, from Eq. 11.59, we get

$$2\sin^2\frac{\theta}{2} = 2\sin^2\frac{\alpha}{2}\sin^2\phi \tag{11.81}$$

Hence the bracket term of Eq. 11.80

$$2\left(1 - \sin^2\frac{\alpha}{2}\sin^2\phi\right) - 1 \tag{11.82}$$

or Eq. 11.80 is given by

$$K\,dx = \frac{2(1 - \sin^2(\alpha/2)\sin^2\phi)\,d\phi}{\sqrt{1 - \sin^2(\alpha/2)\sin^2\phi}} - \frac{d\phi}{\sqrt{1 - \sin^2(\alpha/2)\sin^2\phi}} \tag{11.83}$$

Integrating this, we get

$$Kx = 2E(k, \phi) - F(k, \phi) \tag{11.84}$$

in which $E(k, \phi)$ is the elliptic integral given on p. 123 of Peirce's Tables.
Since

$$Kl = \mathcal{K} \quad \text{(Eq. 11.67)} \tag{11.85}$$

it follows that, from Eq. 11.84,

$$\frac{x}{l} = \frac{1}{\mathcal{K}}[2E(k, \phi) - F(k, \phi)] \tag{11.86}$$

and finally, for $x = h$, see Fig. 11.10;

$$\frac{h}{l} = 2\frac{E}{\mathcal{K}} - 1 \tag{11.87}$$

in which E is also listed on p. 121 of Peirce's Tables.

This completes the analysis of the elastica. For convenience, we list next the important relations which were obtained for this structure.

$$Kl = \mathcal{K} \tag{11.85}$$

$$\frac{\delta}{l} = \frac{2}{\mathcal{K}}\sin\frac{\alpha}{2} \tag{11.74}$$

$$\frac{P}{P_{\text{Euler}}} = \left[\frac{2}{\pi}\mathcal{K}\right]^2 \tag{11.71}$$

$$\frac{x}{l} = \frac{1}{\mathcal{K}}[2E(k, \phi) - F(k, \phi)] \tag{11.86}$$

$$\frac{h}{l} = 2\frac{E}{\mathcal{K}} - 1 \tag{11.87}$$

As an indication of the results obtained by the above theory, Table 7 is prepared.

TABLE 7

α	$\mathcal{K}$	$\frac{P}{P_{\text{Euler}}} = \left(\frac{2}{\pi}\mathcal{K}\right)^2$	$\frac{\delta}{l} = \frac{2}{\mathcal{K}} \text{Sin} \frac{\alpha}{2}$
0	1.5708	1.000	0
10	1.5738	1.004	0.1116
20	1.5828	1.016	0.2193
40	1.6200	1.062	0.4221
60	1.6858	1.152	0.5930
90	1.8541	1.392	0.7625

Thus, for a load P which is 39.2% greater than P_{Euler}, the above indicates a deflected structure as shown in Fig. 11.11. Similar results are obtained for other conditions, all as indicated by the tabular values.†

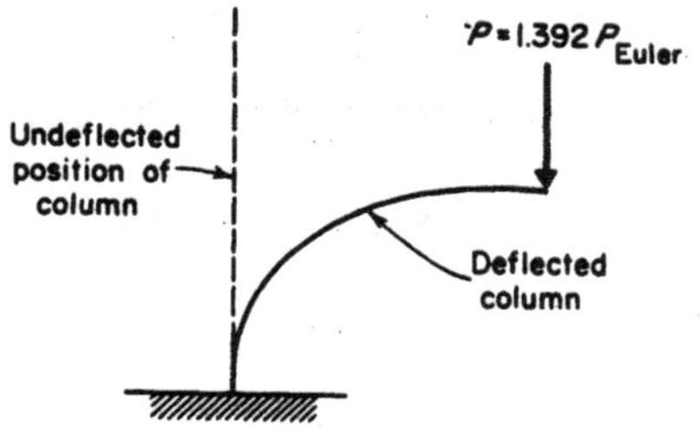

Fig. 11.11

In order to show how the elastica solution applies to a particular column, and more important, in order to indicate most strikingly exactly how the elastica solution compares with the more approximate Euler solution, we solve two special columns using the elastica results.

1. We shall analyze an "engineering-type" column, i.e., a column for which

$$\frac{\text{length}}{\text{least dimension}} = 40 \tag{11.88}$$

2. We shall analyze an "elastica-type" column, i.e., a column for which

$$\frac{\text{length}}{\text{least dimension}} = 400 \tag{11.89}$$

†Although the elastica solution is itself an approximate one (in that it essentially assumes that plane sections before bending are plane after bending) it does represent an improvement on the small deflection Bernoulli-Euler solution. Tests indicate that the elastica solution predicts fairly well the large deflection shape of columns, providing shear effects are unimportant in the deflected structure — as they generally are in elastica-type structures.

Column 1 — the Engineering Column

Given a column of cross-sectional dimensions and length as shown in Fig. 11.12.

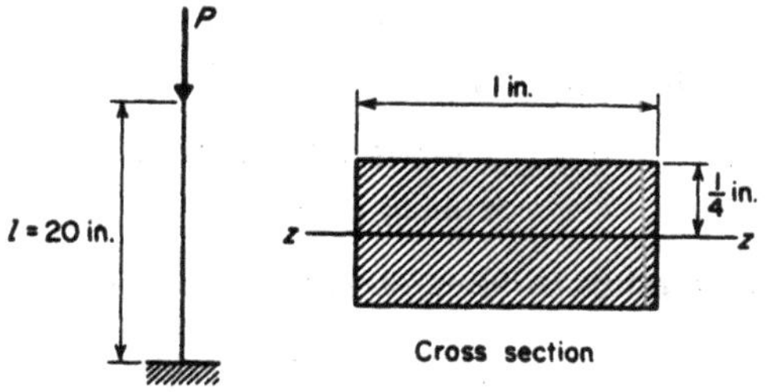

Fig. 11.12

Assume that the material is steel, with $E = 30{,}000{,}000$ psi, and that the allowable normal stress, $\sigma_{\text{allowable}}$, equals 30,000 psi. Then

$$P_{\text{Euler}} = \frac{\pi^2 EI_z}{4l^2} \tag{11.90}$$

$$= \frac{\pi^2(30{,}000{,}000)}{4(20)^2}\left[\frac{(1)(1/2)^3}{12}\right] \tag{11.91}$$

$$= 2000 \text{ lb} \tag{11.92}$$

Therefore, the stress due to direct compression, σ_{direct}, is given by

$$\sigma_{\text{direct}} = \frac{P_{\text{Euler}}}{A} = \frac{2000}{(1)(1/2)} \tag{11.93}$$

$$= 4000 \text{ psi} \tag{11.94}$$

Now assume (see second line of Table 7)

$$P = P_{\text{Euler}} + 0.4\%\ P_{\text{Euler}} \tag{11.95}$$

$$= 2000 + 8 \tag{11.96}$$

$$= 2008 \text{ lb} \tag{11.97}$$

The direct stress is now given by

$$\sigma_{\text{direct}} = \frac{P}{A} = \frac{2008}{(1)(1/2)} \tag{11.98}$$

$$= 4016 \text{ psi} \tag{11.99}$$

Then, from Table 7,

$$\frac{\delta}{l} = 0.11 \tag{11.100}$$

or

$$\delta = (0.11)(20) = 2.2 \text{ in.} \tag{11.101}$$

Hence the maximum bending moment is

$$M_{max} = P\delta = 2008(2.2) \tag{11.102}$$

$$= 4420 \text{ in.-lb} \tag{11.103}$$

Therefore, the maximum bending stress is given by

$$\sigma_{bending} = \frac{M_{max}y}{I_z} \tag{11.104}$$

$$= \frac{(4420)(1/4)}{\left(\frac{(1)(1/2)^3}{12}\right)} \tag{11.105}$$

$$= 106{,}000 \text{ psi} \tag{11.106}$$

and the normal stress, $\sigma_{combined}$, is given by the total of the direct and bending stresses:

$$\sigma_{combined} = \sigma_{direct} + \sigma_{bending} \tag{11.107}$$

$$= 4016 + 106{,}000 \tag{11.108}$$

$$\cong 110{,}000 \text{ psi} \tag{11.109}$$

In other words, for an ordinary "engineering-type" column, a load only very little more than P_{Euler} will cause a stress which is larger than the allowable stress. Stated differently again, *for "engineering type" columns, for all practical purposes, the maximum allowable direct load is the Euler load.*

Now let us consider the "elastica-type" column.

Column 2 - the Elastica Column

Given a column of cross-sectional dimensions and length as shown in Fig. 11.13.

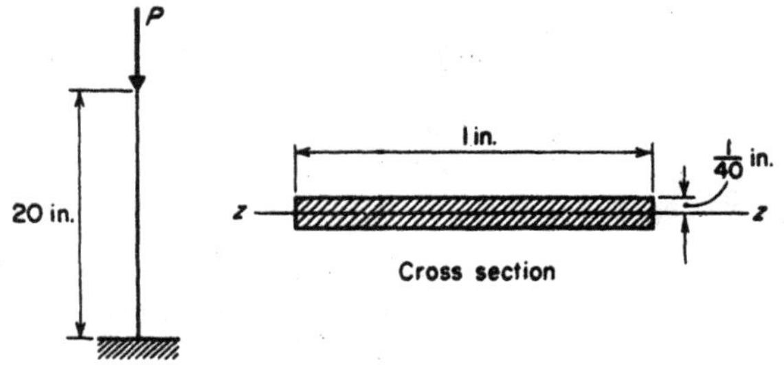

Fig. 11.13

Assume that the material is steel, $E = 30{,}000{,}000$ psi, and that the allowable normal stress, $\sigma_{\text{allowable}}$, equals 30,000 psi.
For this column,

$$P_{\text{Euler}} = \frac{\pi^2 E I_z}{4l^2} = \frac{\pi^2(30{,}000{,}000)}{(4)(20)^2}\left[\frac{1(1/20)^3}{12}\right] \tag{11.110}$$

$$= 2 \text{ lb} \tag{11.111}$$

Therefore, the direct stress due to P_{Euler} is

$$\sigma_{\text{direct}} = \frac{P_{\text{Euler}}}{A} \tag{11.112}$$

$$= \frac{2}{(1)(1/20)} \tag{11.113}$$

$$= 40 \text{ psi} \tag{11.114}$$

If, once more,

$$P = P_{\text{Euler}} + 0.4\%\ P_{\text{Euler}} \tag{11.115}$$

$$\cong 2 \text{ lb} \tag{11.116}$$

and, as before (Table 7),

$$\frac{\delta}{l} = 0.11 \text{ in.} \tag{11.117}$$

or

$$\delta = 2.2 \text{ in.} \tag{11.118}$$

then the maximum bending moment is given by

$$M_{\max} = P\delta = 2(2.2) \tag{11.119}$$

$$= 4.4 \text{ in.-lb} \tag{11.120}$$

Hence the maximum bending stress is given by

$$\sigma_{\text{bending}} = \frac{M_{\max} y}{I_z} \tag{11.121}$$

$$= \frac{(4.4)(1/40)}{\left(\frac{(1)(1/20)^3}{12}\right)} \tag{11.122}$$

$$= 10{,}600 \text{ psi} \tag{11.123}$$

The combined stress, σ_{combined}, is given by the total of the direct and bending stresses:

$$\sigma_{\text{combined}} = \sigma_{\text{direct}} + \sigma_{\text{bending}} \tag{11.124}$$

$$= 40 + 10{,}600 \tag{11.125}$$

$$\cong 10{,}600 \text{ psi} \tag{11.126}$$

In other words, for an ordinary "elastica-type" column, a load larger than P_{Euler} can be sustained by the structure without causing failure or excessive stress. Stated another way, *for "elastic-type" columns, the permissible maximum allowable direct load may be larger than the Euler load.*

The two examples just considered show very clearly that, although the Euler analysis is an approximate one, for the "engineering-type" column it does represent the ultimate compressive load which the member can sustain.

However — it is also important to understand that there are columns (here called "elastica-type" columns) which can sustain loads larger than P_{Euler} (without failure and with stresses below the proportional limit).

11-6 Column Behaviour When Stresses Exceed the Proportional Limit

The Euler column formula for the hinged-hinged bar is given by

$$P_{CR} = \frac{\pi^2 EI_z}{l^2} \tag{11.127}$$

If we recall that

$$I_z = Ar^2 \tag{11.128}$$

in which A is the area of the cross section and r is the radius of gyration with respect to the z-axis, then this formula may be put into the form

$$\frac{P_{CR}}{A} = \frac{\pi^2 E}{(l/r)^2} \tag{11.129}$$

and (see Fig. 11.14) P_{CR}/A is, in effect, the compressive stress acting on the

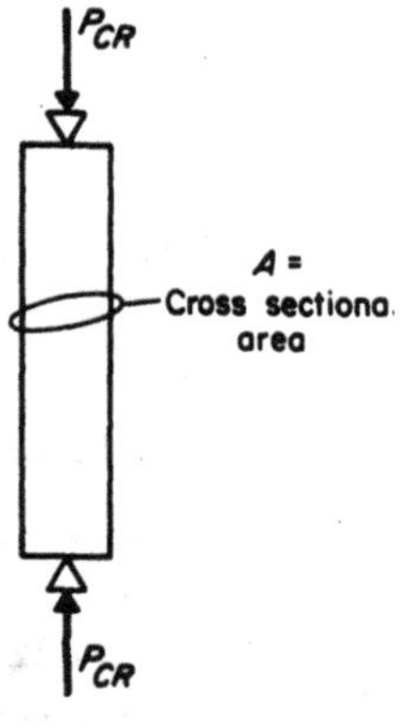

Fig. 11.14

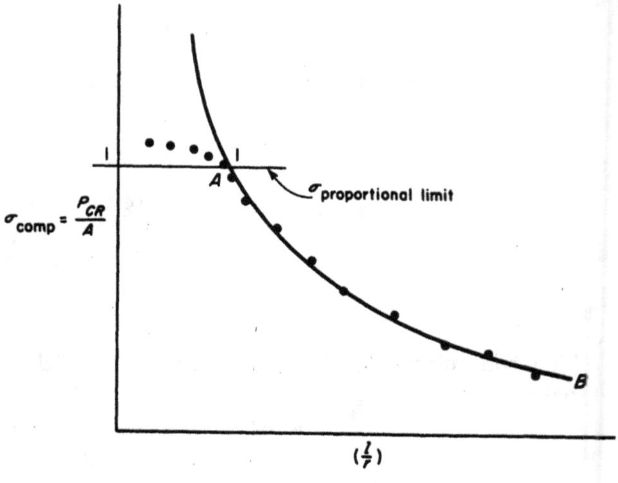

Fig. 11.15

cross section. Thus, a plot of P_{CR}/A vs. l/r would give the curve shown in Fig. 11.15.

If tests are performed on columns (especially chosen for their initial straightness) of varying l/r and these are plotted on the curve, it will be found that, approximately, for a typical metal they fall as shown by the points.

This indicates quite strikingly that the Euler formula accurately predicts buckling phenomena, for portion A-B of the curve, *only when the stress is less than the proportional limit.* This, however, is no more than we ought to expect, since the Euler formula was derived starting with

$$\frac{d^2v}{dx^2} = -\frac{M(x)}{EI_z} \quad \text{(Eq. 11.1)} \tag{11.130}$$

and this formula (see Art. 6.3) assumes a linear relation between stress and strain. That is, this formula holds only when stresses are below the proportional limit (see Fig. 11.16), in which region the modulus of elasticity

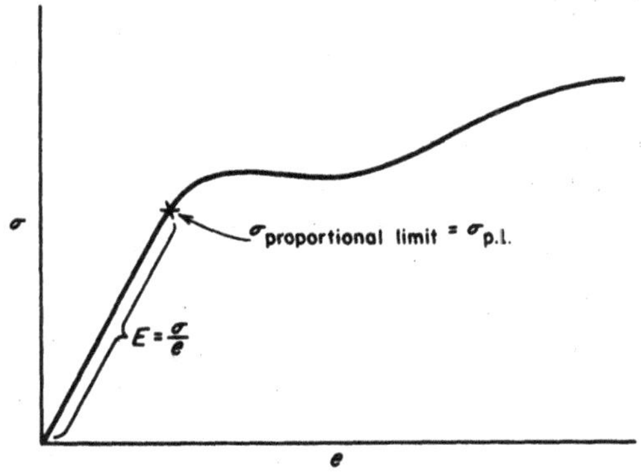

Fig. 11.16

is constant and equal to E. But, in Fig. 11.15, we see that for the lower values of l/r, the stress indicated by the Euler formula is larger than the proportional limit; hence we have no right to expect the Euler formula to accurately predict buckling loads in this region.

There are several different procedures which have been recommended for determining the buckling loads when the stresses are larger than the proportional limit. We shall describe one of these methods which is particularly simple and easy to apply and which is sufficiently accurate for many engineering applications. A more complete discussion of this subject will be found in Ref. 22.

Refer to Fig. 11.17, which shows a typical metal stress-strain curve.

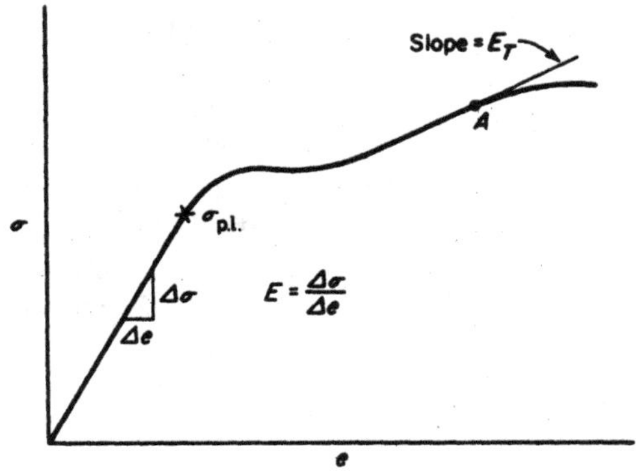

Fig. 11.17

Note that for stresses below the proportional limit the slope of the stress-strain curve is given by

$$E = \frac{\Delta\sigma}{\Delta e} \tag{11.131}$$

This is defined as the "modulus of elasticity."

It is clear that the constant value of E is only maintained below the proportional limit. For stresses greater than the proportional limit, we can still define — locally, at each point, as at "A" — a ratio of $\Delta\sigma/\Delta e$, which corresponds to a so-called "tangent modulus, E_T."

That is, at each point on the curve past the proportional limit, we define the slope to the stress-strain curve as the "tangent modulus of elasticity, E_T."

Then for any given column material having a given stress-strain curve, we determine the value of the tangent modulus, E_T, at each point where

$$\sigma > \text{proportional limit} \tag{11.132}$$

In the Euler formula, see Fig. 11.18, where the l/r indicates a stress greater than the proportional limit, instead of E, we then insert E_T, corresponding to the given stress, or

$$\frac{P_{CR}}{A} = \frac{\pi^2 E_T}{(l/r)^2}, \quad \sigma > \sigma_{p.l.} \tag{11.133}$$

This will give for a complete modified Euler curve a representation as shown in Fig. 11.18.

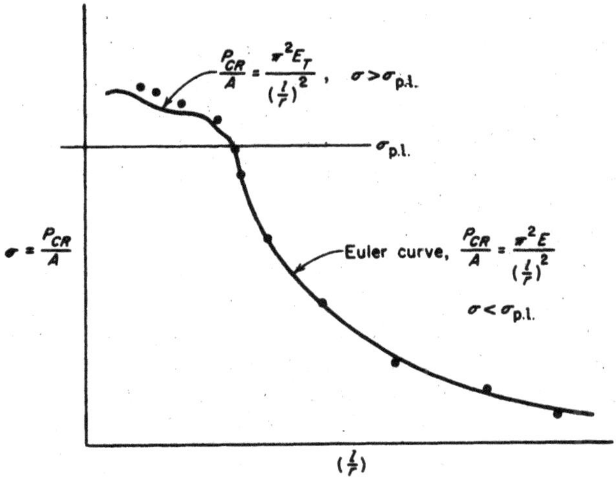

Fig. 11.18

If test values (for columns carefully prepared as to end conditions and specially chosen for initial straightness) are plotted on this curve they appear as shown. Note that the actual P_{CR}/A values are somewhat larger than indicated by the tangent modulus. There are theoretical reasons why this should be so; see Ref. 22. However, for engineering purposes, the tangent modulus gives results which are probably sufficiently accurate.

11-7 The Engineering Form of the Column Formulas

In actual engineering applications of a column formula, consideration must be given to the following facts:

1. The loading on columns is frequently variable within quite wide limits, with a rather uncertain maximum value.

2. Columns used in buildings, bridges and similar structures very often have some initial slight curvature or, as it is called, *eccentricity of loading*. This introduces initial bending stresses which are not considered in the Euler derivation.

3. In the Euler formula, the quantity E, modulus of elasticity, appears. Even for so highly controlled a material as steel, it is found that E may vary between 27,000,000 psi and 32,000,000 psi. Also I, the moment of inertia, varies for the same size beams coming from the same mill due to necessary tolerances in the rolling machines.

4. It is found that the Euler formula gives rather good results, when compared with experiments, for large values of l/r — say for $l/r > 200$. For smaller values of l/r, for the engineering column (as against an ideal,

laboratory-type column which was considered in the results shown in Fig. 11.18) there is some deviation between Euler results and actual results.

In view of the above, it is necessary that some modification be made to the Euler formula. The most important change is that a "factor of safety" must be included to account for the uncertainty of load, material and shape. Secondly, some consideration should be given to the deviation between Euler and actual results for smaller l/r values.

All of these factors are taken care of, in engineering design practice, by the introduction of arbitrary, empirical column formulas. These formulas are the result of much study and experimentation, are based upon the considered engineering judgement of the different groups involved, and represent sound, economical engineering design practice for columns.

A typical column design specification is given by the American Institute of Steel Construction in the *AISC Manual*. We shall use this specification in our work on columns in this chapter. It holds for assumed hinged-hinged columns and makes the further assumption that all columns in ordinary building or bridge construction which are supported at both ends may be considered as hinged-hinged.

We emphasize, however, that although this specification for steel column design is authoritative and very widely used, it is by no means the only specification for steel column design. Also, there are other specifications for aluminum columns, for cast iron columns, for timber columns, for high-strength steel columns and so on for other materials. In all cases, the starting point is the Euler column formula. But practical considerations in all cases introduce the requirement that the Euler formula be modified to a lesser or greater degree.

The AISC specification formulas for steel columns are as follows:

(a) If l/r is not greater than 120, the allowable concentric load, P, on a column is given by

$$\frac{P}{A} = 17{,}000 - 0.485\frac{l^2}{r^2} \tag{11.134}$$

(There is an added stipulation on this formulation which we need not consider here.)

(b) For axially loaded columns with l/r greater than 120 and less than 200,

$$\frac{P}{A} = \frac{18{,}000}{1 + \dfrac{l^2}{18{,}000r^2}} \tag{11.135}$$

In the next section we shall investigate and design typical columns using these specifications.

11-8 The Design and Investigation of Engineering Columns

We can best illustrate the investigation phase of column analysis by investigating a typical steel column, as follows:

EXAMPLE 1 What is the allowable concentric load on a 14 WF 84, 25 ft long, supported at its ends?

SOLUTION We refer to Table 4 on p. 131 showing the properties of WF sections for use in designing.

Note that for the 14 WF 84 the moment of inertia, I_{yy} is the minimum value for the cross section. Hence it governs the allowable load and must be used in the formula for allowable load. Also shown in the table is the radius of gyration, r, for this moment of inertia. Thus,

$$\frac{l}{r_{\min}} = \frac{25(12)}{3.02} \tag{11.136}$$

$$= 99.3 \tag{11.137}$$

This is less than 120. Hence, in accordance with the AISC specification, the allowable load, P, is given by

$$P = A\left(17{,}000 - 0.485\frac{l^2}{r^2}\right) \tag{11.138}$$

in which A (see Table 4) is 24.71 sq in. Thus,

$$P = 24.71[17{,}000 - 0.485(99.3)^2] \tag{11.139}$$

$$= 302{,}000 \text{ lb} \tag{11.140}$$

EXAMPLE 2 (a *design* problem) What size column must be used to support 100 kip if the column is supported at its ends and is 10 ft long? Let us assume the column depth to be 8 in., that is, the column is an 8 WF.

SOLUTION There are various methods which have been recommended for the design procedure. In all cases, however, the technique is one of cut-and-try. We must make some initial assumption. Some engineers prefer an assumption on the stress; other engineers prefer an assumption on the l/r.

Let us proceed on the latter basis and assume $l/r = 100$. Then

$$\frac{P}{A} = 17{,}000 - 0.485\frac{l^2}{r^2} \tag{11.141}$$

$$\frac{P}{A} = 17{,}000 - 4850 \tag{11.142}$$

$$= 12{,}150 \text{ psi} \tag{11.143}$$

Since the actual load is 100,000 lb, this means that, for the assumed l/r, a column is required having an A_{req} given by

$$A_{req} = \frac{P}{12{,}150} \tag{11.144}$$

$$= \frac{100{,}000}{12{,}150} \tag{11.145}$$

$$= 8.23 \text{ sq in.} \tag{11.146}$$

Also, for $l/r = 100$ we have, as a required radius of gyration,

$$r_{req} = \frac{l}{100} = \frac{120}{100} = 1.2 \text{ in.} \tag{11.147}$$

Checking these requirements against the member sizes given in Table 4, we have

		8 WF 28		8 WF 24	
r_{req}	A_{req}	r_{min}	A	r_{min}	A
1.20	8.23	1.62	8.23	1.61	7.06

Thus, although the area for the 8 WF 28 is equal to the required area, its r_{min} is much larger than the required r_{min}. Hence it is too strong and will support much more than the 100 K.

On the other hand, the 8 WF 24 has a smaller area than required, but does have a larger r_{min}. The beam therefore should be checked for allowable load, and it is found that, for 8 WF 24,

$$P = A\left(17{,}000 - 0.485\frac{l^2}{r^2}\right) \tag{11.148}$$

$$= 7.06\left(17{,}000 - 0.485\frac{120^2}{1.61^2}\right) \tag{11.149}$$

$$= 101K \tag{11.150}$$

so that this member is satisfactory.

In connection with this design solution we may mention the following facts.

1. There are other columns besides the 8 WF 24 which can be used for this load and length. Thus, a 6 WF 25-10 ft long will also support 100K. In all cases, there are generally several different wide-flange columns which will be suitable for a particular load and length. Which one is used depends upon several factors, including economy (i.e., weight of member), clearance requirements, rolling mill schedules, and similar considerations.

2. The AISC has a listing of the allowable loads on the sections for various unsupported lengths. Thus, instead of having to design steel columns, an engineer, when using the AISC specifications, need only enter the proper column and row of tables already prepared and choose the size column. If clearances govern the permissible size, he will simply choose the column with the required maximum dimension. Therefore, for rolled columns using the AISC specification, solving the actual steel column design problem is merely a matter of referring to the *AISC Manual*.

11-9 Critical Buckling Loads Obtained by the Finite Difference Method

In this article we shall describe how critical bending buckling loads may be determined using the finite difference method.

As pointed out earlier, the finite difference method is particularly useful because it sets up a solution in a form especially applicable to electronic computer applications. This is true also for its use in the column buckling problem, that is, there are some complicated column problems which can not be solved exactly but which can be approximately analyzed using the finite difference formulation and electronic computers.

We illustrate the finite difference method by solving the same problem in three ways. The first solution will be a very rough (finite difference) approximation, but we shall see (in the finite difference sense) that it does give results which are not too far from the more accurate Euler results. The second solution will represent an improvement on the first (i.e., a smaller finite difference interval) and we shall see that it gives — without too much additional labor — much better results. Finally, we shall generalize the technique and indicate how it is used for a more complex column analysis, illustrating this with a still smaller grid solution. This generalization will also indicate how the higher buckling modes (see Fig. 11.4) may be obtained.

Example 1 Consider the hinged-hinged column shown in Fig. 11.19. Determine P_{CR} using finite difference methods with $\Delta x = l/2$.

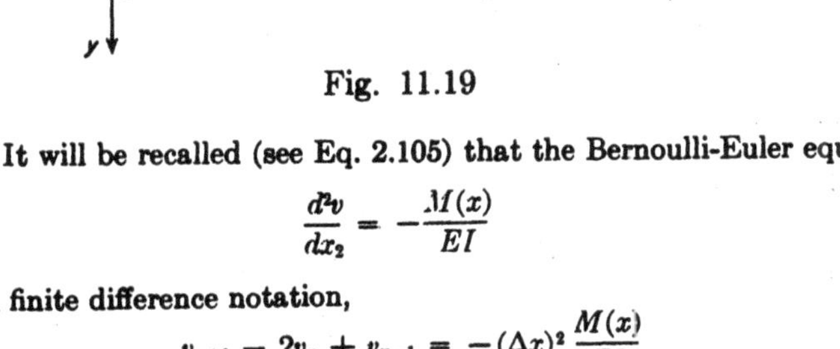

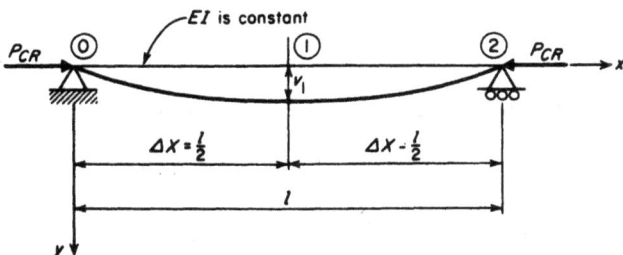

Fig. 11.19

Solution It will be recalled (see Eq. 2.105) that the Bernoulli-Euler equation

$$\frac{d^2v}{dx^2} = -\frac{M(x)}{EI} \tag{11.151}$$

becomes, in finite difference notation,

$$v_{n+1} - 2v_n + v_{n-1} = -(\Delta x)^2 \frac{M(x)}{EI} \tag{11.152}$$

It will be recalled also (see Art. 11-2) that the critical buckling load, P_{CR}, may be thought of as the load which will just hold the column in a slightly deflected position, as shown in Fig. 11.19.

Thus, we consider our column to be in a deflected position and divided into stations — 0 , 1 and 2 , as shown. Then, applying the above relation at point 1 and noting that at this point $M(x) = P_{CR}v_1$, and $\Delta x = l/2$, we have

$$v_0 - 2v_1 + v_2 = -\left(\frac{l}{2}\right)^2 \frac{P_{CR}v_1}{EI} \tag{11.153}$$

Now,

$$v_0 = v_2 = 0 \tag{11.154}$$

so that the equation becomes

$$0 - 2v_1 - 0 = -\frac{l^2}{4}\frac{P_{CR}v_1}{EI} \tag{11.155}$$

and

$$P_{CR} = \frac{8EI}{l^2} \quad \left(\text{vs. } \frac{9.87EI}{l^2}, \text{ Euler}\right) \tag{11.156}$$

Hence, even for the very rough grid chosen, we see that a fair value for the buckling load is obtained.

Example 2 Consider the beam divided as in Fig. 11.20, with $\Delta x = l/3$. Determine P_{CR} using the finite difference method.

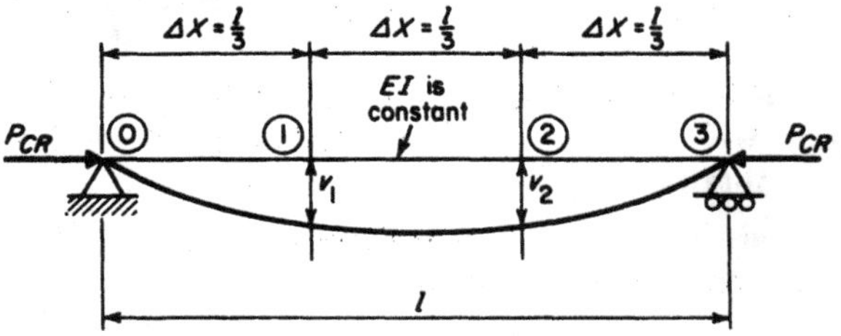

Fig. 11.20

Solution In this case we have at point 1 , considering 0 , 1 and 2 ,

$$v_0 - 2v_1 + v_2 = -(l/3)^2 \frac{P_{CR}v_1}{EI} \tag{11.157}$$

and at point 2 , considering 1 , 2 and 3 ,

$$v_1 - 2v_2 + v_3 = -(l/3)^2 \frac{P_{CR}v_2}{EI} \tag{11.158}$$

In this problem

$$v_0 = v_3 = 0 \tag{11.159}$$

and, due to symmetry,

$$v_1 = v_2 \tag{11.160}$$

Thus, Eqs. 11.157 and 11.158 become

$$-v_1 = -\frac{l^2}{9}\frac{P_{CR}v_1}{EI} \tag{11.161}$$

or

$$P_{CR} = \frac{9EI}{l^2}\left[\text{vs. } \frac{(9.87)EI}{l^2}, \text{ Euler}\right] \tag{11.162}$$

Hence, by making Δx smaller, but utilizing the properties of symmetry, we obtain a more accurate solution with very little additional labor.

In the general case, with many points in the column being chosen for the finite difference formulation, we have, at a typical point,

$$v_{n+1} - 2v_n + v_{n-1} = -(\Delta x)^2 \frac{P_{CR} v_n}{EI} \tag{11.163}$$

or

$$v_{n+1} - \left[2 - (\Delta x)^2 \frac{P_{CR}}{EI}\right] v_n + v_{n-1} = 0 \tag{11.164}$$

At the point $n + 1$, this expression would be

$$v_{n+2} - \left[2 - (\Delta x)^2 \frac{P_{CR}}{EI}\right] v_{n+1} + v_n = 0 \tag{11.165}$$

and at v_{n-1}, it would be

$$v_n - \left[2 - (\Delta x)^2 \frac{P_{CR}}{EI}\right] v_{n-1} + v_{n-2} = 0 \tag{11.166}$$

Thus, we have in this set a series of j homogeneous equations, in terms of j unknown deflections:

$$v_1, v_2 \cdots v_j \tag{11.167}$$

To illustrate our point let us assume we have two such equations, namely,

$$a\,v_1 + b\,v_2 = 0 \tag{11.168}$$

$$c\,v_1 + d\,v_2 = 0 \tag{11.169}$$

Now it is well known that a solution to this pair of equations is given (see Ref. 7, for example) by

$$v_1 = \frac{\begin{vmatrix} 0 & b \\ 0 & d \end{vmatrix}}{\begin{vmatrix} a & b \\ c & d \end{vmatrix}} \tag{11.170}$$

and

$$v_2 = \frac{\begin{vmatrix} a & 0 \\ c & 0 \end{vmatrix}}{\begin{vmatrix} a & b \\ c & d \end{vmatrix}} \tag{11.171}$$

The first column of the top determinant of Eq. 11.170, is obtained by replacing the coefficient of v_1 in each equation by the right-hand side of the equation, and the second column is made up of the coefficients of v_2. The denominator determinant is made up of the coefficients of v_1 and v_2 in each equation. A similar description holds for Eq. 11.171, except that here the top determinant substitution is made for v_2.

The top determinants in both Eq. 11.170 and 11.171 are obviously zero. It follows, therefore, that v_1 *and* v_2 *in these equations can be different from zero only if the denominator determinants are also equal to zero.*

This is the criterion which we apply to Eqs. 11.164, 11.165 and 11.166. That is, the denominator determinant must be zero, or in one form, for a particular problem, the buckling equation or criterion becomes

$$\begin{vmatrix} -\left[2-(\Delta x)^2\dfrac{P_{CR}}{EI}\right] & 1 & 0 & \cdots & 0 \\ 1 & -\left[2-(\Delta x)^2\dfrac{P_{CR}}{EI}\right] & 1 & \cdots & \cdot \\ 0 & 1 & -\left[2-(\Delta x)^2\dfrac{P_{CR}}{EI}\right] & \cdots & \cdot \\ \vdots & \vdots & \vdots & \ddots & \vdots \\ 0 & 0 & 0 & \cdots & -\left[2-(\Delta x)^2\dfrac{P_{CR}}{EI}\right] \end{vmatrix} = 0 \tag{11.172}$$

This will be a jth-order equation in terms of P_{CR}. It will therefore, in general, have j solutions for P_{CR}, corresponding to the various modes of buckling as obtained for the Euler solution in Art. 11-2. The lowest critical value will, in general, be the only one desired, although it is conceivable that the others also will be of interest.

In any event, this represents a procedure, using finite difference methods, for solving complicated column buckling problems which are not capable of exact solutions.

We shall indicate in our next illustrative problem a special application of this more general technique.

Example 3 Consider the column of Fig. 11.21, with $\Delta x = l/4$. By finite difference methods, determine the critical buckling loads, P_{CR_1} and P_{CR_2}, corresponding to the two buckling modes shown in Fig. 11.21.

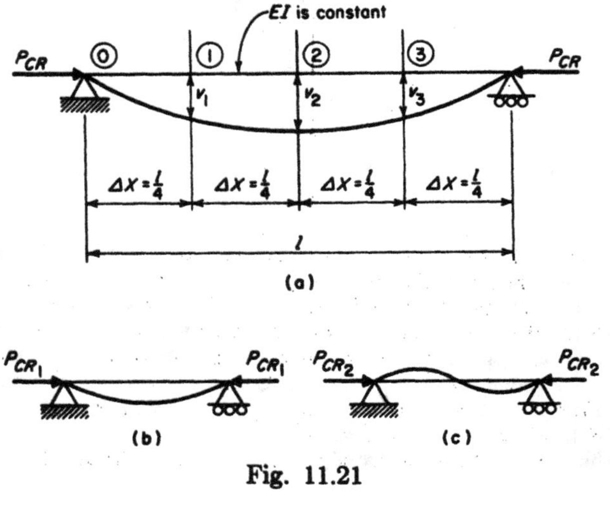

Fig. 11.21

SOLUTION The three finite difference equations, in terms of the general deflections v_1, v_2 and v_3, are (since $v_0 = v_4 = 0$)

$$-2v_1 + v_2 = -\frac{l^2}{16}\frac{P_{CR}v_1}{EI} \tag{11.173}$$

$$v_1 - 2v_2 + v_3 = -\frac{l^2}{16}\frac{P_{CR}v_2}{EI} \tag{11.174}$$

$$v_2 - 2v_3 = -\frac{l^2}{16}\frac{P_{CR}v_3}{EI} \tag{11.175}$$

In general terms, the critical loads are given by

$$\begin{vmatrix} -\left(2 - \frac{P_{CR}l^2}{16EI}\right) & 1 & 0 \\ 1 & -\left(2 - \frac{P_{CR}l^2}{16EI}\right) & 1 \\ 0 & 1 & -\left(2 - \frac{P_{CR}l^2}{16EI}\right) \end{vmatrix} = 0 \tag{11.176}$$

In this problem, however, due to symmetry we can simplify the solution for P_{CR_1} and P_{CR_3} as follows: refer to Fig. 11.21b. Note, for P_{CR_1}, that

$$v_1 = v_3 \tag{11.177}$$

Therefore, Eqs. 11.173, 11.174 and 11.175, are equivalent to

$$-\left(2 - \frac{l^2P_{CR_1}}{16EI}\right)v_1 + v_2 = 0 \tag{11.178}$$

$$2v_1 - \left(2 - \frac{l^2P_{CR_1}}{16EI}\right)v_2 = 0 \tag{11.179}$$

so that

$$\begin{vmatrix} -\left(2 - \frac{l^2P_{CR_1}}{16EI}\right) & 1 \\ 2 & -\left(2 - \frac{l^2P_{CR_1}}{16EI}\right) \end{vmatrix} = 0 \tag{11.180}$$

or

$$\left(2 - \frac{l^2P_{CR_1}}{16EI}\right)^2 - 2 = 0 \tag{11.181}$$

which gives

$$2 - \frac{l^2P_{CR_1}}{16EI} = \pm\sqrt{2} = \pm 1.414 \tag{11.182}$$

The smallest value of P_{CR_1}, which represents the lowest buckling load requires the positive sign before the 1.414, giving

$$2 - \frac{l^2 P_{CR_1}}{16EI} = 1.414 \tag{11.183}$$

or

$$P_{CR_1} = \frac{16(0.586)EI}{l^2} = \frac{9.38EI}{l^2} \left[\text{vs. } \frac{9.87EI}{l^2}, \text{ Euler} \right] \tag{11.184}$$

To determine P_{CR_2}, note from Fig. 11.21c that in this case,

$$v_1 = -v_3 \tag{11.185}$$

Therefore the three equations become

$$-2v_1 + v_2 = -\frac{l^2 P_{CR_2} v_1}{16EI} \tag{11.186}$$

$$-2v_2 = -\frac{l^2 P_{CR_2} v_2}{16EI} \tag{11.187}$$

$$2v_1 + v_2 = +\frac{l^2 P_{CR_2} v_1}{16EI} \tag{11.188}$$

Adding the first and third of these, we conclude that $v_2 = 0$ (as shown on Fig. 11.21c). Therefore, from either the first or third we obtain

$$2 = \frac{P_{CR_2} l^2}{16EI} \tag{11.189}$$

or

$$P_{CR_2} = \frac{32EI}{l^2} \left[\text{vs. } \frac{4\pi^2 EI}{l^2} = \frac{39.48EI}{l^2}, \text{ Euler} \right] \tag{11.190}$$

Thus, we have determined, using finite difference techniques, approximate values for the two lowest critical buckling loads for a column. Higher modes, if desired, can be obtained in an extension of these methods as described previously, and shown in Eq. 11.172.

11-10 Summary

We discussed the bending instability problem from several different points of view, all aimed at giving the student a more thorough understanding of the basic instability phenomenon.

First we derived the classical Euler solution for the different boundary conditions, pointing out the shortcomings and advantages of this approach. Following this, we derived the more accurate, large deflection column instability solution — the one which applies to the elastica.

It was shown that the Euler solution for the engineering-type column is a lower limit of the elastica solution, and hence entirely adequate for the engineering column.

The empirical engineering formulas were then described. In particular we mentioned the formulas recommended by the American Institute of Steel Construction as given in their specifications.

Finally, we discussed in some detail the solution of buckling problems using the finite difference method. It was pointed out that approximate solutions can be obtained for all the modes of buckling of the column. Also, in common with other applications using finite difference techniques, the formulation is one which lends itself particularly to electronic computer programming and solution.

Problems

1. Determine the allowable Euler column load for a 8 WF 24, 25 ft long, using a factor of safety of 2.5.
2. For the column of Prob. 1, determine the allowable load using the AISC specification.
3. A 10 WF 39 is braced at mid-height as shown in the figure. Determine the allowable column load for this structure. Use AISC formula.

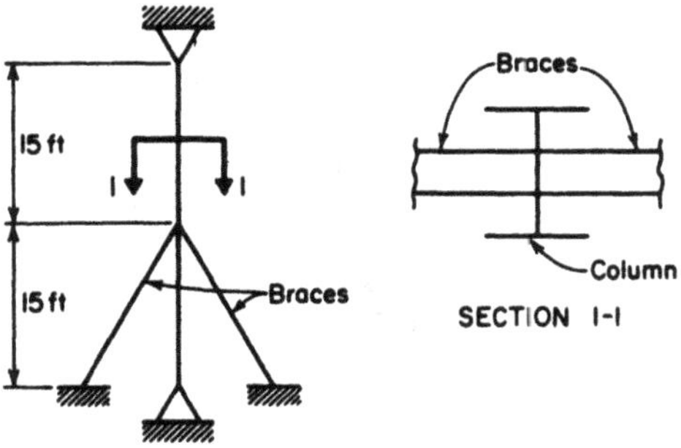

4. What is the allowable Euler column load for a 3 × 6 timber column, 10 ft high? Use a factor of safety of 2.5, $E = 1{,}200{,}000$ psi.
5. A timber column 15 ft high is to support a load of 25,000 lb. Using the Euler formula with a factor of safety of 2.5, what size square column is required? $E = 1{,}500{,}000$ psi. Assume the minimum l/r is 10, the maximum l/r is 50.
6. What is the allowable column load for a built-up steel column 20 ft long, constructed of angles and plate as shown in the figure? Use AISC specifications.

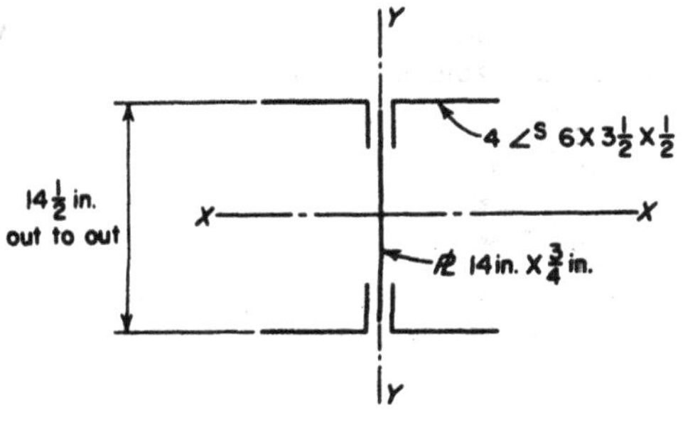

7. What is the allowable AISC column load for the double channel column shown in the figure, if it is 25 ft high? Channels act as a single unit. Lacing merely ties the two members together, is not considered in determining allowable load.

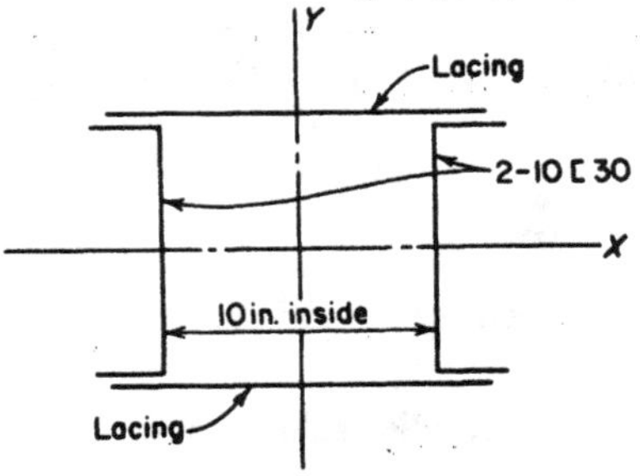

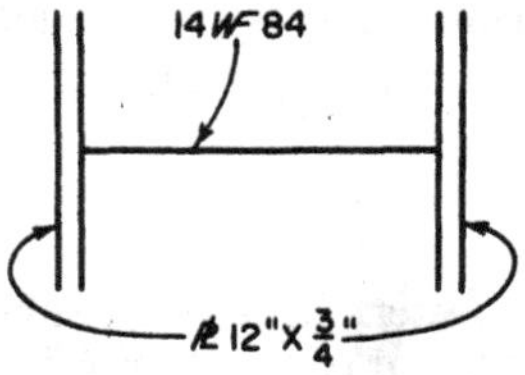

8. What is the allowable AISC column load for the built-up column shown in the figure if it is 30 ft high?

9. What is the allowable AISC column load for the built-up column shown in the figure? It is 20 ft high.

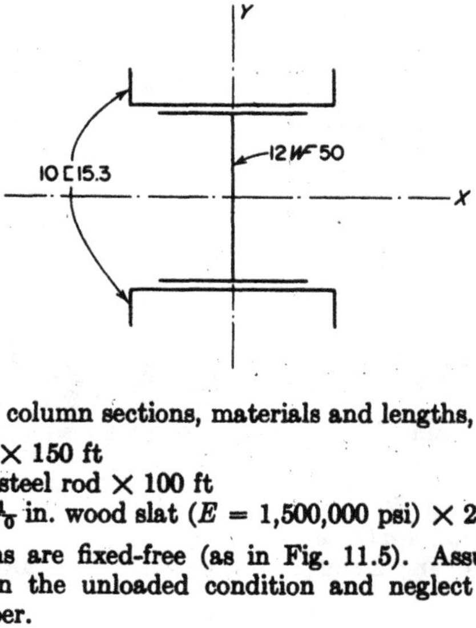

10. Given the following column sections, materials and lengths,

8 WF 24 × 150 ft
1 in. dia steel rod × 100 ft
1 in. × $\frac{1}{20}$ in. wood slat (E = 1,500,000 psi) × 25 ft

Assume the columns are fixed-free (as in Fig. 11.5). Assume also they are perfectly straight in the unloaded condition and neglect the effect of the weight of the member.

(a) Determine P_{Euler} and the compressive stress corresponding to P_{Euler}.
(b) Using the elastica solution, obtain the maximum combined compressive stress for

(1) $P = 1.02\,P_{\text{Euler}}$
(2) $P = 1.10\,P_{\text{Euler}}$
(3) $P = 1.25\,P_{\text{Euler}}$
(4) $P = 1.50\,P_{\text{Euler}}$

(c) Using the elastica solution, obtain the deflected shape of the columns for the different conditions in (b) above.

11. Set up the equations and the solution determinant for obtaining the critical compressive buckling loads (by finite difference methods) for the following columns. Attempt to obtain actual solutions if possible, although in general the algebraic work involved in obtaining a solution will be of such a magnitude that electronic (or similar) computers will be needed. *Hint:* for (c) and (d), vertical reactions are induced.

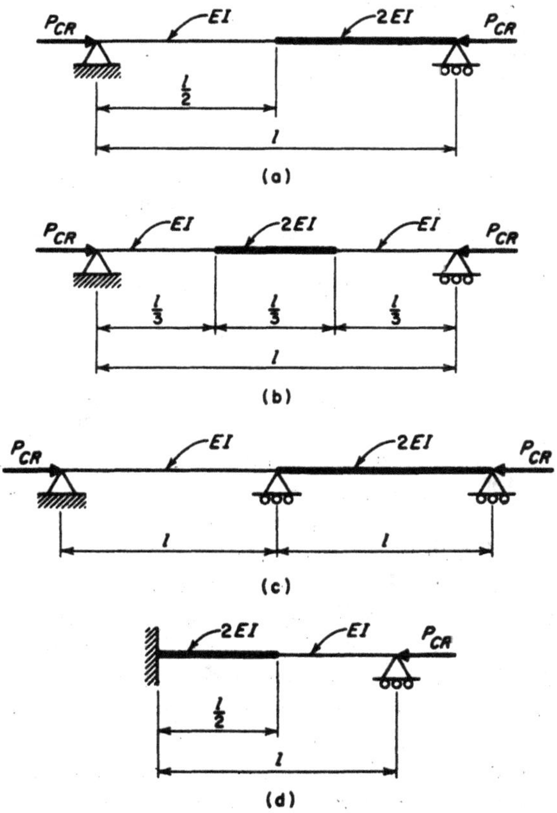

Chapter 12

THE TORSION PROBLEM

12-1 Introduction

In the preceding chapters we studied in some detail problems which arise from a consideration of the stress tensor having the form

$$\begin{pmatrix} \sigma_x & 0 & 0 \\ 0 & 0 & 0 \\ 0 & 0 & 0 \end{pmatrix} \qquad (12.1)$$

These were the direct load problem, the beam-bending problem, and the bending instability problem.

In this chapter we shall consider problems which occur when the stress tensor is assumed to contain only the *shear stress*. We shall see that, for our purposes, two kinds of problems are encountered —

1. the near-trivial pure shear structure;
2. the much more important pure torsion structure.

Problem (1) will be discussed first. Following this we shall present, in rather abbreviated form, the classical St. Venant solution to the torsion problem. We shall discuss the significance of warping in this problem and having the exact general solution, we will develop the solutions for two cross sections of engineering interest — the circle and the ellipse. Following this we shall indicate very briefly how solutions for more complicated cross sections are obtained. The square cross section will then be solved using the finite difference technique.

12-2 The Pure Shear Structure

Consider the two-dimensional stress tensor (see Art. 4–6) —

$$T = \begin{pmatrix} \sigma_x & \tau_{xy} \\ \tau_{yx} & \sigma_y \end{pmatrix} \qquad (12.2)$$

Let us assume that all elements are zero except the two shear stresses, $\tau_{xy} = \tau_{yx}$. Then the stress tensor becomes

$$\begin{pmatrix} 0 & \tau_{xy} \\ \tau_{yx} & 0 \end{pmatrix} \tag{12.3}$$

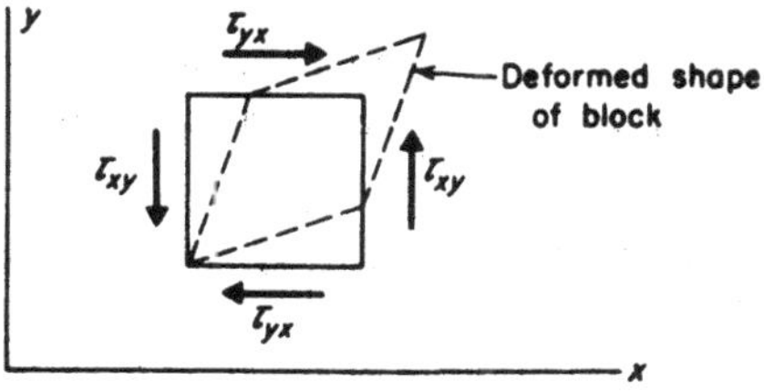

Fig. 12.1

and conditions are as shown in Fig. 12.1.

The equilibrium equations (see Eq. 4.22) require (body forces neglected)

$$\frac{\partial \tau_{xy}}{\partial x} = 0 \tag{12.4}$$

$$\frac{\partial \tau_{yx}}{\partial y} = 0 \tag{12.5}$$

or

$$\tau_{xy} = \tau_{yx} = \text{constant everywhere in the body.} \tag{12.6}$$

Hooke's Law gives

$$\left.\begin{aligned} \gamma_{xy} &= \frac{\tau_{xy}}{G} = \text{constant} \\ \text{and} \qquad \gamma_{yx} &= \frac{\tau_{yx}}{G} = \text{constant} \end{aligned}\right\} \tag{12.7}$$

Obviously the strain compatibility condition is identically satisfied, $0 = 0$.

With reference to the boundary conditions consider first the face perpendicular to the x-axis. Along this face, $l = 1$, $m = 0$, so that the boundary condition equation, Eq. 4.76, becomes

$$(1 \quad 0)\begin{pmatrix} 0 & \tau_{xy} \\ \tau_{yx} & 0 \end{pmatrix} = (0 \quad \tau_{xy}) \tag{12.8}$$

or the surface stress τ_{xy} must act on this face — as, in fact it does, see Fig. 12.1. A similar analysis applies to the other three faces.

Thus we have, essentially, the solution for a block of material subjected to constant, equal, shear stresses acting on the four faces perpendicular to the x- and y-axes and deforming as shown in Fig. 12.1. Although this represents an exact solution in the linearized mathematical theory of elasticity, it is really a near-trivial solution and of rather limited practical interest. We shall not consider it further in this text.

12-3 The Classical St. Venant Torsion Solution

The solution of the torsion problem, as obtained by St. Venant, represents one of the truly great landmarks in the mathematical theory of elasticity. It accounted for many observed experimental phenomena which other, earlier, theories could not predict. Also, it has been the starting point for many additional investigations and extensions of its theories in both elasticity and applied mathematics.

In demonstrating the St. Venant theory we begin by stating that we are seeking a solution to the problem of a pure torsion acting on a cylindrical (not necessarily circular) cross section. This is as shown in Fig. 12.2.

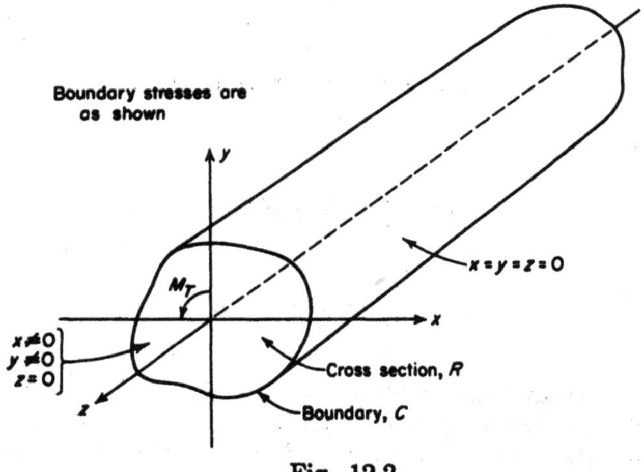

Fig. 12.2

The applied torque is M_T, the length of the bar is along the z-axis, C is the *boundary* of any cross section, and R is the *region*, or area, of any cross section. A balancing torque is applied at the far end of the bar.

Experiments indicate that a torque applied as in Fig. 12.2, develops shear stresses on the cross section. Hence, we assume as our stress tensor

$$\begin{pmatrix} 0 & 0 & \tau_{xz} \\ 0 & 0 & \tau_{yz} \\ \tau_{zx} & \tau_{zy} & 0 \end{pmatrix} \tag{12.9}$$

which means we are assuming stresses acting as shown in Fig. 12.3. Then

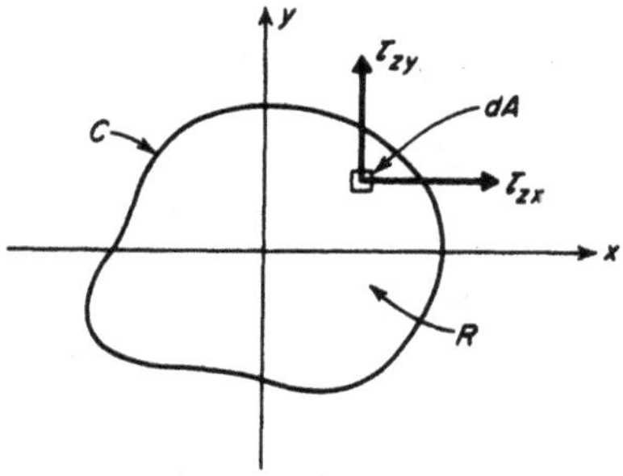

Fig. 12.3

the equilibrium equations (see Eq. 4.22) give, for everywhere in R (body forces neglected),

$$\begin{pmatrix} \dfrac{\partial}{\partial x} & \dfrac{\partial}{\partial y} & \dfrac{\partial}{\partial z} \end{pmatrix} \begin{pmatrix} 0 & 0 & \tau_{xz} \\ 0 & 0 & \tau_{yz} \\ \tau_{zx} & \tau_{zy} & 0 \end{pmatrix} = 0 \qquad (12.10)$$

or

$$\frac{\partial \tau_{zx}}{\partial z} = 0 \qquad (12.11)$$

$$\frac{\partial \tau_{zy}}{\partial z} = 0 \qquad (12.12)$$

$$\frac{\partial \tau_{xz}}{\partial x} + \frac{\partial \tau_{yz}}{\partial y} = 0 \qquad (12.13)$$

The first two of the above imply that

$$\tau_{zx} = \tau_{xz} = \text{independent of } z \qquad (12.14)$$

$$\tau_{zy} = \tau_{yz} = \text{independent of } z \qquad (12.15)$$

Hooke's Law (see Eq. 4.60) is given by

$$\begin{pmatrix} e_x & \frac{1}{2}\gamma_{xy} & \frac{1}{2}\gamma_{xz} \\ \frac{1}{2}\gamma_{yx} & e_y & \frac{1}{2}\gamma_{yz} \\ \frac{1}{2}\gamma_{zx} & \frac{1}{2}\gamma_{zy} & e_z \end{pmatrix} = \frac{1+\nu}{E} \begin{pmatrix} \sigma_x & \tau_{xy} & \tau_{xz} \\ \tau_{yx} & \sigma_y & \tau_{yz} \\ \tau_{zx} & \tau_{zy} & \sigma_z \end{pmatrix}$$

$$- \frac{\nu}{E}(\sigma_x + \sigma_y + \sigma_z) \begin{pmatrix} 1 & 0 & 0 \\ 0 & 1 & 0 \\ 0 & 0 & 1 \end{pmatrix} \qquad (12.16)$$

and for our stress tensor this gives at once

$$\gamma_{zx} = \frac{2(1+\nu)}{E}\tau_{zx} = \frac{\tau_{zx}}{G} = \frac{\partial u}{\partial z} + \frac{\partial w}{\partial x} \tag{12.17}$$

$$\gamma_{zy} = \frac{2(1+\nu)}{E}\tau_{zy} = \frac{\tau_{zy}}{G} = \frac{\partial v}{\partial z} + \frac{\partial w}{\partial y} \tag{12.18}$$

$$e_x = \frac{\partial u}{\partial x} = 0 \tag{12.19}$$

$$e_y = \frac{\partial v}{\partial y} = 0 \tag{12.20}$$

$$e_z = \frac{\partial w}{\partial z} = 0 \tag{12.21}$$

$$\gamma_{xy} = \frac{\partial u}{\partial y} + \frac{\partial v}{\partial x} = 0 \tag{12.22}$$

From the third, fourth and fifth of the above we have

$$u = u(y, z) \tag{12.23}$$

$$v = v(x, z) \tag{12.24}$$

$$w = w(x, y) \tag{12.25}$$

Now, when a torque such as shown in Fig. 12.2, is applied to a cylindrical

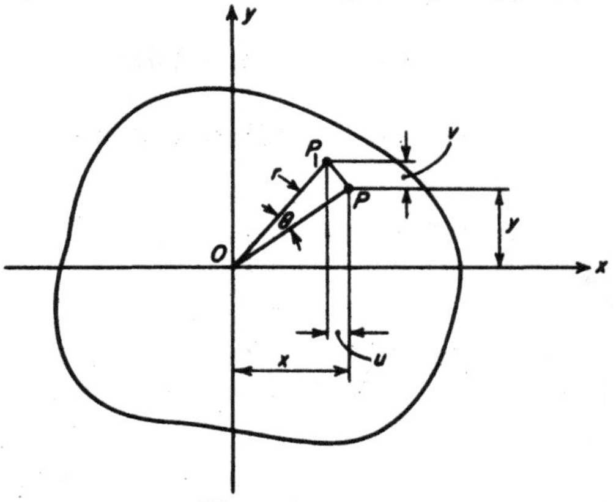

Fig. 12.4

bar, it is found that the cross section rotates about some point, such as O, Fig. 12.4, and the point P moves to P_1. Therefore, from the figure,

$$\overline{PP_1} = r\theta \tag{12.26}$$

and

$$\frac{u}{\overline{PP_1}} = -\frac{y}{r} \tag{12.27}$$

or

$$u = -\frac{\overline{PP_1}y}{r} = -y\theta \tag{12.28}$$

the negative sign being required since u is obviously in a negative x direction.
Similarly,

$$v = +x\theta \tag{12.29}$$

The bar twists through an angle, θ, which varies linearly from the end as

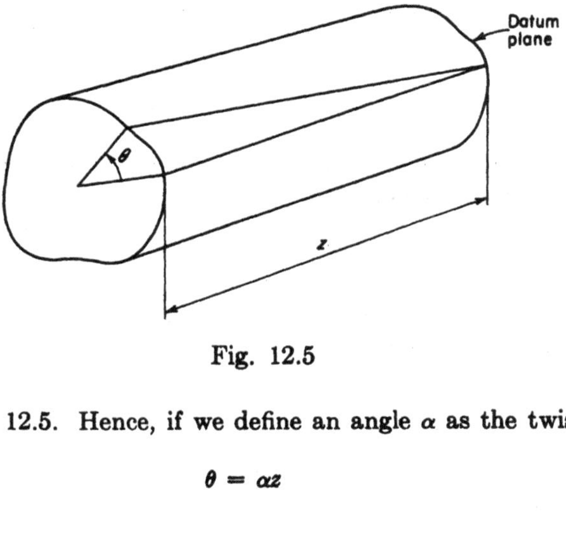

Fig. 12.5

shown in Fig. 12.5. Hence, if we define an angle α as the twist per unit length, then

$$\theta = \alpha z \tag{12.30}$$

and we have

$$u = -\alpha yz \tag{12.31}$$

$$v = \alpha xz \tag{12.32}$$

We must also consider the possibility of a deformation in the z-direction, a *warping*, which is given by w. Tests indicate that for non-circular cross sections, there is a warping of the face when a pure torque is applied. This warping is considered in the St. Venant theory. Indeed, it was the inclusion of this term, and the subsequent solution obtained therewith, that constituted St. Venant's great achievement, since none of the earlier torsion theories could properly account for this term.

Thus, we have

$$u = -\alpha yz \tag{12.33}$$

$$v = \alpha xz \tag{12.34}$$

$$w = w(x, y) \tag{12.35}$$

and we are in agreement with the requirements previously obtained — Eqs. 12.22, 12.23, 12.24, 12.25.

Substituting these values for u, v and w in Eqs. 12.17 and 12.18, we obtain

$$\gamma_{zx} = -\alpha y + \frac{\partial w}{\partial x} \tag{12.36}$$

$$\gamma_{zy} = \alpha x + \frac{\partial w}{\partial y} \tag{12.37}$$

and also

$$\tau_{zx} = G\gamma_{zx} = -G\alpha y + G\frac{\partial w}{\partial x} \tag{12.38}$$

$$\tau_{zy} = G\gamma_{zy} = G\alpha x + G\frac{\partial w}{\partial y} \tag{12.39}$$

It will simplify later expressions if we introduce a new function, the warping function, ϕ, defined by

$$\alpha\,\phi(x, y) = w(x, y) \tag{12.40}$$

Then

$$\tau_{zx} = G\alpha\left(-y + \frac{\partial \phi}{\partial x}\right) \tag{12.41}$$

$$\tau_{zy} = G\alpha\left(x + \frac{\partial \phi}{\partial y}\right) \tag{12.42}$$

Using the above and substituting into Eq. 12.13, we have, for everywhere in R,

$$\frac{\partial}{\partial x}\left[G\alpha\left(-y + \frac{\partial \phi}{\partial x}\right)\right] + \frac{\partial}{\partial y}\left[G\alpha\left(x + \frac{\partial \phi}{\partial y}\right)\right] = 0 \tag{12.43}$$

or

$$\nabla^2\phi = \frac{\partial^2\phi}{\partial x^2} + \frac{\partial^2\phi}{\partial y^2} = 0, \text{ in } R \tag{12.44}$$

Thus the warping function, ϕ, is *harmonic* (or satisfies the Laplace equation) everywhere in R, i.e., throughout the bar.

If the strains, as given by Eqs. 12.19–12.22, 12.36 and 12.37, are substituted in the strain compatibility equations, it will be found that these are identically satisfied, $0 = 0$. Hence the strains are compatible.

Turning now to the boundary conditions, we note that unit normals to the boundary, C, have components (l, m, zero) since the normal is always

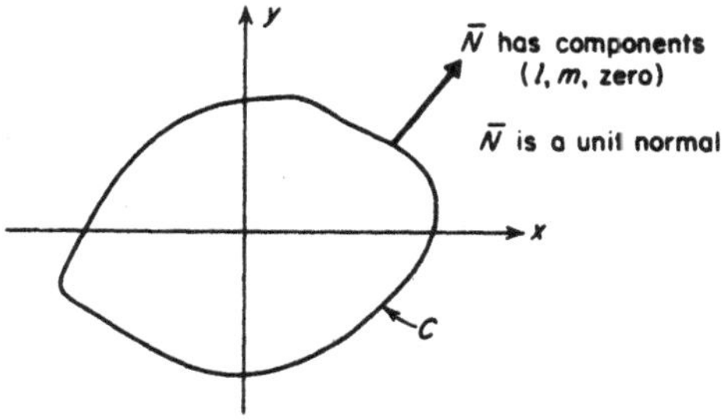

Fig. 12.6

perpendicular to the direction of the z-axis. Hence the boundary condition (see Eq. 4.42)†

$$(l \quad m \quad n)\begin{pmatrix} \sigma_x & \tau_{xy} & \tau_{xz} \\ \tau_{yx} & \sigma_y & \tau_{yz} \\ \tau_{zx} & \tau_{zy} & \sigma_z \end{pmatrix} = (\bar{x} \quad \bar{y} \quad \bar{z}) \tag{12.45}$$

becomes, since $\bar{x} = \bar{y} = \bar{z} = 0$, on C

$$\tau_{xz}l + \tau_{yz}m = 0 \tag{12.46}$$

or, using Eqs. 12.41 and 12.42,

$$\left(-y + \frac{\partial\phi}{\partial x}\right)l + \left(x + \frac{\partial\phi}{\partial y}\right)m = 0, \text{ on } C \tag{12.47}$$

Thus, we have shown that the equations of equilibrium, compatibility conditions, Hooke's Law and boundary conditions for the assumed stress tensor and deformation of Fig. 12.4, require that we find for each bar a function, ϕ, *the warping function*, such that

everywhere in R, $\nabla^2\phi = 0$;

everywhere on C, $\left(-y + \frac{\partial\phi}{\partial x}\right)l + \left(x + \frac{\partial\phi}{\partial y}\right)m = 0$;

and such that, given this ϕ, then

$$\tau_{zx} = G\alpha\left(-y + \frac{\partial\phi}{\partial x}\right) \tag{12.41}$$

$$\tau_{zy} = G\alpha\left(x + \frac{\partial\phi}{\partial y}\right) \tag{12.42}$$

†The boundary condition on the end faces considered as boundaries are also satisfied. The student should verify that on these faces the boundary condition requires that $x = \tau_{zx}$ and that $y = \tau_{zy}$, which are, of course, identically satisfied. See also the footnote on p. 89. A similar statement applies to this problem.

It may be shown (see, for example, Refs. 4, 8) that the above does represent a solution for a constant torque along the length of a cylindrical bar. Furthermore, if we introduce a new function, ψ, which is connected with ϕ by the relation

$$\phi + i\psi = f(z) \tag{12.48}$$

(see Ref. 4),† in which

$$z = x + iy \tag{12.49}$$

then the result can be simplified to the following:

The solution to the constant torque problem for a cylindrical bar requires finding a function, ψ, such that

$$\nabla^2\psi = 0, \text{ everywhere in } R \tag{12.50}$$

$$\psi - \tfrac{1}{2}(x^2 + y^2) = \text{a constant, say } d, \text{ everywhere on } C \tag{12.51}$$

Then,

$$\tau_{zx} = G\alpha\left(\frac{\partial\psi}{\partial y} - y\right) \tag{12.52}$$

$$\tau_{zy} = -G\alpha\left(\frac{\partial\psi}{\partial x} - x\right) \tag{12.53}$$

and

$$M_T = \alpha\left[G\int\int_R\left(x^2 + y^2 - x\frac{\partial\psi}{\partial x} - y\frac{\partial\psi}{\partial y}\right)dA\right] \tag{12.54}$$

Also, having ψ, the warping function, ϕ, may be determined from the fact that

$$\phi + i\psi = f(z), \qquad z = x + iy \tag{12.55}$$

The above results were given without proof, although many of the preliminary steps were shown. The interested student may check Refs. 4, 8, 9, and 10 for a more detailed discussion in which the proofs of all of the above are given in detail. For our purposes, we shall assume that the treatment given herein is sufficiently complete.

In the next section we shall obtain two exact solutions to the equations given above. They will, therefore, represent exact solutions to the torsion problem as formulated in the linearized mathematical theory of elasticity.

12-4 Two Exact Solutions to the Torsion Problem

We repeat here, for convenience, the requirements for an exact solution to the torsion problem as given in the last article.

†The statements of Eqs. 12.48 and 12.49 are the common functional relations of complex-variable theory. The elements of this theory, including the Cauchy-Riemann relations, are utilized in going from Eqs. 12.41 and 12.42 to Eqs. 12.50–12.55.

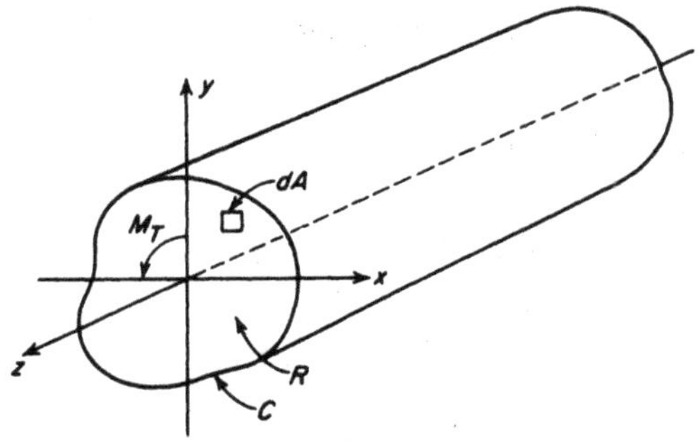

Fig. 12.7

Given any cylindrical bar, having a cross section of any shape, Fig. 12.7. The solution will be obtained if we can find a function, $\psi(x, y)$, such that

$$\nabla^2\psi = 0, \text{ everywhere in } R \tag{12.50}$$

$$\psi - \tfrac{1}{2}(x^2 + y^2) = d, \text{ everywhere on } C \tag{12.51}$$

With function $\psi(x, y)$ determined, then

$$\tau_{zx} = G\alpha\left(\frac{\partial\psi}{\partial y} - y\right) \tag{12.52}$$

$$\tau_{zy} = -G\alpha\left(\frac{\partial\psi}{\partial x} - x\right) \tag{12.53}$$

and

$$\alpha = \text{twist per unit length} = \frac{M_T}{G\iint_R\left(x^2 + y^2 - x\frac{\partial\psi}{\partial x} - y\frac{\partial\psi}{\partial y}\right)dA} \tag{12.56}$$

EXAMPLE 1 — THE CIRCULAR CROSS SECTION To indicate an application of this theory to a problem of practical interest, let us consider the bar with a circular cross section, Fig. 12.8, of radius r. The given applied torque is M_T. Determine the stress, deformation and warping for this bar.

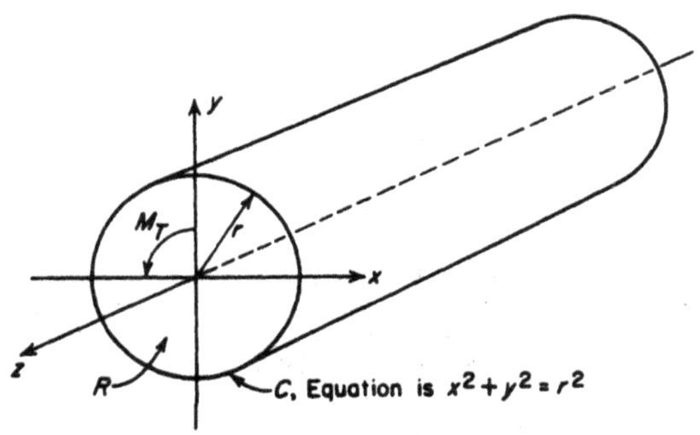

Fig. 12.8

SOLUTION Let us try for ψ, the constant quantity K.†

Then

$$\nabla^2\psi = \nabla^2 K = 0, \text{ as required, everywhere in } R$$

On C, we have

$$\psi - \tfrac{1}{2}(x^2 + y^2) = d \tag{12.57}$$

so that for $\psi = K$,

$$2(K - d) = x^2 + y^2 \tag{12.58}$$

Hence if

$$2(K - d) = r^2 \tag{12.59}$$

we satisfy the boundary condition on C. We choose K and d so that this is so. Then

$$\tau_{zx} = G\alpha\left(\frac{\partial\psi}{\partial y} - y\right) = -yG\alpha \tag{12.60}$$

$$\tau_{zy} = -G\alpha\left(\frac{\partial\psi}{\partial x} - x\right) = xG\alpha \tag{12.61}$$

and we see that stresses τ_{zx} and τ_{zy} correspond to a *resultant* stress, τ_{res} (see Fig. 12.9), such that

1. τ_{res} varies linearly with distance, ρ, from the center of the cross section, O;

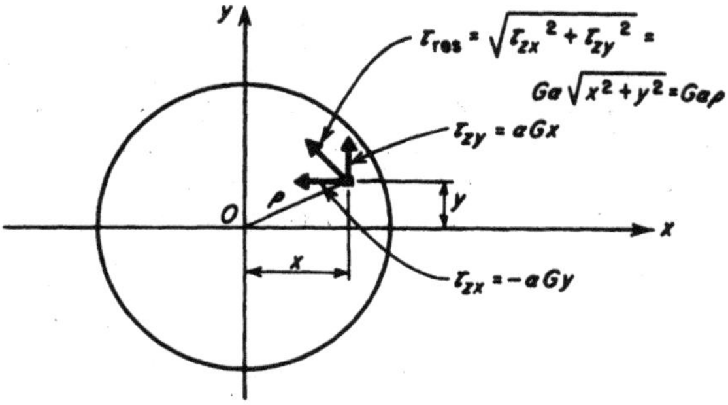

Fig. 12.9

†It may be shown that solutions to the torsion problem are unique (see Ref. 4), that is, a function which satisfies the field equation (Eq. 12.50) and the boundary condition (Eq. 12.51) is the only solution to the problem, *no matter how it was obtained*. Thus, we might justify this choice of ψ on the grounds that it was taken as a trial function and, upon substitution into the torsion problem equations, it was found to give a solution (and hence a *unique* solution) to the circular cross section bar. In reality, this is essentially how it was obtained. The same may be said for the elliptic cross section function of Eq. 12.70.

2. τ_{res} is perpendicular to a radial line from the center, O, to the point in question. Also, α, the twist per unit length, is given by

$$\alpha = \frac{M_T}{G\iint_R \left(x^2 + y^2 - x\frac{\partial\psi}{\partial x} - y\frac{\partial\psi}{\partial y}\right)dA} \tag{12.62}$$

$$= \frac{M_T}{G\iint_R (x^2 + y^2)dA} \tag{12.63}$$

$$= \frac{M_T}{GI_\rho} \tag{12.64}$$

where I_ρ is the polar moment of inertia of the cross section with respect to the origin, O (see Art. 3.2). Furthermore, since (see Fig. 12.9)

$$\tau_{res} = \alpha G\rho \tag{12.65}$$

it follows from the above that

$$\tau_{res} = \frac{M_T\rho}{I_\rho} \tag{12.66}$$

which permits us to find the resultant shear stress on a circular cross section due to an applied torque, M_T. Note that this value will be a maximum at the outside fibre, where ρ is a maximum.

Finally, since ψ is a constant, it follows that ϕ is a constant (see Eq. 12.55), and therefore the *warping* is zero, since this constant may be taken equal to zero (any other constant value for ϕ or w simply represents a rigid body movement of the entire bar and not a distortion of the cross section).

Summarizing the solution for the circular bar, we have M_T, the applied torque, which is given. Then

$$\tau_{res} = \frac{M_T\rho}{I_\rho} = \text{resultant shear stress at } \rho \tag{12.67}$$

$$\alpha = \frac{M_T}{GI_\rho} = \text{twist per unit length} \tag{12.68}$$

$$w = 0 = \text{warping} \tag{12.69}$$

Example 2 — The Elliptical Cross Section We now wish to determine the stresses, angle of twist, and warping for an elliptical cross section bar (Fig. 12.10) subjected to a constant torque, M_T.

Solution Without indicating how ψ is obtained (see Ref. 4), let us assume for ψ the function†

$$\psi = e(x^2 - y^2) \tag{12.70}$$

where e is some constant. Then certainly

$$\nabla^2\psi = \frac{\partial^2\psi}{\partial x^2} + \frac{\partial^2\psi}{\partial y^2} = 0, \text{ in } R \tag{12.71}$$

as required. The student should verify this.

†See footnote on page 238.

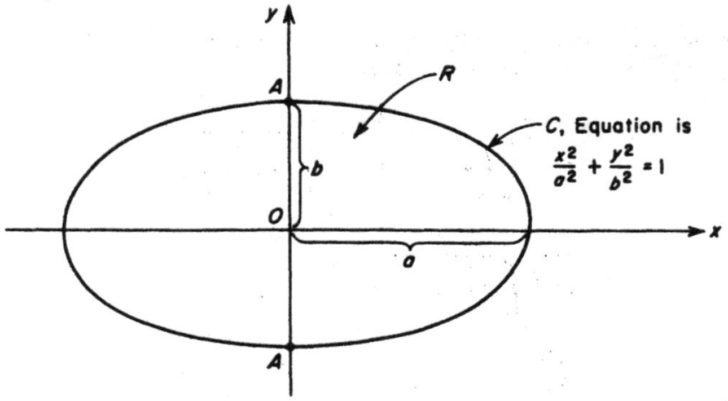

Fig. 12.10

Also, on C we require that

$$\psi - \tfrac{1}{2}(x^2 + y^2) = d \tag{12.72}$$

or that, on C, using the assumed value for ψ

$$e(x^2 - y^2) - \tfrac{1}{2}(x^2 + y^2) = d \tag{12.73}$$

This is equivalent to

$$\frac{x^2}{-2d/(1 - 2e)} + \frac{y^2}{-2d/(1 + 2e)} = 1 \tag{12.74}$$

and for this to be true, i.e., for this to be given by

$$\frac{x^2}{a^2} + \frac{y^2}{b^2} = 1 \tag{12.75}$$

requires that

$$\frac{-2d}{1 - 2e} = a^2 \tag{12.76}$$

and

$$\frac{-2d}{1 + 2e} = b^2 \tag{12.77}$$

which gives

$$e = \frac{a^2 - b^2}{2(a^2 + b^2)} \tag{12.78}$$

Hence, function ψ, which satisfies all requirements in R and on the boundary C is given by

$$\psi = \frac{a^2 - b^2}{2(a^2 + b^2)}(x^2 - y^2) \tag{12.79}$$

The stresses are then given by

$$\tau_{zx} = G\alpha\left(\frac{\partial\psi}{\partial y} - y\right) \tag{12.80}$$

$$= \frac{-2G\alpha a^2}{a^2 + b^2}y \tag{12.81}$$

$$\tau_{zy} = -G\alpha\left(\frac{\partial\psi}{\partial x} - x\right) \tag{12.82}$$

$$= \frac{2G\alpha b^2}{a^2 + b^2}x \tag{12.83}$$

and

$$\alpha = \frac{M_T}{G\iint_R\left(x^2 + y^2 - x\frac{\partial\psi}{\partial x} - y\frac{\partial\psi}{\partial y}\right)dA} \tag{12.84}$$

$$= \frac{a^2 + b^2}{\pi G a^3 b^3}M_T \tag{12.85}$$

Finally, the warping function, ϕ, is given by

$$\phi = -\frac{a^2 - b^2}{a^2 + b^2}xy \tag{12.86}$$

The above represents a complete solution to the problem of an elliptical bar subjected to a constant torque M_T. The student should note the following:

1. The maximum shear stress occurs on the boundary at the *minor* diameter at points A, Fig. 12.10. This is an unexpected result but one which is borne out by experiment.

2. This cross section *warps*. Indeed, the only cross section which will not warp when subjected to a pure torsion is the circular cross section solved earlier in this section.

12-5 Torsion Solutions for Other Cross-Sectional Shapes

It is possible to obtain exact solutions to the torsion problem for many cross sections of practical interest. However, there are many more cross sections for which exact solutions can not be obtained. For such problems we must resort to some approximate method. There are various approximate procedures and techniques which have been developed for solving these problems. Three such techniques are listed here.

1. The membrane analogy. In 1903, Ludwig Prandtl described his torsion-membrane analogy. He pointed out that the differential equation of the torsion problem is the same as the differential equation for small lateral displacements of the surface of a thin membrane subjected to lateral

pressures. By properly correlating the two sets of boundary conditions it is possible to draw an analogy between the torsion and the membrane problems. Soap films have been used in place of membranes, and many irregular shapes have been investigated for torsion in this manner. It is necessary that the membrane be stretched across a hole having a boundary the same shape as that of the bar under investigation and that it be distorted with a slight, lateral pressure. See Ref. 9.

2. There are various energy conservation theorems which can be utilized to give approximate solutions. The student is referred to Ref. 10 for a more detailed discussion of this point.

3. It is possible to solve the problem using finite difference methods and extensions of the techniques described in Chapter 2. See the next article. Further discussion of this will be found in Ref. 6. See also Prob. 8 at the end of this chapter.

The engineering forms of solutions for various conditions met with in practical problems are described in some detail in Ref. 25.

12-6 The Square Cross Section Solved by the Finite Difference Method

We will now indicate an application of finite difference methods in the solution of torsion problems. The problem to be solved will be the square

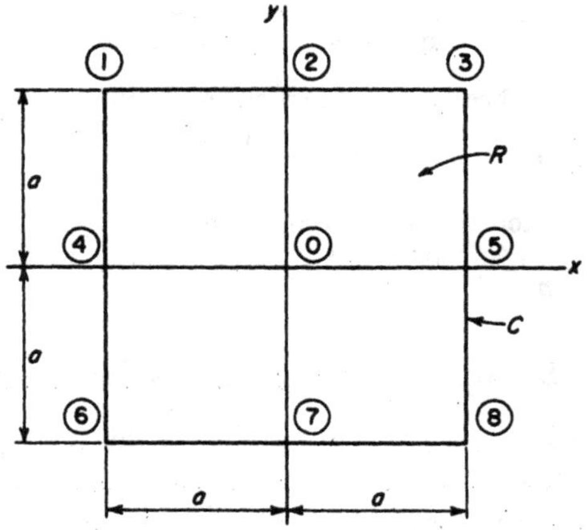

Fig. 12.11

cross section member subjected to a constant torque. The exact solution for this is known (see Ref. 9) and we shall compare our approximate solution to this known, exact solution.

The cross section is shown in Fig. 12.11. Also shown is the net used in the solution. Note particularly that the network chosen is the coarsest one possible. In spite of this, the solution will not be excessively inaccurate.

Referring to Eqs. 12.50 and 12.51, we see that the solution requires finding a function, ψ, such that

$$\nabla^2\psi = 0, \text{ everywhere in } R \tag{12.87}$$

$$\psi - \tfrac{1}{2}(x^2 + y^2) = \text{constant on } C. \tag{12.88}$$

Then

$$\tau_{zx} = G\alpha\left(\frac{\partial\psi}{\partial y} - y\right) \tag{12.89}$$

$$\tau_{zy} = -G\alpha\left(\frac{\partial\psi}{\partial x} - x\right) \tag{12.90}$$

and

$$\alpha = \frac{M_T}{G\iint_R\left(x^2 + y^2 - x\frac{\partial\psi}{\partial x} - y\frac{\partial\psi}{\partial y}\right)dA} \tag{12.91}$$

In finite different notation (see Eq. 2.98), since $\Delta x = \Delta y = a$, the field equation, Eq. 12.87, becomes

$$\frac{\psi_2 - 2\psi_0 + \psi_7}{a^2} + \frac{\psi_4 - 2\psi_0 + \psi_5}{a^2} = 0 \tag{12.92}$$

and on the boundary, taking the constant equal to zero (which may be done without affecting the stresses or other elastic behaviour),† we obtain

$$\left.\begin{aligned}
\psi_1 - \tfrac{1}{2}(a^2 + a^2) &= 0\\
\psi_2 - \tfrac{1}{2}(a^2) &= 0\\
\psi_3 - \tfrac{1}{2}(a^2 + a^2) &= 0\\
\psi_4 - \tfrac{1}{2}(a^2) &= 0\\
\psi_5 - \tfrac{1}{2}(a^2) &= 0\\
\psi_6 - \tfrac{1}{2}(a^2 + a^2) &= 0\\
\psi_7 - \tfrac{1}{2}(a^2) &= 0\\
\psi_8 - \tfrac{1}{2}(a^2 + a^2) &= 0
\end{aligned}\right\} \tag{12.93}$$

Therefore,

$$\psi_1 = \psi_3 = \psi_6 = \psi_8 = a^2 \tag{12.94}$$

and

$$\psi_2 = \psi_4 = \psi_5 = \psi_7 = \frac{a^2}{2} \tag{12.95}$$

†This is so because stresses, deformations and other quantities are given by *derivatives* of ψ. Hence adding a constant to ψ does not change the elastic behaviour.

Substituting these values in Eq. 12.92, we find

$$\frac{a^2/2 - 2\psi_0 + a^2/2}{a^2} + \frac{a^2/2 - 2\psi_0 + a^2/2}{a^2} = 0 \qquad (12.96)$$

or

$$\psi_0 = \frac{a^2}{2} \qquad (12.97)$$

Thus, for the network chosen in Fig. 12.11, we satisfy the field equation (Eq. 12.87) and the boundary conditions (Eq. 12.88) with ψ values as given in Eqs. 12.94, 12.95 and 12.97.

The maximum shear stress will occur at the outer boundary midpoint on each face. Thus, at point 5 we have

$$(\tau_{zy})_5 = -G\alpha\left(\frac{\partial\psi}{\partial x} - x\right) \qquad (12.98)$$

which, in terms of the ψ's just obtained, gives

$$(\tau_{zy})_5 = -G\alpha\left[\frac{\psi_0 - \psi_5}{-a} - a\right] \qquad (12.99)$$

$$= G\alpha a \qquad (12.100)$$

The exact solution for this stress gives (see Ref. 9)

$$(\tau_{zy})_5 = 1.35G\alpha a \qquad (12.101)$$

We shall next determine the value of the integral in Eq. 12.91 defining α, the twist per unit length.

This integral is given, in finite difference notation, by

$$\sum_R \left(x^2 + y^2 - x\frac{\Delta\psi}{\Delta x} - y\frac{\Delta\psi}{\Delta y}\right)\Delta A \qquad (12.102)$$

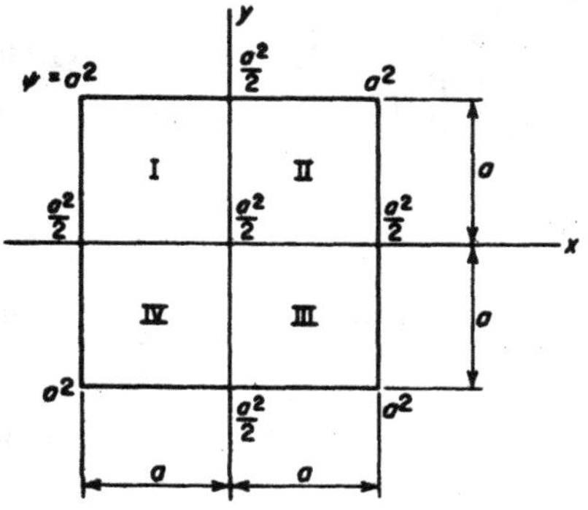

Fig. 12.12

Examining the ψ values at the net points as given in Eqs. 12.94, 12.95, 12.97, and as shown in Fig. 12.12, it is clear that the contributions due to

$$\left(x\frac{\Delta\psi}{\Delta x}\right)\Delta A \qquad (12.103)$$

and

$$\left(y\frac{\Delta\psi}{\Delta y}\right)\Delta A \qquad (12.104)$$

are equal for each of the areas, I, II, III, and IV.

Also, a typical value (for I, as an example) with $\Delta\psi$ averaged over the four corners is

$$x\frac{\Delta\psi}{\Delta x} = -\frac{a}{2}\left(-\frac{a^2 - a^2/2 + a^2/2 - a^2/2}{2}\Big/a\right) \tag{12.105}$$

$$= \frac{a^2}{8} \tag{12.106}$$

so that for the entire area the contribution is eight times this value, multiplied by a^2 (the area of each sub-area).

Similarly (see Eq. 3.13), the contribution due to the first two terms is given by

$$\iint_R (x^2 + y^2)\, dA = 2\frac{(2a)^4}{12} \tag{12.107}$$

$$= \frac{32a^4}{12} \tag{12.108}$$

Hence the value of the integral is given by

$$\sum_R \left(x^2 + y^2 - x\frac{\Delta\psi}{\Delta x} - y\frac{\Delta\psi}{\Delta y}\right)\Delta A = \frac{32a^4}{12} - a^4 \tag{12.109}$$

$$= 1.67a^4 \tag{12.110}$$

The exact value (see Ref. 9) is given by

$$\iint_R \left(x^2 + y^2 - x\frac{\partial\psi}{\partial x} - y\frac{\partial\psi}{\partial y}\right) dA = 2.25a^4 \tag{12.111}$$

12-7 Summary

It was pointed out in this chapter that by considering a single pair of shear stresses only in the stress tensor, the near-trivial shear block solution is obtained.

By considering *two* pairs of shear stresses we obtain the classical St. Venant solution to the pure torsion problem for a cylindrical bar.

We indicated in some detail the steps in this exact solution of the linearized equations of elasticity. Having this general solution, we applied the theory to two cross sections of practical interest: (a) the circular cross section and (b) the elliptical cross section. Complete solutions were obtained for these two cases. It was then noted that there are various approximate techniques available for solving the torsion problem and some of these were briefly described. In order to illustrate the application of finite difference techniques to the torsion problem, the square cross section was solved using these methods.

Problems

1. What is the maximum shear stress and total angle of twist for a steel shaft, 2 in. in diameter, 10 ft long, subjected to a torque of 1000 in.-lb? Use $G = 12{,}000{,}000$ psi.

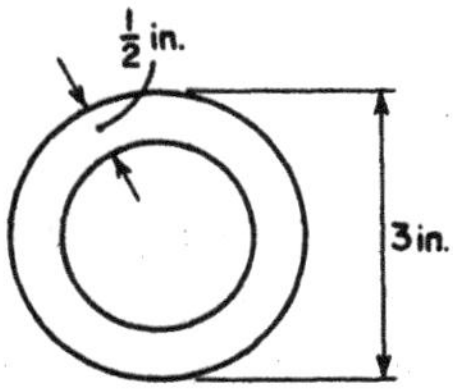

Prob. 2

2. Assuming that the circular cross-section relations, properly modified, apply also to the hollow tube cross section, determine the maximum shear stress and angle of twist for a shaft having the cross section shown, if it is 15 ft long, $G = 1{,}500{,}000$ psi, and applied torque is 10,000 in.-lb.

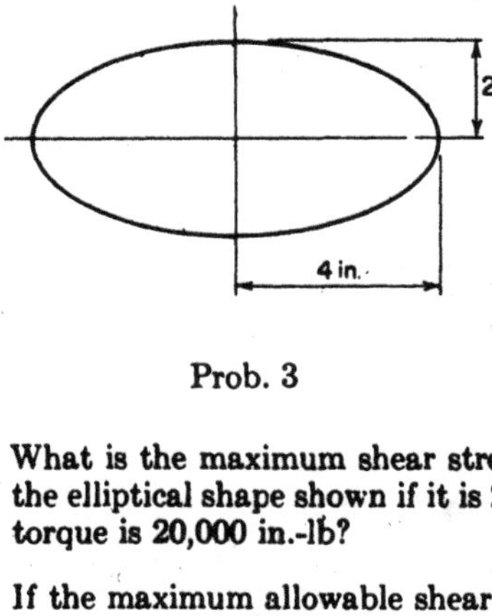

Prob. 3

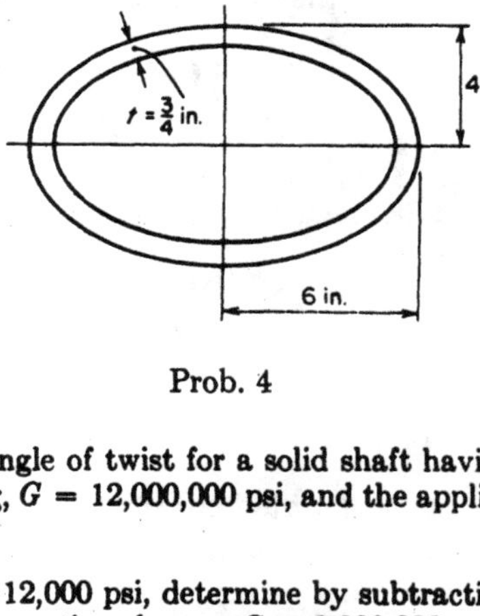

Prob. 4

3. What is the maximum shear stress and angle of twist for a solid shaft having the elliptical shape shown if it is 20 ft long, $G = 12{,}000{,}000$ psi, and the applied torque is 20,000 in.-lb?

4. If the maximum allowable shear stress is 12,000 psi, determine by subtraction the allowable applied torque for the cross section shown. $G = 2{,}000{,}000$ psi.

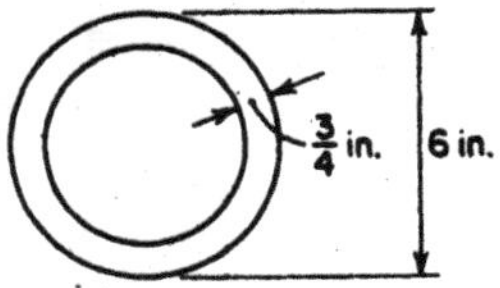

Prob. 5

5. If the allowable shear stress is 12,000 psi, the member is 20 ft long, and $G = 2{,}000{,}000$ psi, determine the permissible torque for the hollow circular bar shown.

6. Show that, if $a = b$, the elliptical torsion formula solutions become the same as the circular solution, with $r = a$.

7. Prove that the maximum shear stress for the elliptical cross section occurs on the boundary at the minor diameter.

8. (a) Using the finite difference method, determine the shear stresses at the mid-points of ail sides and also the twist per unit length for the rectangle shown.
   (b) Is the stress greater at point 1 or point 2? Would you predict this without knowing the correct answer?
   (c) Compare your result with the exact solution:

$$\frac{\tau_{max}}{G\alpha(2a)} = 0.930$$

$$\frac{M_T}{G\alpha(2a)^3(2b)} = 0.229$$

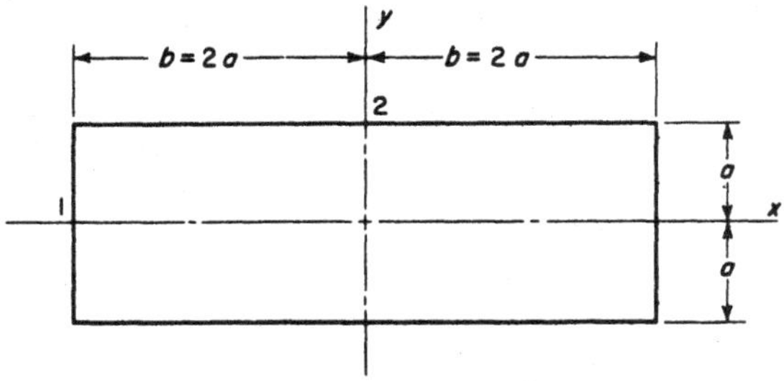

9. Show that if, in the torsion problem, we introduce a new variable, Ψ, defined by

$$\Psi = \psi - 1/2(x^2 + y^2)$$

then the solution is given by

$$\nabla^2\Psi = -2 \text{ in } R$$

$$\Psi = 0 \text{ on } C$$

and

$$\tau_{zx} = G\alpha\frac{\partial\Psi}{\partial y}$$

$$\tau_{zy} = -G\alpha\frac{\partial\Psi}{\partial x}$$

$$\alpha = \frac{M_T}{-G\iint_R\left(x\frac{\partial\Psi}{\partial x} + y\frac{\partial\Psi}{\partial y}\right)dx\,dy} = \frac{M_T}{2G\iint_R \Psi\,dx\,dy}$$

Hint: The last expression for α is obtained by using Green's relation between area and boundary integrals and the boundary condition on Ψ.

10. Solve the square cross section of Art. 12-6 using the finite difference formulation applied to the solution of Prob. 9 above.

Chapter 13

ENGINEERING ELASTICITY ANALYSIS OF SPECIAL STRUCTURES

13-1 Introduction

In this chapter we shall discuss extensions of the various engineering elasticity applications given in the previous chapters. In order to treat these special structures most systematically we shall present a classification system which will emphasize the similarities and differences of the units being discussed.

We shall see that a natural system of classification, based upon increasing order of difficulty, is one which groups structures into pairs as follows:

1. A uni-stress, or single-stress, plane structure (a cord or cable) and
2. A uni-stress, or single-stress, space structure (a membrane).

Another pair of structures naturally classified together is

3. A shear and bending stress straight linear structure (a beam or a rigid frame) and
4. A shear and bending stress straight plane structure (a plate).

A final pair of structures which may be classified together is

5. A shear and bending stress curved plane structure (a curved beam, or arch) and
6. A shear and bending stress curved space structure (a shell).

It is beyond the scope of a textbook such as this to discuss all six of the above structures in detail. We have, of course, treated the beam at great length. We shall also discuss briefly in this chapter the rigid frame — essentially a structure composed of beam elements, and hence one which can, at this point, be described quite readily. We will also discuss at some length the cable and its space counterpart, the membrane. A much briefer treatment will be given of the arch and its space counterpart, the shell.

The solutions of the special paired structures which we shall discuss in this chapter represent essentially engineering-type extensions of the various engineering elasticity solutions given in the preceding chapters of the text.

We shall emphasize what may be called the *generalization technique*, a procedure by means of which a one-dimensional solution may be extended to a two-dimensional solution, or a two-dimensional solution extended to a three-dimensional solution. However, we caution that the generalization technique must be used with care, since very often an increase in the "dimension" of a physical problem introduces effects which are not considered in the lower dimensional form of the solution. Where no new effect of this type is introduced, the generalization technique may be used to give approximate forms of the corresponding solutions. Some examples of this other than those given in this text will be found in Ref. 4.

13-2 The Cable and the Membrane

The prime characteristic of the cable and the membrane is that each of these is a structure which can support only tensile stresses. They can not support compression stresses and they can not support bending stresses. This then is the determining feature in any study of these structures, such as the one we now present.

13-3 The Cable

A cable can, in general, be subject to the following three types of loading: (a) a concentrated load, (b) a uniform load, and (c) its own weight (see Fig. 13.1, a, b and c). We shall analyze the cable for all three types of loadings.

(a) *Cable Supporting a Concentrated Load*

Since the cable can only support tensile stresses, it follows that equilibrium (see Fig. 13.2) requires that

$$\Sigma F_V = 0 \qquad (13.1)$$

or

$$T_{1V} + T_{2V} - P = 0 \qquad (13.2)$$

and

$$\Sigma F_H = 0 \qquad (13.3)$$

or

$$T_{1H} = T_{2H} \qquad (13.4)$$

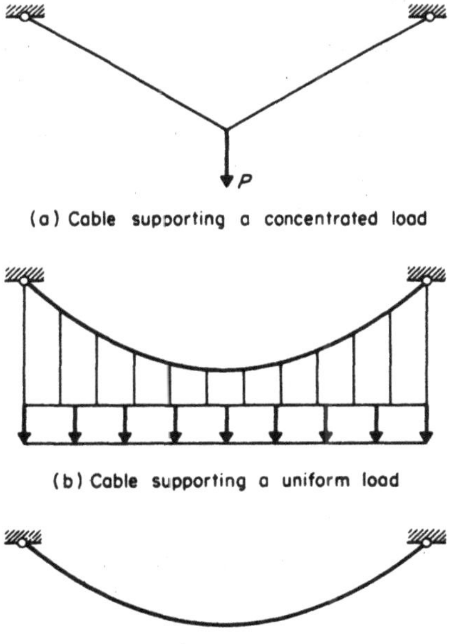

(a) Cable supporting a concentrated load

(b) Cable supporting a uniform load

(c) Cable supporting its own weight

Fig. 13.1

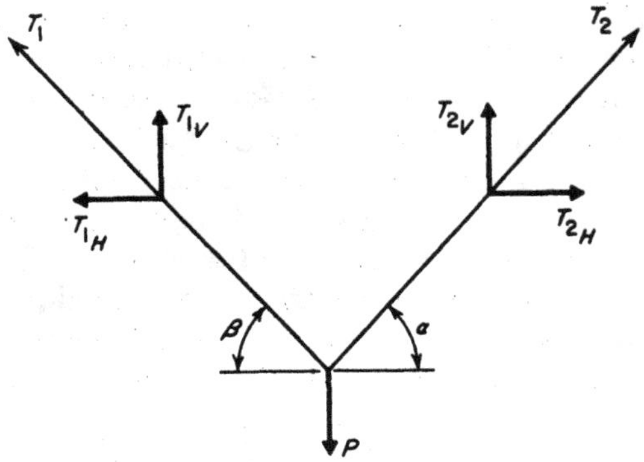

Fig. 13.2

Hence both portions of the cable are straight; and, since

$$\left.\begin{aligned} T_{1V} &= T_1 \sin \beta \\ T_{1H} &= T_1 \cos \beta \end{aligned}\right\} \tag{13.5}$$

and

$$\left.\begin{aligned} T_{2V} &= T_2 \sin \alpha \\ T_{2H} &= T_2 \cos \alpha \end{aligned}\right\} \tag{13.6}$$

we can determine α, β, T_1 and T_2 without difficulty. Note the lengths of each part of the cable are also known, that is, α and β are known.

(b) *Cable Supporting a Uniform Load*

Note, in Fig. 13.3, that the cable is suspended from points A and B. The points of attachment of the uniform load to the cable may be assumed to be

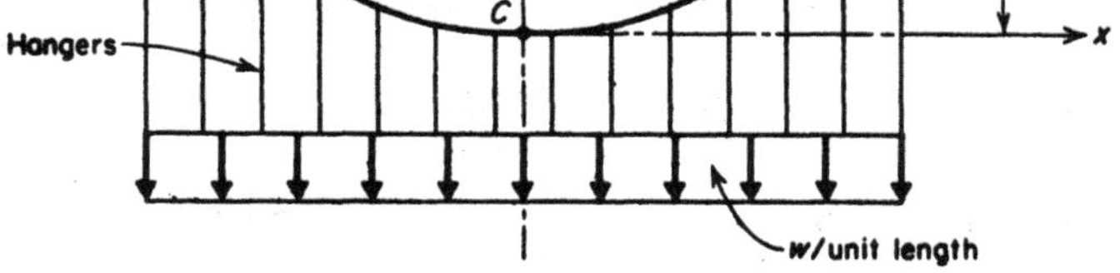

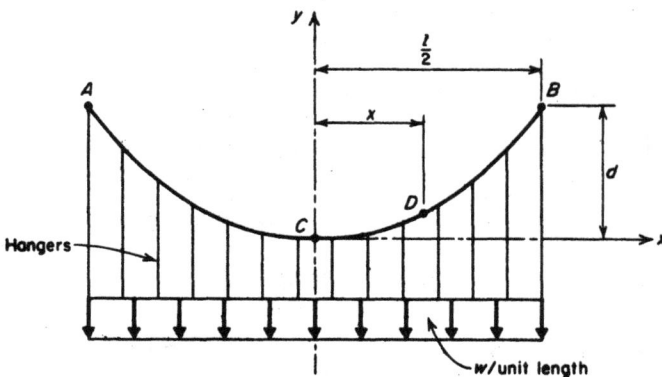

Fig. 13.3

so close together that the cable is loaded with a uniform load per *horizontal* distance.

The resultant tension at any point is in the direction of the cable at the point in question. Consider a portion of the cable CD, Fig. 13.4, where C is

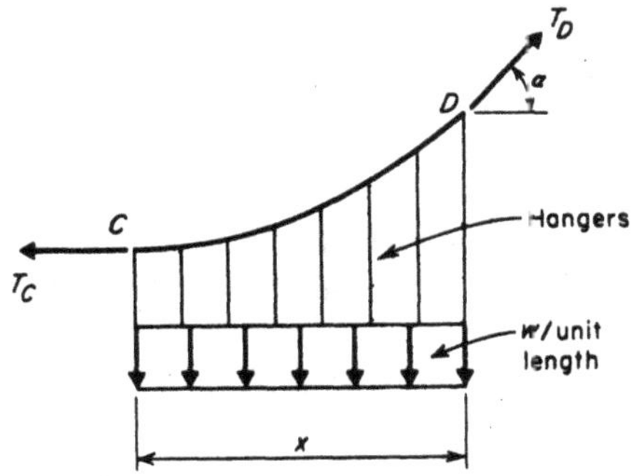

Fig. 13.4

the midpoint (i.e., the lowest point) of the cable and D is any other point, a distance x from C. Then, once again applying the engineering form of the equilibrium equations, we have

$$\Sigma F_V = 0 \tag{13.7}$$

and

$$\Sigma F_H = 0 \tag{13.8}$$

The first of these, $\Sigma F_V = 0$ gives

$$(w)(x) = T_D \sin \alpha \tag{13.9}$$

and the second, $\Sigma F_H = 0$, gives

$$T_C = T_D \cos \alpha \tag{13.10}$$

Dividing Eq. 13.9 by Eq. 13.10 gives

$$\tan \alpha = \frac{wx}{T_C} \tag{13.11}$$

and, since

$$\tan \alpha = \frac{dy}{dx} \tag{13.12}$$

we have

$$\frac{dy}{dx} = \frac{wx}{T_C} \tag{13.13}$$

This may be integrated at once to give

$$y = \frac{wx^2}{2T_C} + \text{constant} \tag{13.14}$$

and, since $y = 0$ when $x = 0$, the constant $= 0$, so

$$y = \frac{wx^2}{2T_C} \tag{13.15}$$

This is the equation of the curve assumed by the cable. Note that it is a parabola. Note also that, if ρ is the radius of curvature at any point, then, approximately,

$$\frac{d^2y}{dx^2} = \frac{w}{T_C} = \frac{1}{\rho} \tag{13.16}$$

or

$$w = \frac{T_C}{\rho} \tag{13.17}$$

We shall point out later the similarity between this equation and the corresponding relation for the space counterpart of the cable, the membrane.

Going back to the cable (Fig. 13.3), we have at once,

$$d = \frac{wl^2}{8T_C} \tag{13.18}$$

where d is the sag of the cable and l is its horizontal projection.

The length, S, of the cable is given by

$$S = \int_A^B ds \tag{13.19}$$

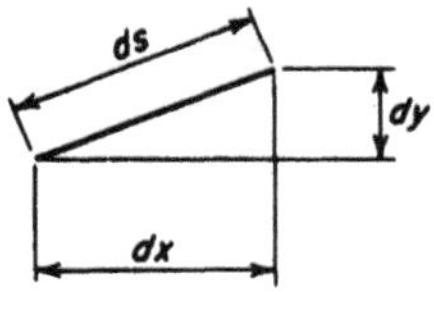

Fig. 13.5

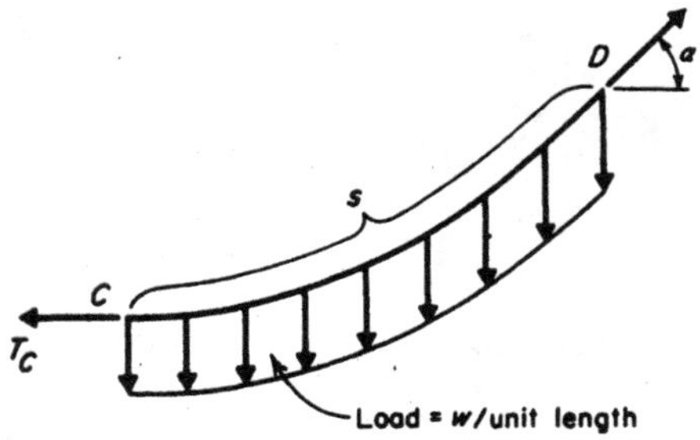

Fig. 13.6

where ds is a differential arc measured along the cable (see Fig. 13.5). Note that

$$ds^2 = dx^2 + dy^2 \tag{13.20}$$

or

$$ds = \sqrt{1 + (dy/dx)^2}\, dx \tag{13.21}$$

and therefore

$$CB = \int_0^{l/2} \sqrt{1 + (dy/dx)^2}\, dx \tag{13.22}$$

Now (see Eq. 13.13)

$$\frac{dy}{dx} = \frac{wx}{T_C} \tag{13.23}$$

or

$$CB = \int_0^{l/2} \sqrt{1 + w^2x^2/T_C^2}\, dx \tag{13.24}$$

which integrates (see Peirce's Tables, Eq. 124) to

$$CB = \frac{wl}{4T_C}\sqrt{T_C^2/w^2 + l^2/4} + \frac{T_C}{2w}\log_e\left(\frac{\frac{l}{2} + \sqrt{T_C^2/w^2 + l^2/4}}{T_C/w}\right) \tag{13.25}$$

Expanding the terms in the parentheses and under the radical, we obtain the following equations (which the student should verify):

$$S = l + \frac{w^2l^3}{24T_C^2} - \frac{w^4l^4}{768T_C^4} + \ldots. \tag{13.26}$$

or

$$S = l + \frac{8d^2}{3l} - \frac{32d^4}{5l^3} + \ldots \tag{13.27}$$

To obtain the maximum tension in the cable (from Eq. 13.10),

$$T_C = T_D \cos\alpha \tag{13.28}$$

or

$$T_D = \frac{T_C}{\cos\alpha} \tag{13.29}$$

so that the smaller $\cos\alpha$ (i.e., larger α) the larger is T_D. This means the maximum tension occurs at the supports.

The problem just considered is essentially that of the suspended bridge in which we neglect the weight of the suspending cable.

(c) *Cable Supporting Its Own Weight*

If the cable is supporting its own weight, then the loading is no longer a uniform one with horizontal distance (see Fig. 13.6). In this case we again have, for the engineering form of the equilibrium equations,

$$\Sigma F_V = 0 \tag{13.30}$$

and

$$\Sigma F_H = 0 \tag{13.31}$$

which leads to the following two expressions:

$$T_C = T_D \cos \alpha \tag{13.32}$$

and

$$ws \doteq T_D \sin \alpha \tag{13.33}$$

Dividing once more, we obtain

$$\tan \alpha = \frac{ws}{T_C} \tag{13.34}$$

or

$$\frac{dy}{dx} = \frac{ws}{T_C} \tag{13.35}$$

Again (see Fig. 13.5), we have

$$ds^2 = dx^2 + dy^2 \tag{13.36}$$

and since

$$dy = \frac{ws\,dx}{T_C} \tag{13.37}$$

The expression for ds becomes

$$dx = \frac{ds}{\sqrt{1 + w^2s^2/T_C^2}} = \frac{T_C}{w} \frac{ds}{\sqrt{T_C^2/w^2 + s^2}} \tag{13.38}$$

This may be integrated at once (see Peirce's Tables, Eq. 126a) between the limits O-S on the right hand side, to give

$$x = \frac{T_C}{w} \log_e \left\{ \frac{S + \sqrt{T_C^2/w^2 + S^2}}{T_C/w} \right\} \tag{13.39}$$

In a similar manner, if we eliminate dx from the equations,

$$\frac{dy}{dx} = \frac{ws}{T_C} \tag{13.40}$$

and

$$ds^2 = dx^2 + dy^2 \tag{13.41}$$

we would obtain

$$y = \sqrt{T_C^2/w^2 + S^2} - \frac{T_C}{w} \tag{13.42}$$

Now, if S is eliminated from the above equations for x and y, we get

$$y + \frac{T_C}{w} = \frac{T_C}{2w} \left(e^{wx/T_C} + e^{-wx/T_C} \right) \tag{13.43}$$

which is an equation of the *catenary* with the origin at the vertex and with the y-axis as the axis of symmetry. Once again we see that the maximum tension occurs at the support.

If the length of the horizontal span is l, the maximum deflection is obtained from Eq. 13.43, by substituting $x = l/2$. The half-length of the cable can be found from Eq. 13.39 by substituting $x = l/2$ and solving for S.

The student should note that Eq. 13.39 can also be written in the form

$$S = \frac{T_C}{2w}\left(e^{wx/T_C} - e^{-wx/T_C}\right) \tag{13.44}$$

and this will generally simplify the computation for S. Similarly, the hyperbolic identities lead to the following forms of Eqs. 13.43 and 13.44:

$$y + \frac{T_C}{w} = \frac{T_C}{w}\cosh\frac{wx}{T_C} \tag{13.45}$$

and

$$S = \frac{T_C}{w}\sinh\frac{wx}{T_C} \tag{13.46}$$

and these expressions are generally simpler to use in numerical computations.

13-4 The Membrane

The membrane, as we noted, is the space counterpart of the plane cable It also, just as is the case for the cable, cannot support compression or bending loads. Equilibrium is maintained solely by means of tensile stresses introduced in the membrane, acting at each point in the direction of the slope of the membrane at the point in question.

Whereas the cable may be used to support concentrated loads, (Fig. 13.1), the membrane is rarely called upon to perform in this manner. No simple theory is available for determining the stresses in the membrane for this loading. As a first approximation which may be suitable for engineering purposes, it is suggested that a form of *strip theory* be utilized (see Fig. 13.7), in which various strips of the membrane are assumed to act as

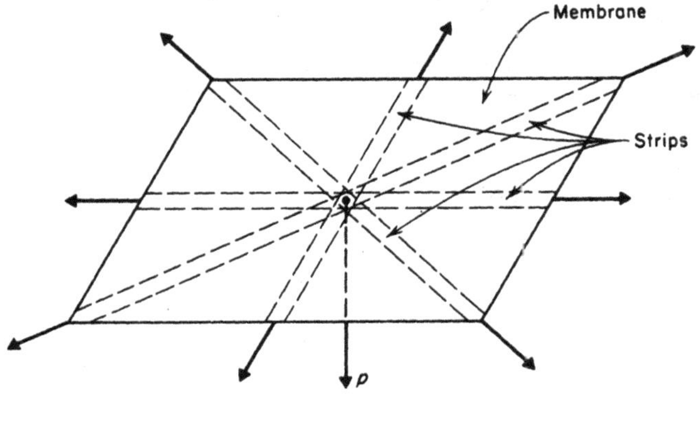

Fig. 13.7

separate cables supporting the load — with the load divided up among all the cables (i.e., strips) which are assumed to be supporting it.

The more usual loading on a membrane is a pressure loading which is normal, at all points, to the membrane. This loading may vary over the membrane. Under this loading, a differential free-body piece of the membrane will be in equilibrium under the edge tensile loads per unit length, t_1 and t_2, as shown in Fig. 13.8. In this figure

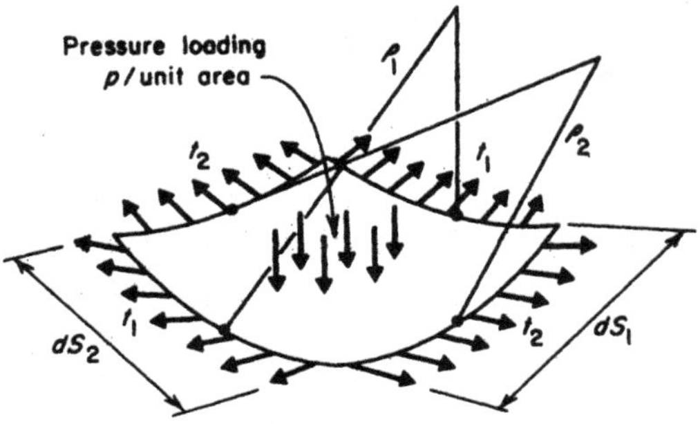

Fig. 13.8

dS_1 and dS_2 are the differential sides
ρ_1 and ρ_2 are the corresponding radii of curvature
t_1 and t_2 are the loads per unit length normal to dS_2 and dS_1, respectively and, if h = thickness of membrane, then $t_1 = h\sigma_1$ and $t_2 = h\sigma_2$, where σ_1 and σ_2 are the corresponding *stresses* in psi.

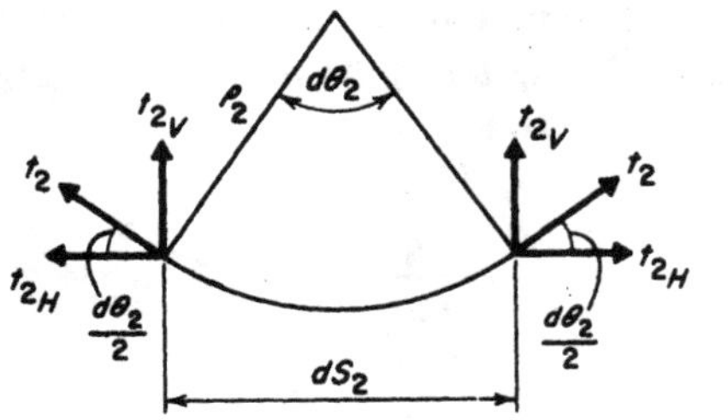

Fig. 13.9

Let us consider conditions in planes parallel to dS_2. These are shown in Fig. 13.9. Note, in this figure, that the net vertical force is given by

$$2t_{2V}dS_1 = 2t_2\frac{d\theta_2}{2}dS_1 \tag{13.47}$$

and, since

$$\rho_2 d\theta_2 = dS_2 \tag{13.48}$$

we have

$$2t_{2V}dS_1 = \frac{t_2}{\rho_2}dS_1dS_2 \tag{13.49}$$

In exactly the same way, for the two stresses normal to the other two sides, we would find

$$2t_{1V}dS_2 = \frac{t_1}{\rho_1}dS_1dS_2 \tag{13.50}$$

If now, for the differential element of Fig. 13.9, we take the engineering equation of equilibrium,

$$\sum F_V = 0 \tag{13.51}$$

we have

$$p dS_1 dS_2 = \frac{t_1}{\rho} dS_1 dS_2 + \frac{t_2}{\rho_2} dS_1 dS_2 \tag{13.52}$$

or, finally,

$$p = \frac{t_1}{\rho_1} + \frac{t_2}{\rho_2} \tag{13.53}$$

The student should compare this with the corresponding relation for the plane structure, the counterpart of the membrane, as given in Eq. 13.17. Note how the three-dimensional form of the equation may be thought of as a generalization of the simpler two-dimensional form, when the latter is applied at a point where the load is normal to the structure, as was assumed for the membrane solution. This is an example of the "generalization technique" described in the introduction to this chapter.

13-5 The Membrane Solution for the Cylindrical Container

As a first application of the membrane formula, Eq. 13.53, we consider a cylindrical container of circular cross section subjected to a uniform internal pressure, Fig. 13.10. Let h equal thickness of shell, hence $t_1 = \sigma_1 h$,

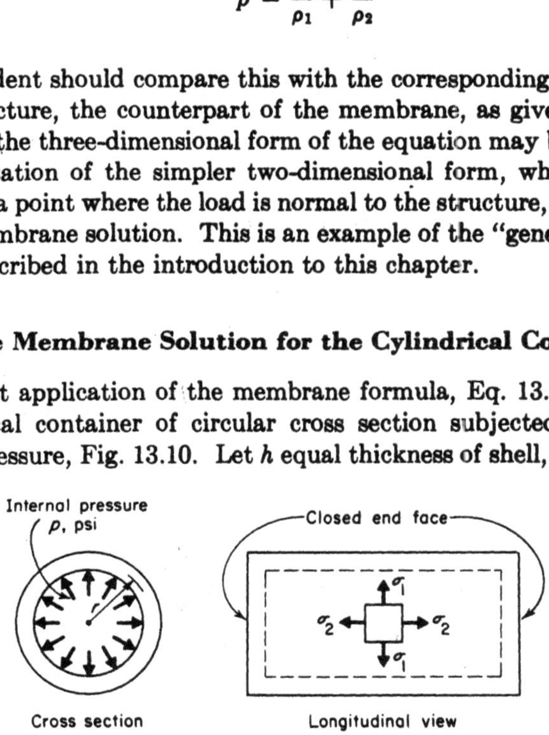

Fig. 13.10

$t_2 = \sigma_2 h$, where σ_1 and σ_2 are the *stresses*. The equation governing the stresses (Eq. 13.53) is

$$p = \frac{h\sigma_1}{\rho_1} + \frac{h\sigma_2}{\rho_2} \tag{13.54}$$

Note that for this case

$$\left.\begin{aligned} \rho_1 &= r \\ \rho_2 &= \infty \end{aligned}\right\} \tag{13.55}$$

Hence we have

$$p = \frac{h\sigma_1}{r} + 0 \tag{13.56}$$

or

$$\sigma_1 = \frac{pr}{h} \tag{13.57}$$

To determine σ_2 we consider a free body of the closed end face plus a small portion of the tank, Fig. 13.11. Then balance of forces in the x direction gives

$$\Sigma F_x = 0 \tag{13.58}$$

$$\sigma_2(2\pi r)h = p(\pi r^2) \tag{13.59}$$

or

$$\sigma_2 = \frac{pr}{2h} \tag{13.60}$$

which is one-half the value of the transverse membrane stress.

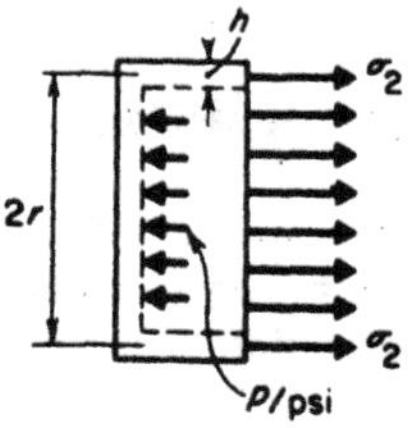

Fig. 13.11

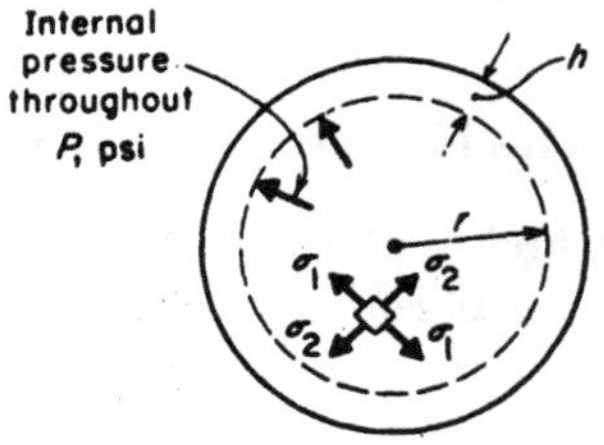

Fig. 13.12

13-6 The Membrane Solution for the Spherical Container

If the internal pressure is uniform, of value p psi (Fig. 13.12), it follows from symmetry that

$$\sigma_1 = \sigma_2 = \sigma \tag{13.61}$$

and

$$r_1 = r_2 = r \tag{13.62}$$

Therefore, the governing equation

$$p = \frac{t_1}{\rho_1} + \frac{t_2}{\rho_2} \tag{13.63}$$

becomes

$$p = \frac{h\sigma}{r} + \frac{h\sigma}{r} \tag{13.64}$$

or

$$\sigma = \sigma_1 = \sigma_2 = \frac{pr}{2h} \tag{13.65}$$

The two examples given in Arts. 13-5 and 13-6 are typical of membrane stress solutions for simple structures. A more detailed discussion will be found in Ref. 20.

13-7 The Rigid Frame and Its Relation to the Continuous Beam

A typical rigid frame and loading is shown in Fig. 13.13. The rigid frame *ABCD* is made up of beam elements *AB*, *BC* and *CD*, rigidly connected at

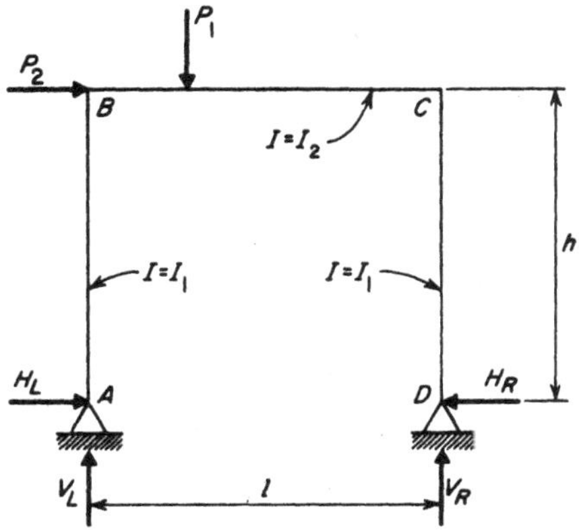

Fig. 13.13

their intersections, *B* and *C*. We may explain more clearly what is meant by "rigidly connected" if we compare a typical rigid frame joint, say *B*, to a typical continuous beam condition at a support, Fig. 13.14.

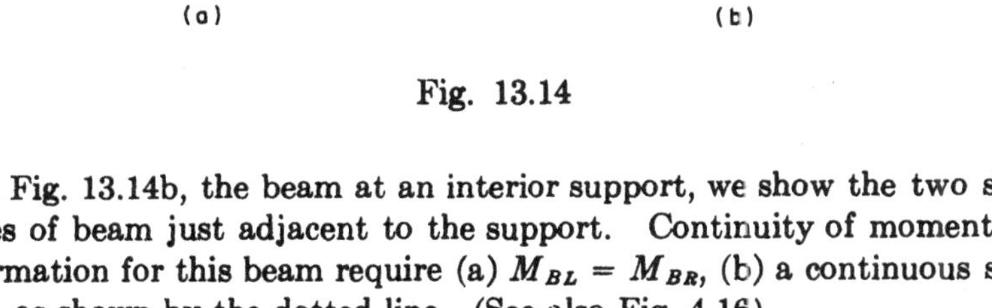

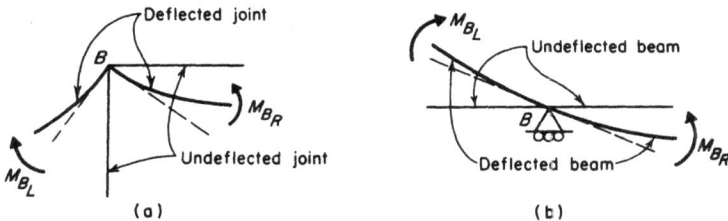

Fig. 13.14

In Fig. 13.14b, the beam at an interior support, we show the two short pieces of beam just adjacent to the support. Continuity of moment and deformation for this beam require (a) $M_{BL} = M_{BR}$, (b) a continuous slope at *B*, as shown by the dotted line. (See also Fig. 4.16).

In exactly the same way, for the rigid frame joint of Fig. 13.14a, continuity of moment and deformation require (a) $M_{BL} = M_{BR}$, (b) the original angle at B, in this case a right angle, must be maintained by the deformed structure, as shown by the pair of dotted tangent lines.

It is clear, therefore, that a rigid frame is very similar to a continuous beam. Thus, just as in Figs. 10.1 and 10.2 we defined two basic statically

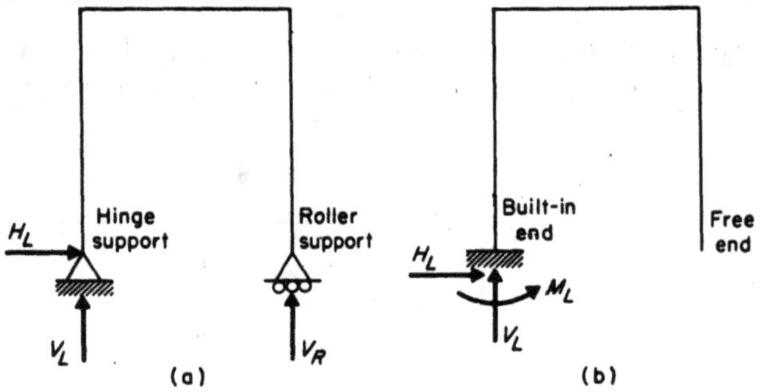

Fig. 13.15

determinate beams (each of which has three unknown reactions), so also we may define two basic statically determinate frames (each of which has three unknown reactions), as shown in Fig. 13.15a and b.†

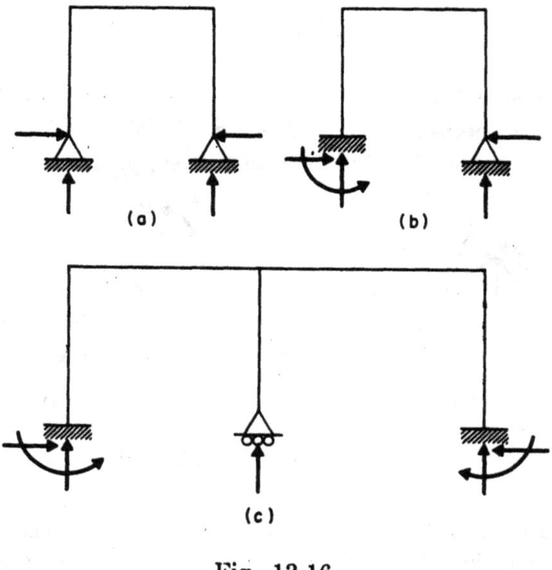

Fig. 13.16

†As in the case of a beam, it is possible in rigid frames to introduce artificial conditions within the structure (such as pins or hinges) which essentially give additional independent equations that permit us to solve for more than three reactions.

Also, just as for beams, we may determine the degree of redundancy of a rigid frame by counting the number of reactions which must be removed to bring the frame to the condition of one or the other of these two basic statically determinate frames.

As examples, consider the frames of Fig. 13.16. In Fig. 13.16a we have one extra reaction, hence one redundant. In Fig. 13.16b there are two redundants, and in Fig. 13.16c there are four redundants.

Because the rigid frame is basically an assembly of beam elements, rigid frames may be analyzed using all of the methods developed for beams. We shall illustrate a typical solution of a rigid frame by means of the conjugate beam method.

13-8 A Rigid Frame Solution Using the Conjugate Beam Method

EXAMPLE Given the rigid frame and loading shown in Fig. 13.17, determine the reactions.

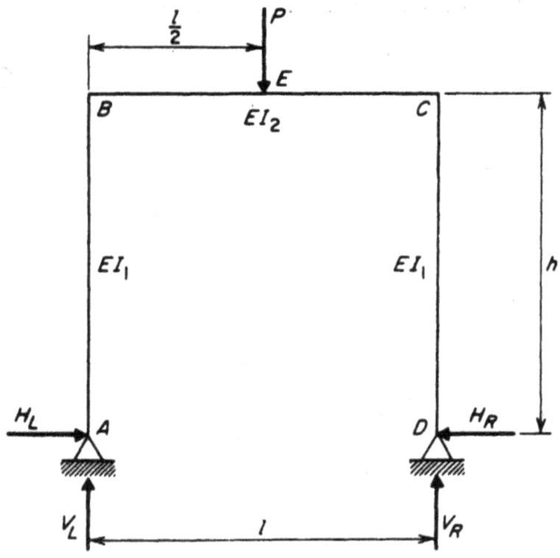

Fig. 13.17

SOLUTION We note first that there is one redundant reaction. Let H_L be this redundant.

Due to symmetry, we need consider only half the structure — say portion ABE. Also due to symmetry, we can say that

$$V_L = V_R = \frac{P}{2} \tag{13.66}$$

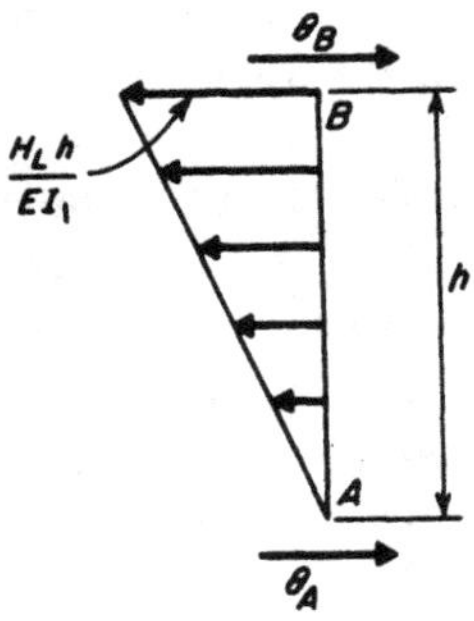

Fig. 13.18

In applying the conjugate beam method to this structure we shall use the sign convention for this procedure as outlined in Art. 9.4. Also, we shall imagine ourselves as stationed *inside* the frame looking outward, in determining the signs of moments. This being so, we have for member AB, Fig. 13.18, an M/EI diagram which is zero at A, varies linearly to its maximum value at B, and acts in the direction shown on the figure. Also shown on the figure are the slopes (i.e., conjugate beam reactions) θ_A and θ_B. Similarly, for portion BE of member BC, we have an M/EI diagram, by parts, as shown in Fig. 13.19. Also shown on this diagram is the slope at the left end, θ_B, and the moment δ_E at the point E (representing the deflection at this point on the actual beam).

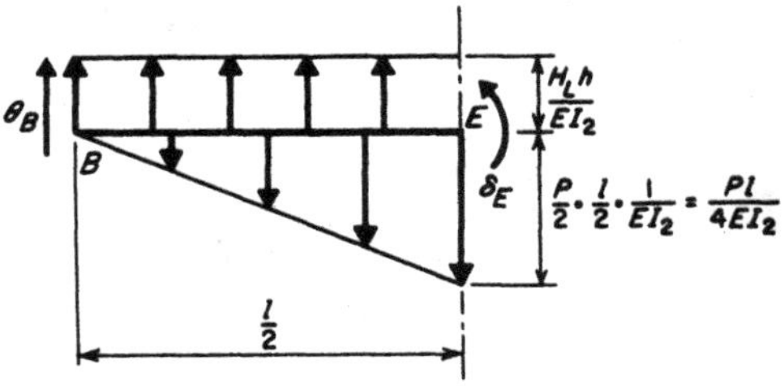

Fig. 13.19

Now, as pointed out in connection with Fig. 13.14, rigid frame action requires that angles at joints be maintained. This implies that θ_B of member AB is equal to θ_B of member BC — in fact, this is the strain compatibility requirement for the form of solution being considered herein.

We obtain θ_B of member AB by taking moments about point A of Fig. 13.18, that is,

$$\theta_B h = \frac{H_L h}{EI_1} \cdot \frac{h}{2} \cdot \frac{2h}{3} \tag{13.67}$$

or

$$\theta_B = \frac{H_L h^2}{3EI_1} \tag{13.68}$$

We get θ_B for member BC by noting that, due to symmetry, the summation of vertical forces on BE of Fig. 13.19 must be zero, that is,

$$\theta_B - \frac{Pl}{4EI_2} \cdot \frac{l}{4} + \frac{H_L h}{EI_2} \cdot \frac{l}{2} = 0 \tag{13.69}$$

or

$$\theta_B = -\frac{H_L hl}{2EI_2} + \frac{Pl^2}{16EI_2} \tag{13.70}$$

Now, equating these two values of θ_B, we obtain

$$\frac{H_L h^2}{3EI_1} = -\frac{H_L hl}{2EI_2} + \frac{Pl^2}{16EI_2} \tag{13.71}$$

and, solving for H_L, we get finally

$$H_L = \frac{\frac{3}{8}\frac{Pl^2}{I_2}}{\frac{3hl}{I_2} + \frac{2h^2}{I_1}} \tag{13.72}$$

Having found H_L and V_L, we can determine the shear and moment at any point — that is, the problem is solved.

The conjugate beam method, which was used in the example to solve a comparatively simple rigid frame, can also be used to solve the more complicated frames. The same general technique is applied, namely,

1. prepare an M/EI diagram for each member of the frame, in terms of the redundants and loading;
2. apply the necessary strain compatibility conditions; there will always be as many of these as there are redundants.

13-9 The Flat Plate

We pointed out in the introduction to this chapter that the flat plate may be thought of as the three-dimensional counterpart of the two-dimensional beam or rigid frame. Also, we suggested there that frequently a *generalization technique* can be applied to paired problems of this type. This shows up very clearly in the case of the beam and plate.

In Art. 6.4, it was shown that for a beam subjected to a transverse loading $p(x)$ the governing differential equation is

$$\frac{d^4v}{dx^4} = \frac{p(x)}{EI} \tag{13.73}$$

or

$$\frac{d^2}{dx^2}\left(\frac{d^2v}{dx^2}\right) = \frac{p(x)}{EI} \tag{13.74}$$

in which EI is the beam stiffness and v is the deflection at any point, x, along the beam.

If now we consider a thin plate, with a transverse loading $p(x, y)$, (see

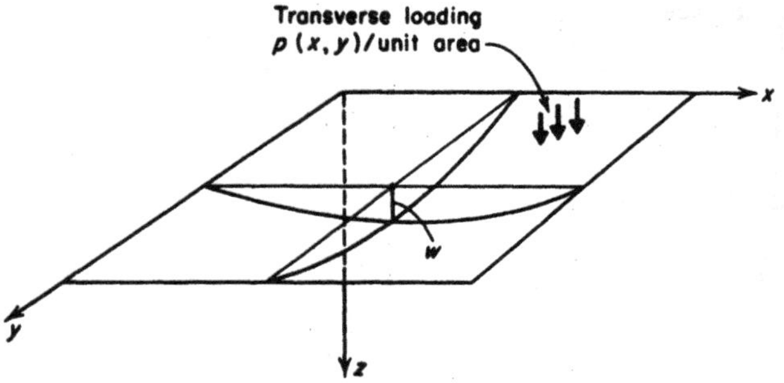

Fig. 13.20

Fig. 13.20) then it may be shown (see Ref. 4) that if $w(x, y)$ is the deflection, the governing differential equation is

$$\nabla^2(\nabla^2 w) = \frac{p(x, y)}{D} \qquad (13.75)$$

in which

$$\nabla^2 = \frac{\partial^2}{\partial x^2} + \frac{\partial^2}{\partial y^2} \qquad (13.76)$$

so that

$$\nabla^2\nabla^2 = \nabla^4 = \frac{\partial^4}{\partial x^4} + 2\frac{\partial^4}{\partial x^2 \partial y^2} + \frac{\partial^4}{\partial y^4} \qquad (13.77)$$

and

$$D = \text{plate stiffness}$$

$$= \frac{Eh^3}{12(1 - \nu^2)}$$

$$\nu = \text{Poisson's ratio}$$

$$h = \text{plate thickness}$$

The student should note particularly how the generalization technique applies in the preceding problem, which in this case involves the substitutions

$$\frac{\partial^2}{\partial x^2} + \frac{\partial^2}{\partial y^2} \text{ or } \nabla^2 \text{ for } \frac{d^2}{dx^2}$$

and

$$D \text{ for } EI$$

We shall not discuss the plate problem further here. There are various ways for obtaining solutions to the different plate problems. Some of these are exact, closed form solutions; others are approximate, numerical solutions. The interested student will find a more detailed discussion of these in Refs. 4, 13, 20, 21, and 25.

13-10 The Curved Beam, or Arch

Up to this point we have not considered structures subjected to bending which are not straight. There are, however, two very important classes of engineering structures which are curved and subjected to bending effects:

1. curved beams, or arches;
2. shells.

In this article we shall very briefly touch upon the analysis of the curved beam, or arch. Much more detailed discussions will be found in Refs. 25 and 26.

A typical curved beam or arch is shown in Fig. 13.21. In this figure AB

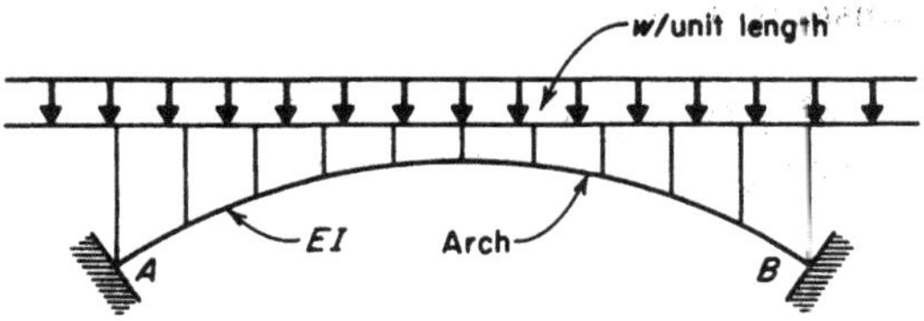

Fig. 13.21

is the arch, with abutments or supports at A and B, $w(x)$ is the loading, and EI is the arch stiffness.

We know that the governing differential equation for a straight beam is

$$\frac{d^2v}{dx^2} = -\frac{M(x)}{EI} \tag{13.78}$$

It may be shown (see Ref. 25) that if S is the distance along the arch, the differential equation governing arch action is

$$\frac{d^2v}{dS^2} + \frac{v}{\rho^2} = -\frac{M(S)}{EI} \tag{13.79}$$

in which

v = radial deflection at any point

ρ = radius of curvature at this point

Methods which may be used to solve this problem are described in Refs. 25 and 26.

13-11 The Shell

The three-dimensional counterpart of the two-dimensional arch is the shell. There are several methods of shell analysis currently being used. All are approximate to some degree or other. A rather detailed discussion of this and related points is given in Ref. 25. In this article we shall not touch further on this subject. Rather, we shall derive still another approximate shell relation using the generalization technique previously described in Arts. 13-1 and 13-9.

We assume that the equation for the shell can be given as a generalization of the two-dimensional arch equation, Eq. 13.79,

$$\frac{d^2v}{dS^2} + \frac{v}{\rho^2} = -\frac{M(S)}{EI} \tag{13.80}$$

We proceed as follows: take d^2/dS^2 of both sides of the above equation, assuming ρ to be constant; this gives

$$\frac{d^2}{dS^2}\left(\frac{d^2v}{dS^2}\right) + \frac{1}{\rho^2}\frac{d^2v}{dS^2} = -\frac{d^2M(S)}{EIdS^2} \tag{13.81}$$

We assume further that for an arch, just as for a straight beam, it will be approximately true that (see Eqs. 6.55 and 6.57)

$$-\frac{d^2M(S)}{dS^2} = w(x) \tag{13.82}$$

Therefore, the above equation becomes

$$\frac{d^2}{dS^2}\left(\frac{d^2v}{dS^2}\right) + \frac{1}{\rho^2}\frac{d^2v}{dS^2} = \frac{w(x)}{EI} \tag{13.83}$$

We now generalize this to be the three-dimensional structure, the shell, by noting (see Art. 13-9) that in going from arch to shell

$$\frac{d^2}{dS^2} \rightarrow \nabla^2 = \frac{\partial^2}{\partial x^2} + \frac{\partial^2}{\partial y^2} \tag{13.84}$$

and

$$EI \rightarrow D = \frac{Eh^3}{12(1-\nu^2)} \tag{13.85}$$

With these substitutions, the shell equations become

$$\nabla^2\nabla^2 v + \frac{1}{\rho^2}\nabla^2 v = \frac{w(x, y)}{D} \tag{13.86}$$

or

$$\frac{\partial^4 v}{\partial x^4} + \frac{2\partial^4 v}{\partial x^2 \partial y^2} + \frac{\partial^4 v}{\partial y^4} + \frac{1}{\rho^2}\left(\frac{\partial^2 v}{\partial x^2} + \frac{\partial^2 v}{\partial y^2}\right) = \frac{w(x, y)}{D} \tag{13.87}$$

In shell analyses it is generally more convenient to present the equations in terms of polar coordinates — x, ρ, ϕ (see Fig. 13.22).

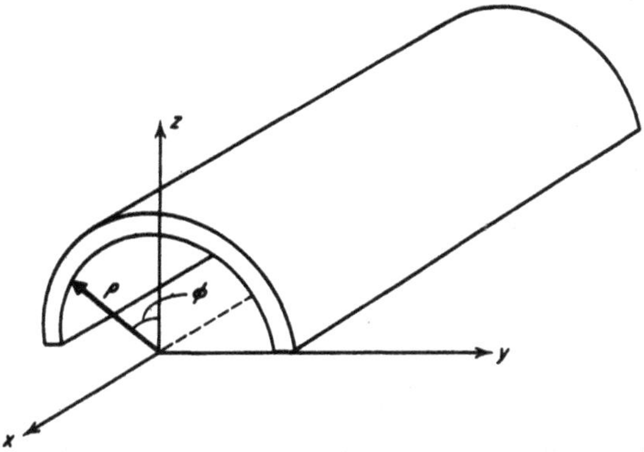

Fig. 13.22

Equation 13.87 may be put in polar form without difficulty by noting that the expression ∇^2, in terms of ρ, ϕ and x, is given (see Ref. 4) by

$$\nabla^2(\ ) = \left(\frac{\partial^2}{\partial\rho^2} + \frac{1}{\rho}\frac{\partial}{\partial\rho} + \frac{1}{\rho^2}\frac{\partial^2}{\partial\phi^2} + \frac{\partial^2}{\partial x^2}\right)(\ ) \tag{13.88}$$

so that Eq. 13.87 is given by

$$\left(\frac{\partial^2}{\partial\rho^2} + \frac{1}{\rho}\frac{\partial}{\partial\rho} + \frac{1}{\rho^2}\frac{\partial^2}{\partial\phi^2} + \frac{\partial^2}{\partial x^2}\right)\left(\frac{\partial^2}{\partial\rho^2} + \frac{1}{\rho}\frac{\partial}{\partial\rho} + \frac{1}{\rho^2}\frac{\partial^2}{\partial\phi^2} + \frac{\partial^2}{\partial x^2}\right)v + \frac{1}{\rho^2}\left(\frac{\partial^2}{\partial\rho^2} + \frac{1}{\rho}\frac{\partial}{\partial\rho} + \frac{1}{\rho^2}\frac{\partial^2}{\partial\phi^2} + \frac{\partial^2}{\partial x^2}\right)v = \frac{w(x, \rho, \phi)}{D} \tag{13.89}$$

In shell analyses it is customary to assume that the radial deflection, v, does not vary with ρ. In other words, all derivatives with respect to ρ may be assumed equal to zero. Therefore Eq. 13.89 becomes

$$\left(\frac{1}{\rho^2}\frac{\partial^2}{\partial\phi^2} + \frac{\partial^2}{\partial x^2}\right)\left(\frac{1}{\rho^2}\frac{\partial^2}{\partial\phi^2} + \frac{\partial^2}{\partial x^2}\right)v + \frac{1}{\rho^2}\left(\frac{1}{\rho^2}\frac{\partial^2}{\partial\phi^2} + \frac{\partial^2}{\partial x^2}\right)v = \frac{w(x, \phi)}{D} \tag{13.90}$$

The above equation represents an approximate solution to the curved shell equation — obtained as a generalization of the two-dimensional counterpart of the three-dimensional shell. It would hold best for a constant radius shell as shown in Fig. 13.22.

13-12 Summary

We discussed in a systematic manner certain paired special structures which may be solved using extensions of the engineering elasticity solutions given in previous chapters.

The structures discussed, and the manner in which they were paired, were

1. cable
2. membrane

3. rigid frame or beam
4. flat plate

5. arch
6. shell

A generalization technique† was described and an application of it was shown for the arch and the shell.

Problems

1. The suspension bridge shown has a 1000-ft span and a sag of 100 ft. Weight of the floor is 3000 lb per horizontal foot. Find the equation of the cable and the maximum and minimum tension in the cable. If the allowable stress in the cable is 60,000 psi, what area of steel cable is needed? What is the length of the cable AB?

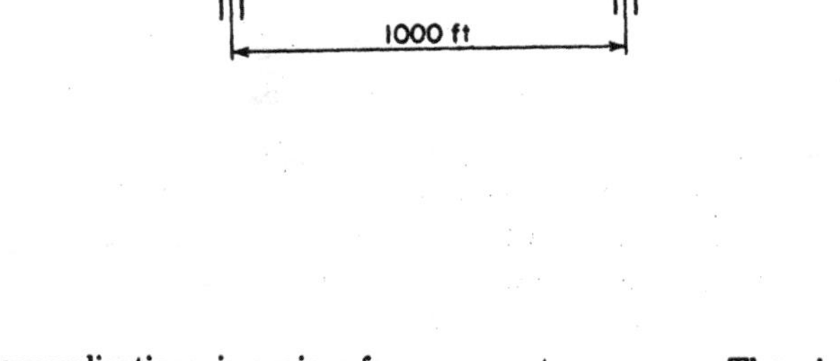

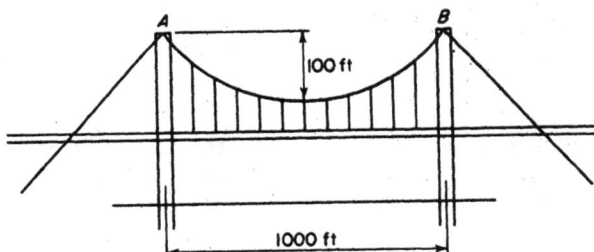

†Such generalizations, in various forms, are not uncommon. Thus, in studying visco-elastic problems (the exact theory of which has not as yet been completely formulated) several authors have arbitrarily generalized to three dimensions the one-dimensional stress-strain law deducible from models. Other examples may be cited in different fields. In every case, of course, the generalized relation remains an unproved postulate unless it predicts results which can be verified experimentally.

2. A wire weighing 1 lb/ft is supported by two piers 300 ft apart. If the horizontal pull on the cable is 1000 lb, what is the center sag and the length of the cable?

3. What pull is required for the wire of Prob. 2 in order that the center sag not exceed 12 in? What is the length of wire for this case?

4. In Prob. 2, what load is applied to the cable at the piers?

5. Determine the reactions for the rigid frame and loading shown.

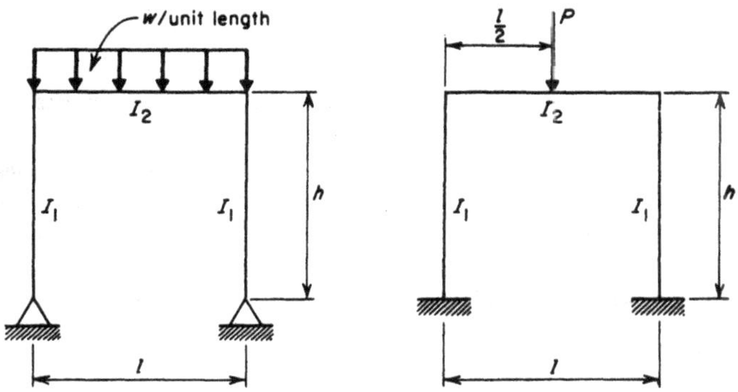

6. Draw the shear and moment diagram for the rigid frame and loading shown.

7. A cylindrical tank with hemispherical ends is fabricated of steel, $\frac{3}{8}$ in. thick. The inside diameter of the tank is 48 in. What is the maximum internal pressure which the tank can stand if the allowable tensile stress is 20,000 psi? Neglect discontinuity stresses at the junction of the cylindrical and hemispherical portions. That is, consider membrane effects only.

8. In the previous problem the student was told to neglect discontinuity stresses at the junction of the spherical and cylindrical portions of the tank. Explain why these arise and how they are resisted by the tank. For this structure, can the material be in a state of pure membrane stress only?

9. Determine the finite difference form of the approximate shell equation, Eq. 13.87.

REFERENCES

Here are listed all references mentioned in the book. When a particular reference number is cited, that reference will be found in this list.

Discussions of matrix-tensor theory which are related to our presentation will be found in

(1) Murnaghan, Francis D., *Introduction to Applied Mathematics*, Wiley, New York, 1948.

(2) Page, L., *Introduction to Theoretical Physics*, Van Nostrand, Princeton, N. J., 1935.

(3) Gibbs, J. W., ed. by E. B. Wilson, *Vector Analysis*, Yale Univ. Press, New Haven, Conn., 1943.

(4) Borg, S. F., *Matrix Tensor Methods in Continuum Mechanics*, D. Van Nostrand Co., Inc., Princeton, N. J., 1963.

Discussions related to the field of finite difference methods will be found in the following texts:

(5) Kárman, Th. von, and Biot, M. A., *Mathematical Methods in Engineering*, McGraw-Hill, New York, 1940.

(6) Allen, D. N. de G., *Relaxation Methods*, McGraw-Hill, New York, 1954.

(7) Scarborough, J. B., *Numerical Mathematical Analysis*, 4th ed., Johns Hopkins Press, Baltimore, Md., 1958.

The modern classic in the mathematical theory of elasticity is

(8) Love, A. E. H., *A Treatise on the Mathematical Theory of Elasticity*, Cambridge Univ. Press, 1927.

The standard American texts in this field are probably

(9) Timoshenko, S., and Goodier, J. N., *Theory of Elasticity*, McGraw-Hill, 1951.

(10) Sokolnikoff, I. S., *Mathematical Theory of Elasticity*, McGraw-Hill, 1946.

The more recent texts from abroad on the subject are

(11) Green, A. E., and Zerna, W., *Theoretical Elasticity*, Oxford Univ. Press, 1954.

(12) Landau, L. D., and Lifshitz, E. M., *Theory of Elasticity*, Pergamon, 1959.

(13) Biezano, C. B., and Grammel, R., *Engineering Dynamics*, Vol. I, *Theory of Elasticity*, Blackie and Son, Ltd., London, Glasgow, 1955.

(14) Swainger, K., *Analysis of Deformation*, Vol. I, *Mathematical Theory*, Chapman and Hall Ltd., London, 1954, and Vol. II, *Experiment and Applied Theory*, Macmillan, New York, 1955.

Additional American textbooks on elasticity in roughly decreasing order of mathematical complexity are

(15) Murnaghan, Francis D., *Finite Deformations of an Elastic Solid*, Wiley, New York, 1951.

(16) Westergaard, H. M., *Theory of Elasticity and Plasticity*, Harvard monograph in applied science, No. 3, Harvard Univ. Press, Cambridge, Mass., 1952.

(17) Sechler, E. E., *Elasticity in Engineering*, Wiley, New York, 1952.

(18) Wang, Chi-teh., *Applied Elasticity*, McGraw-Hill, New York, 1953.

(19) Timoshenko, S., and Lessells, J. M., *Applied Elasticity*, Westinghouse Tech. Night Sch. Press, East Pittsburgh, Pa., 1925.

In the field commonly called strength of materials, the standard American reference is

(20) Timoshenko, S., *Strength of Materials*, Vol. I and Vol. II, Van Nostrand, Princeton, N. J., 1956.

Typical additional references in this field in roughly decreasing order of mathematical complexity, are

(21) Den Hartog, J. P., *Advanced Strength of Materials*, McGraw-Hill, New York, 1952.

(22) Shanley, F. R., *Strength of Materials*, McGraw-Hill, New York, 1957.

(23) Seely, F. B., and Smith, J. O., *Advanced Mechanics of Materials*, Wiley, New York, 1952.

(24) Timoshenko, S., and MacCullough, G. H., *Elements of Strength of Materials*, Van Nostrand, Princeton, N. J., 1949.

In the area of applied strength of materials, i.e., structural engineering, the following listing is representative of those volumes covering the theory or analysis portion of the field.

(25) Borg, S. F., and Gennaro, J. J., *Advanced Structural Analysis*, Van Nostrand, Princeton, N. J., 1959.

(26) Parcel, J. I., and Moorman, R. B., *Analysis of Statically Indeterminate Structures*, Wiley, New York, 1955.

(27) Timoshenko, S., and Young, D. H., *Theory of Structures*, McGraw-Hill, New York, 1945.

(28) Wilbur, J. B., and Norris, C. H., *Elementary Structural Analysis*, McGraw-Hill, New York, 1948.

INDEX

APPENDIX

Appendix to Chapters 1 through 4

In this section are discussed various topics covered in Chapters 1 through 4 of the text. In general, additional or different points of view, derivations and explanations are given for the topics considered. It is hoped that these will clarify and offer additional insights insofar as the subjects discussed are concerned.

The material is identified by referring to the text page that it refers to.

p. 10 — Symmetry and Anti-Symmetry

A simple physical analog to the Eq. 2.27 occurs in the static loading of a simply supported beam, Fig. A.1.

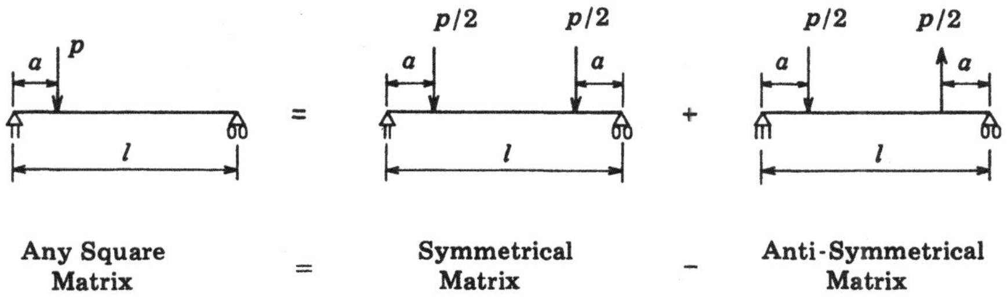

Fig. A.1

p. 20 — Tensor Invariants

We may verify the two invariants of the 2×2 tensor, Eqs. 2.90 and 2.91 by simply performing the indicated operations on the transformation equations 2.82 and 2.83.

p. 40 — The Strain Tensor

The linearized strain tensor may be generated as follows:

$$\text{deformation tensor} = \begin{pmatrix} \dfrac{\partial}{\partial x} \\ \dfrac{\partial}{\partial y} \\ \dfrac{\partial}{\partial z} \end{pmatrix} (u \quad v \quad w) = \Delta\delta$$

$$= \begin{pmatrix} \dfrac{\partial u}{\partial x} & \dfrac{\partial v}{\partial x} & \dfrac{\partial w}{\partial x} \\ \dfrac{\partial u}{\partial y} & \dfrac{\partial v}{\partial y} & \dfrac{\partial w}{\partial y} \\ \dfrac{\partial u}{\partial z} & \dfrac{\partial v}{\partial z} & \dfrac{\partial w}{\partial z} \end{pmatrix} \tag{A.1}$$

The deformation tensor being a square tensor may be represented by the sum of a symmetric tensor and an anti-symmetric tensor (see p. 10) or

$$\text{deformation tensor} : \begin{pmatrix} \dfrac{\partial u}{\partial x} & \dfrac{1}{2}\left(\dfrac{\partial u}{\partial y} + \dfrac{\partial v}{\partial x}\right) & \dfrac{1}{2}\left(\dfrac{\partial u}{\partial z} + \dfrac{\partial w}{\partial x}\right) \\ \dfrac{1}{2}\left(\dfrac{\partial v}{\partial x} + \dfrac{\partial u}{\partial y}\right) & \dfrac{\partial v}{\partial y} & \dfrac{1}{2}\left(\dfrac{\partial v}{\partial z} + \dfrac{\partial w}{\partial y}\right) \\ \dfrac{1}{2}\left(\dfrac{\partial w}{\partial x} + \dfrac{\partial u}{\partial z}\right) & \dfrac{1}{2}\left(\dfrac{\partial w}{\partial y} + \dfrac{\partial v}{\partial z}\right) & \dfrac{\partial w}{\partial z} \end{pmatrix}$$

$$+ \begin{pmatrix} 0 & \dfrac{1}{2}\left(\dfrac{\partial u}{\partial y} - \dfrac{\partial v}{\partial x}\right) & \dfrac{1}{2}\left(\dfrac{\partial u}{\partial z} - \dfrac{\partial w}{\partial x}\right) \\ \dfrac{1}{2}\left(\dfrac{\partial v}{\partial x} - \dfrac{\partial u}{\partial y}\right) & 0 & \dfrac{1}{2}\left(\dfrac{\partial v}{\partial z} - \dfrac{\partial w}{\partial y}\right) \\ \dfrac{1}{2}\left(\dfrac{\partial w}{\partial x} - \dfrac{\partial u}{\partial z}\right) & \dfrac{1}{2}\left(\dfrac{\partial w}{\partial y} - \dfrac{\partial v}{\partial z}\right) & 0 \end{pmatrix}$$

$$= \eta + \frac{1}{2}\Omega \tag{A.2}$$

in which η = the linearized strain tensor
Ω = the rotation tensor which may also be represented by its vector equivalent

$$(\Omega_x\ \Omega_y\ \Omega_z) = \left(\frac{\partial w}{\partial y} - \frac{\partial v}{\partial z}\quad \frac{\partial u}{\partial z} - \frac{\partial w}{\partial x}\quad \frac{\partial v}{\partial x} - \frac{\partial u}{\partial y}\right)$$

p. 43 Footnote — Sign Convention for Stresses

The sign convention given in the footnote on p. 43 is a "mathematical" sign convention. There is another sign convention which follows from the equilibrium of force (Newton's) relation which requires that the *net* force on a body in static equilibrium be equal to zero. Thus, when this convention applies, the sign of a *force* is determined by its *direction* — say positive to the right and negative to the left, although both may be tensile forces.

p. 51 — Invariant Form of the Equilibrium Equation

Note that Eq. 4.22, the equilibrium equation and one of the fundamental relations of elasticity theory is given in terms of *tensors* — i.e., in invariant form, independent of axes.

p. 53 — Invariant Form of the Boundary Conditions

Equation 4.42, another set of fundamental relations in elasticity theory may also be given in invariant form, as follows:

$$(\bar{X}) = \bar{N}T$$

in which $\bar{X}$ = the surface stress vector. Thus, all terms in this equation are tensors.

p. 54 — The Strain Compatibility Conditions

In addition to the discussion of the compatibility equations given in Refs. 4 and 15, one may also analyze these equations from the point of view that since they are fundamental equations of the theory of elasticity, they must be given in invariant form. This last requirement is not at all clear from the usual representations as given in Eqs. 4.48 — 4.53.

That these six equations can, in fact, be given in an invariant form will now be demonstrated and in doing this we shall start from first principles.

As noted above the fundamental equations in any discipline should be given in invariant form (i.e., independent of the coordinate system and also of the axial orientation). Thus, Eq. 4.22 stated the three equations of static equilibrium in terms of tensors and a scalar, Eq. 4.42, stated the boundary condition on the stresses in invariant tensor form and Eq. 4.61 stated Hooke's Law in terms of invariant tensor and isotropic scalar quantities.

One should expect therefore that the six equations of compatibility also can be stated in invariant form and it will now be shown that this is, in fact, the case.

We begin by pointing out that the following two facts have been established insofar as the strain compatibility equations of linearized elasticity theory are concerned. These are

1. There are six equations required.
2. The six equations are given in two entirely different sets of three each, with the equations in each set obtainable from each other by cyclical interchange.

In the following analyses it will be shown

1. Why there are six equations required.
2. Why they are given in two sets of three, markedly different from each other.
3. They will be derived in a concise invariant form.

Fundamental to each of the above will be the requirement that "rotations must be considered as of equal importance to the strains" in the compatibility analyses.

Insofar as the term "invariance" is concerned, as used in this discussion, invariance of an expression or equation can be represented in either of two forms:

1. The equation is given entirely in terms of *tensors*. Thus, η is the tensor representation of the complete nine-term strain expression. η by itself is entirely independent of any coordinate system and is thus an "invariant". Hooke's Law, for example is given in terms of η and also T, the stress tensor and is therefore, in invariant form.
2. The equation is given in terms of the "invariants" of the tensors. These terms are invariant under a rotation of axes and may be transformed into any orthogonal curvilinear coordinate system using well-known techniques. In Hooke's Law, for example, in addition to the stress and strain tensors, one term includes an invariant of the stress (or strain) tensor.

The strain tensor in linearized elasticity theory may be generated in the following manner. See the Appendix discussion relating to p. 40. We begin by considering deformations, represented by the first order tensor, i.e., vector, as either

$$\begin{pmatrix} u \\ v \\ w \end{pmatrix} \quad \text{or} \quad (u \quad v \quad w) \tag{A.3}$$

To introduce the small deformation theory, we have the differential operator, also a first order tensor

$$\Delta = \begin{pmatrix} \dfrac{\partial}{\partial x} \\[2ex] \dfrac{\partial}{\partial y} \\[2ex] \dfrac{\partial}{\partial z} \end{pmatrix} \quad \text{or} \quad \left(\frac{\partial}{\partial x} \quad \frac{\partial}{\partial y} \quad \frac{\partial}{\partial z} \right) \tag{A.4}$$

We may obtain a second order tensor from A.3 and A.4 by performing the operation

$$\begin{pmatrix} \dfrac{\partial}{\partial x} \\[2ex] \dfrac{\partial}{\partial y} \\[2ex] \dfrac{\partial}{\partial z} \end{pmatrix} (u \quad v \quad w) = \begin{pmatrix} \dfrac{\partial u}{\partial x} & \dfrac{\partial v}{\partial x} & \dfrac{\partial w}{\partial x} \\[2ex] \dfrac{\partial u}{\partial y} & \dfrac{\partial v}{\partial y} & \dfrac{\partial w}{\partial y} \\[2ex] \dfrac{\partial u}{\partial z} & \dfrac{\partial v}{\partial z} & \dfrac{\partial w}{\partial z} \end{pmatrix} \tag{A.5}$$

This tensor may be further decomposed into a symmetrical and skew-symmetrical term, each of which is also a second order tensor as follows:

$$\begin{pmatrix} \dfrac{\partial u}{\partial x} & \dfrac{\partial v}{\partial x} & \dfrac{\partial w}{\partial x} \\[2ex] \dfrac{\partial u}{\partial y} & \dfrac{\partial v}{\partial y} & \dfrac{\partial w}{\partial y} \\[2ex] \dfrac{\partial u}{\partial z} & \dfrac{\partial v}{\partial z} & \dfrac{\partial w}{\partial z} \end{pmatrix}$$

[continued on p. 282]

$$= \begin{pmatrix} \dfrac{\partial u}{\partial x} & \dfrac{1}{2}\left(\dfrac{\partial v}{\partial x} + \dfrac{\partial u}{\partial y}\right) & \dfrac{1}{2}\left(\dfrac{\partial w}{\partial x} + \dfrac{\partial u}{\partial z}\right) \\ \dfrac{1}{2}\left(\dfrac{\partial u}{\partial y} + \dfrac{\partial v}{\partial x}\right) & \dfrac{\partial v}{\partial y} & \dfrac{1}{2}\left(\dfrac{\partial v}{\partial z} + \dfrac{\partial w}{\partial y}\right) \\ \dfrac{1}{2}\left(\dfrac{\partial u}{\partial z} + \dfrac{\partial w}{\partial x}\right) & \dfrac{1}{2}\left(\dfrac{\partial v}{\partial z} + \dfrac{\partial w}{\partial y}\right) & \dfrac{\partial w}{\partial z} \end{pmatrix} \tag{A.6a}$$

$$+ \begin{pmatrix} 0 & \dfrac{1}{2}\left(\dfrac{\partial v}{\partial x} - \dfrac{\partial u}{\partial y}\right) & \dfrac{1}{2}\left(\dfrac{\partial w}{\partial x} - \dfrac{\partial u}{\partial z}\right) \\ \dfrac{1}{2}\left(\dfrac{\partial u}{\partial y} - \dfrac{\partial v}{\partial x}\right) & 0 & \dfrac{1}{2}\left(\dfrac{\partial w}{\partial y} - \dfrac{\partial v}{\partial z}\right) \\ \dfrac{1}{2}\left(\dfrac{\partial u}{\partial z} - \dfrac{\partial w}{\partial x}\right) & \dfrac{1}{2}\left(\dfrac{\partial v}{\partial z} - \dfrac{\partial w}{\partial y}\right) & 0 \end{pmatrix}$$

$$= \eta + \frac{1}{2}\Omega \tag{A.6b}$$

in which η is just the symmetrical strain tensor and Ω is the curl or rotation tensor which may be represented more compactly and usefully for our purposes as

$$\Omega = (\Omega_x \ \Omega_y \ \Omega_z) = \left(\frac{\partial v}{\partial z} - \frac{\partial w}{\partial y} \quad \frac{\partial w}{\partial x} - \frac{\partial u}{\partial z} \quad \frac{\partial u}{\partial y} - \frac{\partial v}{\partial x}\right) \tag{A.7}$$

Two significant facts emerge from an examination of Eqs. (A.6a) and (A.6b):

1. It is clear that the rotation components are as important as the strain components in the separation of the fundamental differential deformation tensor, Eq. (A.5) into its symmetrical and skew-symmetrical parts.
2. It is obvious that there are *nine* independent strain and rotation terms (*six* strains and *three* rotations) and these are given in terms of *three* deformations. Therefore, there are *six* additional independent equations necessary among the strain and rotation elements.

The introduction of rotation is essential. It explains why *six* additional compatibility equations or conditions are required. And in addition, as will be shown, it is an essential element in the derivation of one invariant set of these compatibility conditions. So far as the writer knows, only Ref. 6 had previously introduced the rotations as an essential part of the derivation of the compatibility conditions.

A third fundamental tensor, in elasticity theory, is the ∇^2 tensor, obtained as follows:

$$\nabla^2 = \begin{pmatrix} \dfrac{\partial}{\partial x} \\[2ex] \dfrac{\partial}{\partial y} \\[2ex] \dfrac{\partial}{\partial z} \end{pmatrix} \begin{pmatrix} \dfrac{\partial}{\partial x} & \dfrac{\partial}{\partial y} & \dfrac{\partial}{\partial z} \end{pmatrix}$$

$$= \begin{pmatrix} \dfrac{\partial^2}{\partial x^2} & \dfrac{\partial^2}{\partial x\,\partial y} & \dfrac{\partial^2}{\partial x\,\partial z} \\[2ex] \dfrac{\partial^2}{\partial y\,\partial x} & \dfrac{\partial^2}{\partial y^2} & \dfrac{\partial^2}{\partial y\,\partial z} \\[2ex] \dfrac{\partial^2}{\partial z\,\partial x} & \dfrac{\partial^2}{\partial z\,\partial y} & \dfrac{\partial^2}{\partial z^2} \end{pmatrix} \quad \text{(A.8)}$$

Note, this also is a symmetrical second order tensor, if the order of differentiation is immaterial.

The First Set of Compatibility Conditions

It will now be shown that there are two separate invariant sets (three equations in each) of compatibility equations. The first set will be given in terms of the invariants of the two-dimensional forms of the strain tensor and the ∇^2 tensor, these being, for example, in terms of x, y.

$$\eta = \begin{pmatrix} e_x & \dfrac{1}{2}\gamma_{xy} \\[2ex] \dfrac{1}{2}\gamma_{yx} & e_y \end{pmatrix}, \qquad \nabla^2 = \begin{pmatrix} \dfrac{\partial^2}{\partial x^2} & \dfrac{\partial^2}{\partial x\,\partial y} \\[2ex] \dfrac{\partial^2}{\partial y\,\partial x} & \dfrac{\partial^2}{\partial y^2} \end{pmatrix} \quad \text{(A.9)}$$

Obviously, the two tensors can be given, in two-dimensional form, entirely independent of the third dimension, z in this case.

The invariants, trace, of ∇^2 and η are

$$I_{\nabla^2} = \left(\frac{\partial^2}{\partial x^2} + \frac{\partial^2}{\partial y^2}\right) \quad , \quad I_\eta = (e_x + e_y) \tag{A.10}$$

also

$$\nabla^2 \eta = \begin{pmatrix} \dfrac{\partial^2}{\partial x^2} & \dfrac{\partial^2}{\partial x\, \partial y} \\ \dfrac{\partial^2}{\partial y\, \partial x} & \dfrac{\partial^2}{\partial y^2} \end{pmatrix} \begin{pmatrix} e_x & \dfrac{1}{2}\gamma_{xy} \\ \dfrac{1}{2}\gamma_{yx} & e_y \end{pmatrix} \tag{A.11a}$$

$$= \begin{pmatrix} \dfrac{\partial^2 e_x}{\partial x^2} + \dfrac{1}{2}\dfrac{\partial^2 \gamma_{yx}}{\partial x\, \partial y} & \dfrac{1}{2}\dfrac{\partial^2 \gamma_{xy}}{\partial x^2} + \dfrac{\partial^2 e_y}{\partial x\, \partial y} \\ \dfrac{\partial^2 e_x}{\partial y\, \partial x} + \dfrac{1}{2}\dfrac{\partial^2 \gamma_{yx}}{\partial y^2} & \dfrac{1}{2}\dfrac{\partial^2 \gamma_{xy}}{\partial y\, \partial x} + \dfrac{\partial^2 e_y}{\partial y^2} \end{pmatrix} \tag{A.11b}$$

is a second order tensor, and its trace invariant is

$$I_{\nabla^2\eta} = \frac{\partial^2 e_x}{\partial x^2} + \frac{1}{2}\frac{\partial^2 \gamma_{yx}}{\partial x\, \partial y} + \frac{1}{2}\frac{\partial^2 \gamma_{xy}}{\partial y\, \partial x} + \frac{\partial^2 e_y}{\partial y^2} \tag{A.12}$$

If, now it is assumed that the order of differentiation is immaterial (as it is for finite, single valued continuous functions) and noting, that for these tensors,

$$I_{\nabla^2} I_\eta = I_{\nabla^2\eta} \tag{A.13}$$

we obtain at once from Eqs. A.10, A.12, and A.13,

$$\frac{\partial^2 e_x}{\partial y^2} + \frac{\partial^2 e_y}{\partial x^2} = \frac{\partial^2 \gamma_{xy}}{\partial x\, \partial y} \tag{A.14}$$

which is the first equation of the first set of three (the other two are obtained by cyclical interchange).

Equation A.13 is the invariant form of the first three equations of compatibility.

If the three-dimensional forms of η and ∇^2 are used, we have, again, the invariant identity,

$$I_{\nabla^2} I_\eta = I_{\nabla^2 \eta} \tag{A.15}$$

and we obtain a single equation which is just the sum of the three equations A.14 above. Obviously, this single equation, while correct, is a much weaker statement than the three separate equations A.14.

The Second Set of Compatibility Conditions

The second set of three required equations is of an entirely different form from that of Eq. A.14. For one thing, each equation includes the three-dimensional terms (i.e., elements in x, y and z). It is not surprising that this is so, since this second set is, fundamentally, a statement involving the rotation elements.

We begin by repeating Eq. A.7

$$\Omega = (\Omega_x \ \Omega_y \ \Omega_z) = \left(\frac{\partial v}{\partial z} - \frac{\partial w}{\partial y} \quad \frac{\partial w}{\partial x} - \frac{\partial u}{\partial z} \quad \frac{\partial u}{\partial y} - \frac{\partial v}{\partial x} \right) \tag{A.16}$$

and point out the well-known fact that

$$\nabla \cdot \Omega = 0 \tag{A.17}$$

Statement A.17 is a single equation. We are looking for *three* equations among the Ω components. These are obtained by taking the gradient of A.17, i.e.,

$$\nabla \nabla \cdot \Omega = \nabla^2 \Omega = 0 \tag{A.18}$$

Equation A.15 is the invariant form of the second set of three compatibility conditions. In matrix-tensor notation, it is given by

$$\begin{pmatrix} \dfrac{\partial^2}{\partial x^2} & \dfrac{\partial^2}{\partial x\, \partial y} & \dfrac{\partial^2}{\partial x\, \partial z} \\[2ex] \dfrac{\partial^2}{\partial y\, \partial x} & \dfrac{\partial^2}{\partial y^2} & \dfrac{\partial^2}{\partial y\, \partial z} \\[2ex] \dfrac{\partial^2}{\partial z\, \partial x} & \dfrac{\partial^2}{\partial z\, \partial y} & \dfrac{\partial^2}{\partial z^2} \end{pmatrix} \begin{pmatrix} \Omega_x \\[2ex] \Omega_y \\[2ex] \Omega_z \end{pmatrix} = 0 \tag{A.18a}$$

In expanded form, Eq. A.18a is *three* equations, each of which equals zero, or

$$\frac{\partial}{\partial x}\left[\frac{\partial \Omega_x}{\partial x}+\frac{\partial \Omega_y}{\partial y}+\frac{\partial \Omega_z}{\partial z}\right]=0 \tag{A.19a}$$

$$\frac{\partial}{\partial y}\left[\frac{\partial \Omega_x}{\partial x}+\frac{\partial \Omega_y}{\partial y}+\frac{\partial \Omega_z}{\partial z}\right]=0 \tag{A.19b}$$

$$\frac{\partial}{\partial z}\left[\frac{\partial \Omega_x}{\partial x}+\frac{\partial \Omega_y}{\partial y}+\frac{\partial \Omega_z}{\partial z}\right]=0 \tag{A.19c}$$

These are just the second set of three compatibility conditions, in terms of rotations.

If we want these equations in terms of the strains, we introduce the connections between the γ and Ω terms, namely:

$$\frac{1}{2}\Omega_x = \frac{1}{2}\gamma_{yz} - \frac{\partial w}{\partial y} \tag{A.20a}$$

$$\frac{1}{2}\Omega_y = \frac{1}{2}\gamma_{zx} - \frac{\partial u}{\partial z} \tag{A.20b}$$

$$\frac{1}{2}\Omega_z = \frac{1}{2}\gamma_{xy} - \frac{\partial v}{\partial x} \tag{A.20c}$$

Substituting these in, for example, A.19a we obtain

$$\frac{\partial}{\partial x}\left[\frac{1}{2}\frac{\partial \gamma_{yz}}{\partial x}-\frac{\partial^2 w}{\partial x\,\partial y}+\frac{1}{2}\frac{\partial \gamma_{zx}}{\partial y}-\frac{\partial^2 u}{\partial y\,\partial z}+\frac{1}{2}\frac{\partial \gamma_{xy}}{\partial z}-\frac{\partial^2 v}{\partial x\,\partial z}\right]=0 \tag{A.21}$$

Collecting terms assuming the order of differentiation is immaterial, this becomes,

$$2\frac{\partial^2 e_x}{\partial y\,\partial z}=\frac{\partial}{\partial x}\left[\frac{\partial \gamma_{xy}}{\partial z}+\frac{\partial \gamma_{zx}}{\partial y}-\frac{\partial \gamma_{yz}}{\partial x}\right] \tag{A.22}$$

with the other two, Eqs. A.19b and A.19c obtained by cyclical interchange. Equation A.22 will be recognized as the familiar form in which the compatibility equation is given, Eq. 4.52.

Conclusion

It was first demonstrated that the strain tensor, the rotation tensor, and the ∇^2 tensor are fundamental in the analysis of the compatibility conditions in linearized elasticity.

From a study of these, it became clear that *six* compatibility equations are needed, in order that the *nine* strain and rotation elements be expressible in terms of *three* finite, continuous, single-valued displacements.

These six compatibility conditions in concise, invariant form are

$$I_{\nabla^2} I_\eta = I_{\nabla^2 \eta} \tag{A.15}$$

which is three equations, and

$$\nabla^2 \Omega = 0 \tag{A.18}$$

which is three equations.

p. 56 — Derivation of Hooke's Law

We may *derive* Hooke's Law by referring to the simple tensile test described in Fig. 4.5.

Thus, due to a force in the y direction we have

$$\begin{aligned} e_y &= \frac{\sigma_y}{E} \\ e_x &= -\frac{\nu \sigma_y}{E} \\ e_z &= -\frac{\nu \sigma_y}{E} \end{aligned} \tag{A.23}$$

If now all three forces (in the x, y, and z directions) are applied simultaneously then (for small deformations)

$$e_x = \frac{\sigma_x}{E} - \frac{\nu\sigma_y}{E} - \frac{\nu\sigma_z}{E}$$

$$e_y = \frac{\sigma_y}{E} - \frac{\nu\sigma_z}{E} - \frac{\nu\sigma_x}{E} \tag{A.24}$$

$$e_z = \frac{\sigma_z}{E} - \frac{\nu\sigma_x}{E} - \frac{\nu\sigma_y}{E}$$

(Note that the second and third of these equations are obtained from the first by cyclical interchange.)

Then, clearly

$$e_x = \frac{\sigma_x}{E} - \frac{\nu\sigma_y}{E} - \frac{\nu\sigma_z}{E} + \frac{\nu\sigma_x}{E} - \frac{\nu\sigma_x}{E}$$

$$e_y = \frac{\sigma_y}{E} - \frac{\nu\sigma_z}{E} - \frac{\nu\sigma_x}{E} + \frac{\nu\sigma_y}{E} - \frac{\nu\sigma_y}{E} \tag{A.25}$$

$$e_z = \frac{\sigma_z}{E} - \frac{\nu\sigma_x}{E} - \frac{\nu\sigma_y}{E} + \frac{\nu\sigma_z}{E} - \frac{\nu\sigma_z}{E}$$

which, in *matrix* form becomes

$$\begin{pmatrix} e_x & 0 & 0 \\ 0 & e_y & 0 \\ 0 & 0 & e_z \end{pmatrix} = \frac{1+\nu}{E} \begin{pmatrix} \sigma_x & 0 & 0 \\ 0 & \sigma_y & 0 \\ 0 & 0 & \sigma_z \end{pmatrix}$$

$$- \frac{\nu}{E}(\sigma_x + \sigma_y + \sigma_z) \begin{pmatrix} 1 & 0 & 0 \\ 0 & 1 & 0 \\ 0 & 0 & 1 \end{pmatrix} \tag{A.26}$$

or, in *tensor* notation

$$\eta = \frac{1+\nu}{E} T - \frac{\nu}{E} \mathscr{I}_1 \mathscr{E}_3 \tag{A.27}$$

in which,

$$
\begin{aligned}
1, \nu, E, \mathscr{I}_1 &= \text{invariants} \\
\eta &= \text{linearized strain tensor} \\
T &= \text{stress tensor} \\
\mathscr{I}_1 &= \text{trace invariant of } T \\
\mathscr{E}_3 &= \text{unit tensor}
\end{aligned}
$$

Note particularly that η and T, because they are symmetrical tensors may be given in diagonal form (see Eq. 2.85).

Also, because the above equation is a *tensor* equation, it holds with respect to *all* axial orientations, even those for which η and T are *not* diagonal. Also the invariant (i.e., scalar) terms do not change when the axes are rotated. Thus, if the axes are rotated the equation becomes, in unexpanded matrix form

$$
\begin{pmatrix} e_x & \frac{1}{2}\gamma_{xy} & \frac{1}{2}\gamma_{xz} \\ \frac{1}{2}\gamma_{yx} & e_y & \frac{1}{2}\gamma_{yz} \\ \frac{1}{2}\gamma_{zx} & \frac{1}{2}\gamma_{zy} & e_z \end{pmatrix} = \frac{1+\nu}{E}\begin{pmatrix} \sigma_x & \tau_{xy} & \tau_{xz} \\ \tau_{yx} & \sigma_y & \tau_{yz} \\ \tau_{zx} & \tau_{zy} & \sigma_z \end{pmatrix} - \frac{\nu}{E}(\sigma_x + \sigma_y + \sigma_z)\begin{pmatrix} 1 & 0 & 0 \\ 0 & 1 & 0 \\ 0 & 0 & 1 \end{pmatrix} \tag{A.28}
$$

And this equation now *predicts* a result *independent* of the original test, namely

$$
\frac{1}{2}\gamma_{xy} = \frac{1+\nu}{E}\tau_{xy}
$$

or

$$
\tau_{xy} = \frac{E}{2(1+\nu)}\gamma_{xy} \tag{A.29}
$$

the pure *shear* behavior of a stressed block. All the terms in the expression are measurable and experiments verify the accuracy of the prediction, thus verifying the validity of the linearized form of Hooke's Law and hence of the derivation.

www.ingramcontent.com/pod-product-compliance
Lightning Source LLC
Chambersburg PA
CBHW050636190326
41458CB00008B/2289
* 9 7 8 9 8 1 0 2 0 1 6 4 7 *